DROUGHT COUNTRY

The Dry Times That Have Shaped Australia

ROBERT GODFREE

© CSIRO 2025

All rights reserved. Except under the conditions described in the *Australian Copyright Act 1968* and subsequent amendments, no part of this publication may be reproduced, stored in a retrieval system or transmitted in any form or by any means, electronic, mechanical, photocopying, recording, duplicating or otherwise, without the prior permission of the copyright owner. Contact CSIRO Publishing for all permission requests.

Robert Godfree asserts his right to be known as the author of this work.

 A catalogue record for this book is available from the National Library of Australia

ISBN: 9781486314041 (pbk)
ISBN: 9781486314058 (epdf)
ISBN: 9781486314065 (epub)

How to cite:
Godfree R (2025) *Drought Country: The Dry Times That Have Shaped Australia*. CSIRO Publishing, Melbourne.

Published by:

CSIRO Publishing
36 Gardiner Road, Clayton VIC 3168
Private Bag 10, Clayton South VIC 3169
Australia

Telephone: +61 3 9545 8400
Email: publishing.sales@csiro.au
Website: www.publish.csiro.au
Sign up to our email alerts: publish.csiro.au/earlyalert

Front cover: Sand dune engulfs a tree between Ivanhoe and Menindee, NSW. Photo by Robert Godfree.
Back cover: Fence remnant at the Jumping Sandhills between Booligal and Mossgiel, NSW. Photo by Holly Godfree.

Edited by Joy Window
Cover design by Cath Pirret
Typeset by Envisage Information Technology
Index by Indexicana
Printed in China by 1010 Printing International Ltd

CSIRO Publishing publishes and distributes scientific, technical and health science books, magazines and journals from Australia to a worldwide audience and conducts these activities autonomously from the research activities of the Commonwealth Scientific and Industrial Research Organisation (CSIRO). The views expressed in this publication are those of the author(s) and do not necessarily represent those of, and should not be attributed to, the publisher or CSIRO. The copyright owner shall not be liable for technical or other errors or omissions contained herein. The reader/user accepts all risks and responsibility for losses, damages, costs and other consequences resulting directly or indirectly from using this information.

CSIRO acknowledges the Traditional Owners of the lands that we live and work on across Australia and pays its respect to Elders past and present. CSIRO recognises that Aboriginal and Torres Strait Islander peoples have made and will continue to make extraordinary contributions to all aspects of Australian life including culture, economy and science. CSIRO is committed to reconciliation and demonstrating respect for Indigenous knowledge and science. The use of Western science in this publication should not be interpreted as diminishing the knowledge of plants, animals and environment from Indigenous ecological knowledge systems.

The paper this book is printed on is in accordance with the standards of the Forest Stewardship Council® and other controlled material. The FSC® promotes environmentally responsible, socially beneficial and economically viable management of the world's forests.

Sept24_01

Foreword

In 1994, drought gripped 80 per cent of New South Wales and 40 per cent of Queensland. Midway through that year, I went with a small Prime Ministerial party to see the effects at a place called Dingo, near Blackwater in Queensland. We drove through blackened scrub and woodland, no green leaf or blade of grass, no animal or bird. Near the farmhouse, in a scene of utter desolation, stood half a dozen emaciated beasts, breeding stock the exhausted farmers were trying to keep alive.

In Australia the ever-widening gap between the city and the country grows grotesquely wide in droughts. I would not wonder if now and then a manicured Prime Minister and his well-fed staff do not drift into that farming couple's memory of those forlorn and bitter scenes.

But that drought left no mark on the *nation's* memory. It was not the first time Australians had seen pictures of barren landscapes, withered crops, dry lagoons and dams, of despairing farmers slaughtering their starving sheep and cattle, and it would not be the last. Unlike the Federation Drought or the Millennium Drought, it has no name: just the 1994 drought, like the 1981–82 drought, or the 2018 drought, or scores of others. It was just another drought.

Flood and fire have their days, but drought is the great ogre of Australia. When it is not actually present, it lies in wait. Some droughts creep up slowly, and take hold 'like a cancer', as one farmer said. Others come out of nowhere, do their worst, and leave just as quickly. Though they may vary in length and spread, in Australia droughts are cruelly and ruinously normal. They have long been embedded in the country's literature, art and myths. They are often credited with planting a sort of doleful irony in the national character and making persistence, fatalism and scepticism mainstays of the country's ethos. There is no Australia without drought.

That trip to Dingo produced $164 million in aid to farmers and communities. It also generated conversations about better ways to deal with drought: from 'drought proofing' farms through wiser farming practices and water conservation, to combating salination, dealing with climate change and providing income-contingent loans to stricken farmers instead of emergency aid. They were far from new debates, but they gained more currency in the century's last decade and became part of the national conversation in the drought-afflicted new one.

Robert Godfree had been thinking deeply about such things for years when I met him at the end of the Millennium Drought in 2010. Soon after our first meeting he told me he was planning to write a book on the Federation drought. With the fabled persistence and ingenuity of all good scientists and farmers, he has pushed way beyond his original conception and given us this mighty, mind-altering epic.

Don Watson

For my wife, Holly, a fountain of wisdom in the driest of times,
and
for my family, who have given me much to strive for.

Contents

Foreword iii
Preface viii
Acknowledgements x
Cultural sensitivity warning xi
Introduction: a land fertile for drought xii

PART 1: BIG FELLER DRY   1

CHAPTER 1: Legends 2
CHAPTER 2: Calamity on the *Coquun* 10
CHAPTER 3: Vanishing waters 16

PART 2: AN INGRATIOUS COUNTRY   25

CHAPTER 4: Unsettling times 26
CHAPTER 5: The summer of discontent 40
Colour plates 47
CHAPTER 6: An uncommon and tedious drought 63
CHAPTER 7: Bass and the South Coast drought 68

PART 3: A SKY UNTROUBLED BY CLOUDS   73

CHAPTER 8: The poverty of confinement 74
CHAPTER 9: The stillness of death 91
CHAPTER 10: Depressed times 109
CHAPTER 11: Transitions 127

PART 4: HELL'S HALF CENTURY   149

CHAPTER 12: The Titan's grip 150
CHAPTER 13: The living drought 154

CHAPTER 14: Niobe's ruin	168
CHAPTER 15: Crucified	182
CHAPTER 16: The desert of dry bones	196

PART 5: LIFE IN THE ANTHROPOCENE — 213

CHAPTER 17: River of tears	214
CHAPTER 18: The great divide	230
CHAPTER 19: What lies beneath	238
CHAPTER 20: Day zero	252
CHAPTER 21: Hung out to dry	265

| Obscured horizons | 275 |
| Index | 278 |

Note for readers: References have been abbreviated in the endnotes for space reasons. Full reference details are available at: https://www.publish.csiro.au/book/7989#supplementary

If there were the sound of water only
Not the cicada
And dry grass singing
But sound of water over a rock
Where the hermit-thrush sings in the pine trees
Drip drop drip drop drop drop drop
But there is no water.

– *TS Eliot*, 'The Wasteland'

This desert upon which so many have been broken is vast and calls for largeness of heart ...

– *Cormac McCarthy*, Blood Meridian

Preface

I was lucky enough to grow up at the foot of Mount Kaputar in north-central New South Wales, Australia, raised by parents who, after arriving from New Zealand, saw not a country left barren from a century of sheep grazing but a beautiful part of the world where it was possible, with much work, to carve out a largely independent life. Here, the pioneering days did not seem so far away, and so they purchased a small block of land and set about building a log house out of cypress pine and ironbark using hand tools. There was no electricity or plumbing; in the evenings we read by the light of a kerosene lantern, and by day carted water in 44-gallon (200 L) drums from Bullawa Creek, which flowed steadily out of the Nandewar Ranges. And it rained. Year after year, it rained.

The country was much wetter than my parents had expected, but it was around this time that an old grazier who lived down the road, Tom Gunter, remarked that we just hadn't seen a drought yet. But it kept raining, through 1976, 1977, and 1978. It was torrential, the sound deafening on the corrugated iron roof; rivers flooded, the grass grew waist-high, and insects were so numerous that they blocked radiator screens and overheated cars. But then, in 1979, it stopped. At first it was a little dry, then *really* dry, and then, in 1982, it could no longer be described in terms of dryness. Great clouds of dust rolled in from the western plains, river red gums died, and we excavated the bed of Bullawa Creek to get the last of the water below, which then too disappeared. It was my first drought.

I did not seek a life of studying such events. When I departed from Australia in 1996 for the United States with my future wife after completing my undergraduate degree, I did so with the intention of pursuing my childhood dream of a career in conserving rare mammal species. I was fortunate enough to be admitted into the PhD program in biology at Portland State University, Oregon with a teaching scholarship, and the opportunity to work with two fantastic ecologists, Bob Tinnin and Dick Forbes, in a welcoming and friendly environment sealed the deal. I enrolled, set up a desk in the postgraduate room and began to think of what sort of project I might want to do.

Portland lies in the Willamette Valley, nestled between the Coast Range to the west and, to the east, the Cascades, where tantalising accounts of the presence of the beautiful and reclusive Canada lynx led me to apply for a grant to study the size and demographic structure of these remnant populations. But, as fate would have it, a change in government dried up funding for this research work – a common problem for ecologists – and I found myself looking for another project. It was at this point that my supervisors suggested that I try my hand at forest ecology. Trees have the great advantage of being sessile and require little specialised equipment or funding to study, perfect for an impecunious graduate student. I set out for central Oregon in a blue Volvo loaned to me by my mother-in-law, sporting little more than a tent, a species list and field survey sheets. Three-and-a-half years later I received my doctorate, and my career as an ecologist had begun.

Twenty-two years later, in May 2022, I found myself puzzling over the identity of several different trees and the habitats in which they grew, this time in the Pilliga Forest in north-west New South Wales, about an hour's drive west of where I grew up. In this case the species in question were the grey boxes, a famously difficult taxonomic group, the separation of which requires a detailed understanding of the morphology of their buds, fruit ('gum nuts'), juvenile leaves and growth. This may seem arcane, but the reason for my interest lay above: towering over me the trees stood dead, their branches stripped of their leaves, bark, and ultimately their life by one of the most brutal droughts of our time. Only the hardiest species, or the luckiest, had survived, and if I had any chance of predicting how these forests might change in the future I needed to know why.

This book stems from a lifetime of witnessing the destructive impact of drought on Australians and the landscapes and ecosystems that they inhabit. Its content is naturally rooted in the study of environmental history and ecology, but I have also battled drought as a cattle farmer since 2003 when my wife and I purchased a property east of Narrabri in what I thought, based on my experience in the 1980s and 1990s, would be a drought-breaking year. The promise of wet years beckoned, but it was not to be, for the Millennium Drought was only in its early stages, and since then there have been far more dry years than good ones. I have spent much time dragging livestock out of dams, shooting and then burying starving wildlife, carting water, and watching paddocks blow away.

Over the past three centuries Australia has been struck by dozens of droughts, and by necessity I have focused on only a small subset of those that I think have been the most disastrous or which presided over major transitions in Australian landscapes or society. I have also biased the work towards those places, times, people and themes with which I am most familiar. My apologies to those that I have missed out. Much of this book was written when COVID-19 was sweeping the globe, and so it has not been possible to collect as many first-hand contemporary accounts and recollections as I would have liked. This is especially true of the incredibly rich and diverse history of Aboriginal people living with drought. Rather than leave this out I have relied on admittedly imperfect and culturally myopic newspaper accounts and scholarly research to fill in the blanks, and any associated shortcomings should be viewed in this light. The text of this book was written without assistance from generative artificial intelligence and any errors in it are my own.

This book is built on a skeleton of memories scavenged from time. In it you will find little of cheer, but then droughts have always been the physical and metaphorical embodiment of material and spiritual poverty. For the reader who perseveres it is my hope that in the stories of those who have faced down drought and survived you will find green shoots growing among the cracks, and perhaps a source of future hope and prosperity. All the signs are that we are going to need it.

Acknowledgements

I would like to thank all the many people who have contributed to the creation of this book in small or large part. Thanks to my colleague Nunzio Knerr for his support, feedback on content, and help with coding, Brendan Lepschi for assistance with taxonomic questions and for many helpful discussions over decades of working together, and Cecille Gueidan for her encouragement. Thanks to the many people who generously provided permission to use images, without which the book could not have been completed, and to the staff at libraries and archives across Australia, and particularly at the National Library of Australia and Mitchell Library, all of whom were extremely encouraging and supportive of my work. Thank you also to the many landholders and members of the public that generously provided information, historical accounts and access to their properties. I also appreciate the opportunity to talk with members of Indigenous communities about their experiences with drought in Australia. I thank the staff at CSIRO Publishing, especially Mark Hamilton, Tracey Kudis and Eloise Moir-Ford, for being supportive and above all patient throughout the writing process, and Joy Window for editing the manuscript. I would like to thank my parents Bob and Denise, my children Alex and Julia, my sister Amanda, my mother-in-law Cynthia and my extended family for their unconditional support over the years, and finally my wife, Holly, without whose love, encouragement and effort this book would not have been possible.

Cultural sensitivity warning

Readers are warned that there may be words, descriptions and terms used in this book that are culturally sensitive, and which might not normally be used in certain public or community contexts. While this information may not reflect current understanding, it is provided by the author in a historical context.

This publication may also contain quotations, terms and annotations that reflect the historical attitude of the original author or that of the period in which the item was written, and may be considered inappropriate today.

Aboriginal and Torres Strait Islander peoples are advised that this publication may contain the names and images of people who have passed away.

Introduction: a land fertile for drought

> *Il faut faire parler les silences de l'historie*
> The silences of history must be made to speak
> – Jules Michelet (Journal entry, 30 January 1842)

Throughout human history drought has been associated with death, famine and evil. By depriving us of water, that element most fundamental to life, droughts cut to the very essence of the human being, in body and in spirit. Unlike other natural disasters, droughts last for months, years or even decades. With no clear beginning and an end only visible in retrospect, they become grinding periods of uncertainty, exercises in anguish. As anyone who has had to manage a farm, care for animals, or generate food and income from parched soils through such times can attest, droughts are much more than a mere problem of practicality – they represent the slow attrition of ambition, a great temporal desert that must be crossed before life may begin anew.

The greatest droughts force us to confront deep existential questions about the impermanence of life itself. Those that last for many years threaten to extinguish ecosystems, economies, and, by undercutting the production of food, the entire structure of civil society. Indeed, there is a growing body of evidence that drought was at least partly responsible for the decline or destruction of many civilisations. Perhaps the most famous example is that of the Lowland Classic Maya civilisation in 800–1000 CE;[1] others include the Akkadian Empire and other societies from the Mediterranean to the Indus Valley after 2200 BCE, eastern Mediterranean civilisations including the Hittite Empire during the 'Late Bronze Age collapse' after 1200 BCE, and western and Mississippian Native American cultures in the 11th to 13th centuries.[2] More recently, severe drought in 2007–10 is thought to have contributed to the Syrian uprising in 2011 and subsequent civil war.[3] One of the many lessons that these studies reveal is that societal stability can be susceptible to droughts in which rainfall declines are either sharp but brief, or modest but protracted.[4]

Yet societal (or individual) failure is not inevitable, for some of the world's most celebrated feats of courage, leadership and faith have also emerged in times of great water poverty. Many are centred on heroic figures like the mythological king Aeacus and the Israelites Joseph and Elijah, all of whom rescued their people from the ravages of drought.[5] It is difficult not to be inspired by the constitution of Alexander the Great, who, in solidarity with his army crossing the Gedrosian Desert, poured a precious drink of water presented to him onto the ground despite his own terrible thirst,[6] nor that of Mbuya Nehanda, said to have ended a great drought in Zimbabwe in the 1890s while leading a rebellion against British colonialists.[7] But many more no less heroic endeavours, especially those involving broader community-based mobilisation, have passed with scarcely any fanfare or recognition.

Australia has been fertile ground for such stories, for perhaps no other country can claim to have been so persistently afflicted by drought and irreversibly shaped by its ravages. Here, in the volatile antipodean climate, drought has been an evil visitant lurking behind major nation-building moments from European settlement[8] at Port Jackson to Federation, the displacement of many Aboriginal societies in the era of modern agricultural expansion, and more recently to the onset of the Anthropocene Epoch, in which humans have come to dominate the continent and the planet. To engage with this ecological and cultural history is to engage with the best and worst of human experience: the failures tragic, the triumphs prised grudgingly from the most difficult of circumstances. But the impact of these droughts has ramified far deeper into the building blocks of the nation, into its flora and fauna, and ultimately into the very land itself. For those who care to look, the wounds are still there, some as scars, partly healed by time and rain, but others still raw.

My intent in this book is to take you on a journey through the worst Australian droughts of the past three centuries, which is the timeframe in which there are written, eyewitness records of their existence. This allows us to reconstruct their impacts on the environment and society across a diverse set of themes extracted from a rich body of historiographic and scientific sources. There have been, of course, countless droughts before this, many identified using scientific research or remembered in the oral histories of Aboriginal peoples, and equally worthy of investigation. Even within this scope I have covered only my pick of the most devastating droughts, and, within that selection, an even smaller fraction of the vast body of information that is available for each. I make no pretence of comprehensiveness: each drought is examined through the lens of specific cultural and environmental themes particular to that era, all, it is to be hoped, underpinned by a foundation of natural history and science and liberally peppered with insights informed by my career as a scientist and a childhood in rural Australia.

Before we begin, however, we must briefly set the stage by investigating the nature of drought itself and identify the major periods in Australian history that might attract our interest.

ANATOMY OF A DROUGHT

Droughts come in many guises. The classical definition, a condition of dryness, lack of moisture, or of general aridity,[9] differs from the colloquial usage that typically means a period of low rainfall in which significant impacts are observed in agricultural or natural ecosystems. While the latter is a practical sensibility, it lacks objectivity. For what constitutes a drought, as opposed to a run-of-the-mill dry period, depends very much on your perspective. I have seen 'droughts' in New Zealand that would barely raise an eyebrow in Australia, and barren landscapes in western New South Wales in which no rain has fallen for many months that did not, by local estimation, even pass muster as a notable dry spell. In a very meaningful sense one's definition of drought is a deeply personal matter, a state of mind entered upon breaching a certain threshold of water scarcity. Nevertheless, we scientists tend to dislike subjectivity and have established a variety of criteria to define different kinds of drought and to measure their severity.

We must first start with the concept of *meteorological drought*, which, *sensu stricto*, is a period of low precipitation. The simplest method of measuring the severity of a meteorological drought is to determine the amount that rainfall has deviated from the long-term average (normally several decades to centuries), although there are also many more complex indices that integrate a range of meteorological factors into estimation of drought severity.[10] When plotted over time, rainfall deviations can be used to determine the duration (or length) and dryness (or depth) of drought. Prior to the 1870s–80s rainfall data were comparatively scant, particularly in sparsely settled areas, but from 1889 until the present, rainfall data from the SILO climate database exist for the entire continent.[11] Some rainfall records contained in this database have not been exhaustively cleaned to remove errors, and so models based on them must be taken with a certain grain of salt. Nevertheless, it remains one of the best meteorological datasets on the globe, and later you will encounter many drought maps generated from these data.

While quantifying meteorological drought is a crucial element of climate science, from a practical perspective people are usually more interested in 'downstream' impacts on secondary systems like rivers, forests and human societies. Typically, rainfall deficiencies gradually progress into *soil drought*, which refers to low water supply in the soil itself, usually relative to plant requirements for growth or survival, and *hydrological drought*, which is the loss of water in surface (e.g. rivers, wetlands, reservoirs) or subsurface (groundwater) water bodies. As these conditions intensify, they usually begin to manifest as *ecological drought* in (usually natural) ecosystems and *agricultural drought* in farming and grazing systems, both of which can eventually lead to broader *socioeconomic drought*. As human dominion over fundamental earth system processes have increased, the concept of *anthropogenic drought*, or the process in which water scarcity is an emergent phenomenon involving feedbacks between humans and nature, has found growing applicability.[12]

Each of the droughts that we will examine began as a prolonged period of below average rainfall that then developed into events that had devastating consequences for secondary *socioecological* systems. There is no hard-and-fast rule to how long a drought must be for this to happen because the resilience of human societies and the landscapes they use varies over time in response to environmental, technological and social change. In Australia, some droughts that fit into this category lasted for only 1, 2 or 3 years, but in the coming pages I will argue that the most damaging were either *quasi-decadal* (7–13 years) or *semi-decadal* (4–6 years) in duration, although this may be changing as the global climate warms. Another term to consider is that of the *megadrought*, which normally refers to a multi-decade period of chronically low rainfall. In this book we will encounter and explore of all these forms of drought.

DRY TIMES

Now that we have established this foundation, we can take a brief tour of the main droughts covered in this book. Table i.1 contains a list of these droughts and the themes that I investigate to build a deeper understand of them as phenomena bound by time and place. Figure i.1 contains maps of droughts before 1890 and Plates 1 and 2 contain rainfall maps of the droughts that occurred between 1891 and 2019. We will refer to these throughout the book.

Table i.1: Australian droughts investigated in each part of this book, with dates, main areas affected, and major themes. Dates separated by a forward slash indicate variation in the timing of the onset or termination of drought depending on source and/or location.

Part	Drought name	Dates	Areas affected	Themes
1	Pre-settlement	?mid–late 1700s	Hunter Valley, coastal and tableland NSW, King Island (?); timeframe unclear; historical evidence patchy	Myth-making, oral traditions, pre-European impacts on Aboriginal people, lake levels as record of drought
1	Settlement	1790–93	Sydney Basin; proxy records indicate more extensive	Settlement hardship, starvation, disease, convict life, Aboriginal conflict with European settlers, heatwaves and animal death
2	Bass's	1797–99	Sydney Basin, South Coast NSW; contemporary accounts sparse, includes explorer diaries	Settlement hardship, impacts on Aboriginal people
2	1800s	1802–03/05	Sydney region; intermittent dryness, at times fairly severe	Crop and livestock impacts
2	Blaxland's	1813–15	Sydney Basin, central west NSW; well documented from Sydney to Bathurst	Nature of Australian landscapes, overstocking and livestock population growth, opening of frontier
3	Sturt's	1824/26–29	Central and northern NSW, Vic., south-east Qld; may have lasted longer in northern areas; widespread from proxy records	Drought and inland river systems in pre-European era, smallpox
3	Depression	1835/36–39/42	NSW, Vic., eastern SA, south-east Qld, Tas.; probably affected much of continent	Spread of pastoralism, economic depression, tyranny of distance, degradation of grazing country, wildlife decline, forest dieback, hydrological impacts
3	Mid-1800s	1846–50	NSW, including Darling River country, Vic., southern Qld, eastern deserts; varying in length and timing depending on location	Decline of inland rangelands, vegetation impact, Riverina ecosystems, livestock death, Aboriginal impacts, wildlife decline
3	1850s	1858–59	NSW, SA, Darling River country, probably south-western Qld	Rangeland decline, habitat destruction, probable forest dieback
3	Goyder's	1864–66/68	Flinders Ranges and northern SA pastoral districts, western NSW; widespread in semi-arid to arid zone	Destruction of native fauna, native shrublands and Aboriginal society, limits to western agriculture
4	1870s	1875–77	Southern Darling River to Cooper Creek, Qld, SA, northern NSW	Stock losses, rabbit spread

Table i.1: (Continued)

Part	Drought name	Dates	Areas affected	Themes
4	Great	1890/91–92	Extremely severe in north-west WA, especially north-west pastoral, NT, SA, part western Qld and NSW	Agriculture in semi-arid zone, exploration of central Australia
4	Federation	1895–1902/03	Continental, except Pilbara and Kimberley; possibly drought with highest overall impact, preceded by rabbit plague in south-eastern Australia	Agricultural and ecological collapse, hydrological impacts, landscape degradation, onset of 'dust bowl' era, Aboriginal impacts, heatwave
4	World War I	1911/12–15	Continental, except parts of WA and NT	Aboriginal displacement, Murray River regulation and water extraction, agricultural impact, 'dust bowl' in south-east Australia
4	Central I and II	1922–29 and 1931–38	Back-to-back droughts in central Australia but periodically affecting most of continent	Aboriginal impacts, 'dust bowl' and degradation in central Australia, river levels
4	World War II	1937/41–46	Continental, especially eastern and western Australia	Wind erosion, dust storms, sand drift, emergence of political and scientific response, heatwave
4	Northern Australian	1951–54	Most severe in northern Australia	Drought in tropical areas, Aboriginal impacts
4	Central III and Eastern	1958–65 and 1964/65–67	Central Australia, western Qld and NSW, northern Australia, eastern WA and eastern Australia	Landscape degradation in semi-arid and arid zones, pastoral science and development
5	Western Australian	1975–79/80	South-western WA, Wheatbelt, Gascoyne, Murchison, Goldfields, deserts	Water scarcity, hydrological change, population growth
5	Late 1970s–early 1980s	1979–82/83	Most severe in eastern Australia, Tas., also WA	Degradation of Murray–Darling Basin
5	Early to mid-1990s	1991–95/96	Patchy, worst in northern NSW, southern and eastern Qld	Ecological impacts in tropics, emergence of societal concerns over mental health
5	Millennium	1997/2001–09	Eastern Australia including Tas. and far western Australia; early onset in Vic., record severity in south	Re-emergence of major drought impacts, ecological collapse in Murray–Darling Basin, climate change
5	Big Dry (including 'Black Summer')	2017–19	Continental, except western deserts from Nullarbor to Kimberley	Anthropocene, climate change, carbon emissions, water use, forest dieback, Darling River

Introduction: a land fertile for drought

As we travel through this fascinating history, we will first encounter legends of a great drought passed down from the dim obscurity of pre-history, a mythical event that requires detective work that takes us to places as far-flung as King Island. We then encounter the accounts of malnutrition and conflict of the first European settlers who, knowing nothing of the ferocity of the Australian climate, endured two severe droughts and a brutal heatwave within 10 years of arriving at Port Jackson. Fifteen years later another drought caused graziers who had outgrown the Sydney Basin to break open the frontier in the search of greener pastures, and the era of expansion had begun.

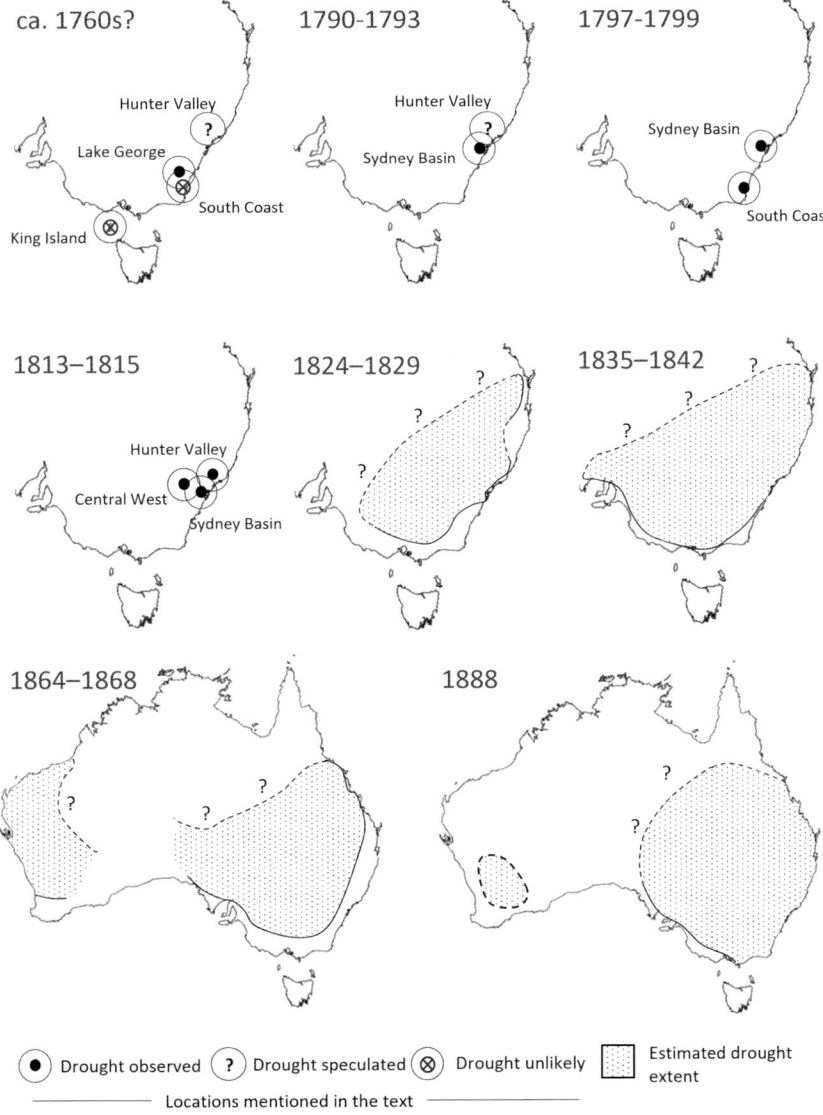

Fig. i.1: Eight major droughts in Australia, ca. 1760s to 1888. Major locations mentioned in the text, and the estimated extent of each drought (solid line = probable based on contemporary observations; dashed line = unknown, limited data), are shown.

xvii

We then join paths with Sturt, Mitchell, Cunningham and other early explorers who saw the drought-stricken Murray–Darling Basin and the Aboriginal people that lived there before it had fallen under the hoof, followed by the accounts of those reduced to poverty by the drought of 1836–42, which tell us much about the economic hazards of life in the early colony. Soon after we witness the spread of squatters across the inland and the destruction of soil, vegetation, and native animals, first in the pastoral country of western New South Wales, and then on a massive scale during the 1864–66 drought of South Australia, the first to cause governments to seriously confront the absolute limits imposed on western agriculture by the climate of the great southern land.

But worse was to come, for in the 1890s the Australia climate entered an epoch of persistently low rainfall that was to last, in some places, for 80 years, a desperate time when great droughts loomed over Federation, two world wars, and the final stages of European settlement of central Australia. Here we encounter a 'living drought', the formation of a 'dust bowl', a 'crucifixion', and hardship on a scale scarcely believable today, and learn how precarious our hold on the land can be when faced with climate extremes. We then finally turn attention to the new millennium, the return of continental-scale drought to a people who had forgotten these lessons, and the culmination of the 2017–19 drought in a 'Black Summer' that left our forests and thoughts as black as coal.

If you haven't already guessed, this is not a cheerful journey, but, if nothing else, it is an honest one. Let us begin.

Endnotes

[1] The terms 'CE' and 'BCE' stand for 'common era' and 'before common era' and are the secular equivalent of the terms 'AD' and 'BC'. I use these interchangeably.

[2] Evans NP *et al.* (2018) *Science* **361**, 498–501; Marshall M (2022) *Nature* **601**, 498–501; Lawler A (2008) *Science* **320**, 1281–1283; Benson *et al.* (2007) *Quaternary Science Reviews* **26**, 336–350; Kaniewski D, Guiot J, Van Campo E (2015) *WIREs Climate Change* **6**, 369–382.

[3] Kelley CP *et al.* (2015) *PNAS* **112**, 3241–3246; for an alternative view see Selby J *et al.* (2017) *Political Geography* **60**, 232–244.

[4] Medina-Elizalde M, Rohling E (2012) *Science* **335**, 956–959; Manning SW *et al.* (2023) *Nature* **614**, 719–724.

[5] Aeacus, in Greek mythology, was the king of the island of Aegina who prayed to Zeus to end a prolonged drought. The story of Joseph and the 7-year famine is told in Genesis chapters 41–47, and that of Elijah in 1 Kings chapters 17–18.

[6] Arrian of Nicomedia. *The anabasis of Alexander*, translated by EJ Chinnock. Hodder and Stoughton, London; book 6, chapter 26. There are a great many variants of this story, but the message is clear.

[7] Mbuya Nehanda, or Nehanda Charwe Nyakasikana, was one of the leaders of the Chimurenga, a revolt against the colonisation of Tanzania in the 1880s–1890s.

[8] The terms 'settlement' and 'settlers' are used throughout this book to identify settlers who came to Australia from overseas. It should be noted that some settlers came from countries outside of Europe, including from Asia and the United States, but 'European' is specified where possible.

[9] As defined by the *Oxford English dictionary*.

[10] Askarimarnani SS, Kiem AS, Twomey CR (2020) *International Journal of Climatology* **41** (Suppl. 1), E912–E934.

[11] Available from https://www.data.qld.gov.au/dataset/silo-climate-database.

[12] AghaKouchak A *et al.* (2021) *Reviews of Geophysics* **59**, e2019RG000683.

Part 1

Big feller dry

1
Legends

there have been droughts, and severe ones too in the dim distance of Time. The blacks in Blackall used to tell of big feller 'dry'
— The Western Champion and General Advertiser for the Central-Western Districts *(Barcaldine), 10 July 1915*

IN 1902 AUSTRALIA WAS BEING RAVAGED BY WHAT HAD BECOME THE MOST protracted and severe period of dryness experienced by Europeans since the time of their first settlement. As the great drought that had presided over the nation's federation dragged on into its eighth and most brutal year, a series of stories began to appear in newspapers that described a tremendous drought that occurred in the mid-1700s, decades before the arrival of the first European settlers aboard the First Fleet. With evocative titles like 'The demon drought' and 'Droughts of the past' (Fig. 1.1), these articles were written by people who were looking to historical experience to make sense of the devastation.[1] The drought broke up in 1903–04, but it ushered in a half-century of much lower rainfall across eastern Australia, and similar stories periodically reappeared during the run of severe droughts that occurred from the 1910s

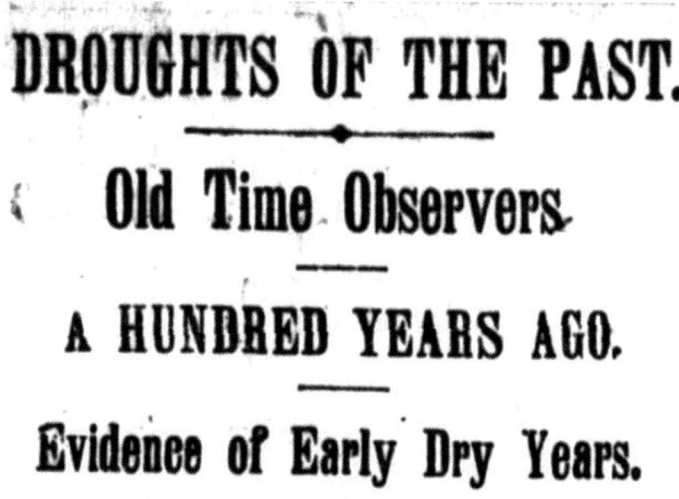

Fig. 1.1: Newspaper articles describing great historical droughts have often appeared during periods of severe drought. This one was published in *The Australian Star* (Sydney) on 14 June 1902 at the height of the Federation Drought.

to 1930s. These culminated in an article published in the *Mudgee Guardian and North-Western Representative* on 5 December 1940, entitled 'Australia's greatest drought. About the year 1760. All vegetation destroyed'.[2] The basic elements of this story later appeared in JC Foley's authoritative 1957 work on Australian droughts.[3]

This series of articles is particularly interesting for two reasons. First, from a cultural perspective, they vividly demonstrate the process of climate myth-building in action, and how, in the absence of a clear scientific or ethnographic record, subsequent interpretations of historical droughts begin to suffer from cumulative error. What develops is often a murky amalgamation of factual historical account, oral tradition, embellishment and speculation, the pieces of which can be formidably difficult to unravel, particularly when unrelated events or cultural traditions become embedded in the narrative. But such histories played an important role in helping drought-afflicted people gain a deeper understanding of their place within the Australian climate, providing not only a source of solace but also an avenue for planning for similar events in the future.

Second, and more importantly, they contain some of the only textual accounts of a great drought that must have been more severe than anything observed in post–European times, one so brutal that entire forests were killed and Aboriginal people in coastal areas died of starvation. Although, as we will see, the source material is obscure, several are also sufficiently detailed that we may place them under scientific and historical scrutiny. The *Mudgee Guardian and North-Western Representative* article provides us with several interesting leads to pursue. It reads:

> *Well over a century ago an explorer sent out by Governor King to examine the coastal country south of Sydney reported over a great expanse of country the giant trees were all dead and surrounded by a new forest growth seemingly about ten years' old. So about the year 1760 a great drought must have terminated after killing the deepest rooted trees. The stories of the blacks also pointed to a time of phenomenal drought about the same epoch, when river beds were overgrown with scrub, and a few surviving wild animals dragged their shrunken bodies to the scant water-holes in an unnatural amity. The blacks themselves perished in great numbers.*

No references were provided, but the main source of information apparently appeared decades earlier in *The Gundagai Independent and Pastoral, Agricultural and Mining Advocate* in 1915.[4] Versions printed in the early 1900s[5] covered much of the same material, and can be traced directly to two seminal works on Australian climate written by Henry Chamberlain Russell, one of Australia's greatest scientists (Box 1.1). Appointed Government Astronomer in 1870, Russell proposed as early as 1871 that rainfall in the colony followed a 19-year cycle, although insufficient data existed at the time to rigorously test the validity of this contention.[6] He quickly set about remedying this, first publishing records of rainfall and drought for Sydney[7] and then by producing 'Meteorological periodicity', which was read before the Royal Society of Sydney on 11 October 1876.

Box 1.1: Henry Chamberlain Russell (1836–1907)

Much of what we know about early Australian climatic history we owe to the efforts of HC Russell. Born on March 17, 1836 in West Maitland in the Hunter Valley of NSW, Russell graduated from the University of Sydney in 1859 and soon began work at the Sydney Observatory. After becoming Government Astronomer in 1870 he embarked on a remarkable scientific career in which he pioneered research in astronomy, natural history, hydrology, meteorology and atmospheric circulation. In 1877 he published *Climate of New South Wales: Descriptive, historical, and tabular*, which together with his paper 'Meteorological periodicity' remain central works for those investigating the climatic history of Australia. His efforts to increase the number of active weather stations and observers greatly enhanced the availability of meteorological data available at the time, and his newspaper articles on contemporary climatic events, especially droughts, were eagerly read across the state. Like most scientists, however, he may have occasionally felt undervalued: after publishing one such article a subsequent letter to the editor published on 23 December 1875 in *The Sydney Morning Herald* read: 'We are indebted to Mr. H. C. Russell for the very interesting tables showing periods of drought in the colony of New South Wales from the year 1840 to the present day. Our gratitude would be a thousandfold to him who could point out a remedy for the great evil we are now suffering under.'

The Bulletin (Sydney), 9 October 1880

In this expansive paper, later incorporated into his 1877 book *Climate of New South Wales: Descriptive, historical, and tabular*, Russell discussed evidence for meteorological cycles using a painstakingly compiled list of Australian droughts and floods and other climatic and solar phenomena from the early colonial period to 1875.[8] Of particular interest are two key pieces of information that he provides as evidence that a great drought might have occurred on the east coast in the mid-1700s: explorer records of tree death along the south-eastern coast from 1802, and Aboriginal oral history from the central Hunter Valley provided to European settlers in 1822. These accounts appear to form the basis of the newspaper article above, albeit in a barely recognisable form.

FOREST DEATH ON THE EAST COAST

When the blacks became able to communicate their ideas they informed the whites that 'big pheller dry' killed all the trees.
– Lachlander and Condobolin and Western Districts Recorder, *16 October 1907*

A central piece of evidence in the 1700s 'great drought' narrative is Russell's statement that a compelling account of mass mortality of trees along the south-eastern coastline was provided by a 'keen observer' aboard a voyage from Sydney to Port Phillip in 1802, directed by Governor Philip Gidley King. Certainly, if tree death had occurred on this scale, it would place the offending drought alongside the most extreme that have occurred anywhere in Australia since European settlement, and probably the worst to strike the region. The statement reads:

> *All the great gum trees were dead in every place I visited, and especially on Elephant Island [King Island], here I saw enormous dead trees, 5 to 6 feet in diameter, surrounded by a dense forest of young trees from 6 to 18 inches in diameter, these were only two or three feet apart, while of the old big trees there were only about 20 to the acre.*

From this he reasonable argued that the growth of the young trees implies that the 'great drought' occurred some 40–50 years prior, around 1750–60.[9] Russell did not appear to know the identity of the observer and may have been working from secondary documents, especially since he later implied that the observations were made on voyages conducted by George Bass, which does not seem to be the case.[10] I have not been able to find the primary source of the specific quote, but the information appears to be based on journal entries provided by the botanist James Flemming in *The voyage of His Majesty's colonial schooner 'Cumberland,' from Sydney to King Island and Port Phillip in 1802–3*.[11] This expedition, led by the Acting Surveyor-General of New South Wales, George Grimes, and under command of Lieutenant Charles Robbins, visited King Island in December 1802–January 1803, and matches the dates of King's governorship.

I can find no mention among Flemming's notes that he saw large numbers of dead trees along the coastline south of Sydney, but while on King Island, he observed evidence of widespread mortality among huge gum trees, and, in some places, their replacement by younger saplings. For example, on 14 December 1802 he noted the remains of some very large gum trees, all of which were rotten, in the vicinity of Sea Elephant River on the eastern side of the island. On 20 December he noted, 'in every place I have seen the large trees are decaying and fresh ones are springing up', and on 31 December 'There have been very large trees, most of which have fallen'. We can be quite confident that Flemming's accounts are accurate because they were corroborated both by Grimes and by Lieutenant Charles Robbins, who was in command of the *Cumberland* at the time,[12] and independently by the botanist Robert Brown, who visited King Island in April 1802 while aboard the *Investigator* with Mathew Flinders.[13] Charles Grimes's map of King's Island (Fig. 1.2) also show numerous places across the island where dead trees were seen.[14]

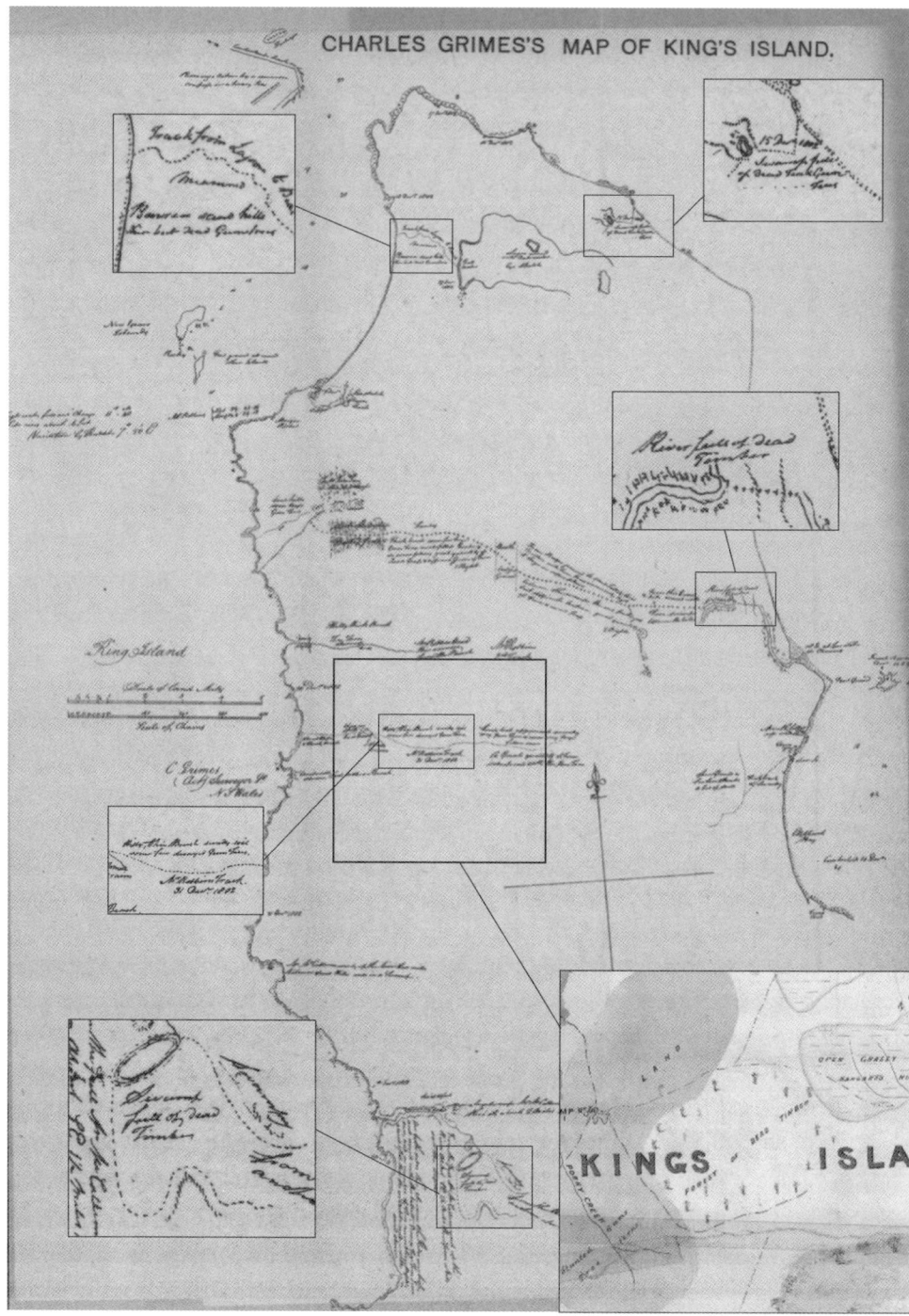

Fig. 1.2: Vegetation map of King Island drawn by Charles Grimes in 1802. Some sites around the island with dead trees are highlighted. National Library of Australia, nla.obj-232245306. The right lower inset is a detail from John Brown's August 1887 vegetation map, which shows a dead forest in the same location as noted by Grimes. Note the presence of dead forests in the south and centre of the island. Tasmanian Archive and Heritage Office, Libraries Tasmania, AF396/1/808.

The key passage that was subsequently used to reconstruct the timing of the 'great drought' that apparently killed the trees was a journal entry made by Flemming on Monday, 10 January 1803:

> *Started at four o'clock in the morning. Met with some fine land, the soil more strong and of a loamy nature. Went over some miles of it – some of a reddish, other of a grey cast. The trees are very high. I measured a gum tree that appeared lately fallen; 100 yards clear stem, and between five and six feet diameter. As near as I could judge there are about 20 per acre. The others small and straight. The poles are only two or three feet apart, very high and straight, from six to eighteen inches in diameter, of various sorts; high fern and sword-grass; a fine deep black soil. Passed a spot of moory ground, where we saw some badgers and kangaroos.*

The dimensions of the trees match those provided in Russell's account, and the species referred to is the Tasmanian blue gum, *Eucalyptus globulus* subsp. *globulus*, which is found across parts of eastern and southern King Island.[15] Here, this giant species can reach at least 40–70 m tall on fertile, well-drained soils, and before their widespread clearing may have grown even taller. *Eucalyptus viminalis* and *E. brookeriana*, which can also grow to 40 m tall, are also present on heavier, poorly drained soils. But is it true that Flemming's passages support the conclusion that some widespread mass mortality event had occurred before their expedition – a conjecture that later became central to the mid-1700s drought narrative?

The island is itself an unlikely place to suffer drought extreme enough to kill mature trees, since it has a cool maritime climate and a moderately high annual rainfall of 800–900 mm. Indeed, in recent years farmers have been moving to the Bass Strait islands from the mainland to escape the ravages of drought and rising temperatures associated with climate change. A drought of this severity would have lasted for years and likely also have affected the south-east mainland and possibly Tasmania. But contrary to claims in later accounts,[16] I am aware of no journal entries by either George Bass or Matthew Flinders on any of their multiple voyages south of Sydney between 1797 and 1801 in which they refer to large areas of drought-killed trees. This includes Bass's epic whaleboat voyage which took place between 3 December 1797 and 25 February 1798, in which he describes in considerable detail the impact of drought on local water sources and local Aboriginal people.[17] Major works published in the mid-1800s including Jevons's (1859) *Some data concerning the climate of Australia and New Zealand* fail to mention Flemming's observations, and it is interesting that Samuel Bennett also dismissed the evidence as 'being capable of explanation on other grounds' in *The history of Australian discovery and colonisation* (1867), 10 years before Russell published 'Meterological periodicity'.

Furthermore, Flemming did not specifically mention that all the huge trees in the location mentioned above were dead, and on two occasions noted that the tops, rather than the whole trees, had died back. Elsewhere he saw apparently live blue gums that were 1–3 feet (30–90 cm) in diameter and from 60–100 feet (18–30 m) tall. Nevertheless, the stands of trees observed by Flemming were clearly in different stages of development, with live trees evidently ranging from 6 inches to 3 feet (15–90 cm) in diameter. Tasmanian blue gum is an extremely fast-growing species that has been planted in many parts of the world for wood production, and saplings can reach reaching more than 20 m tall at 6 to 8 years of age and 50 m high and 1–2 m across at 50 years under silvicultural conditions.[18] Although growth in natural stands is likely to be slower, especially if crowded, the regrowth observed could have been significantly less than 50 years old, and Flemming himself did not venture an estimate as to their age. Thus, dating of the death of the trees is difficult; about all we can say is that the disturbance that pre-dated the larger regrowth likely occurred sometime in the two or three decades that preceded the arrival of the First Fleet.

A critical piece of evidence that supported Bennett's argument was added in 1887 by the surveyor John Brown and then shortly after by the Victorian Field Naturalists Club, who mounted a scientific expedition to the island that featured many of Australia's most eminent naturalists. A map of King Island produced by Brown in August 1887 for the Tasmanian government clearly shows large forests of dead standing timber both in the south of the island to the west of Mount Stanley, and near the centre of the island (Fig. 1.2), where they were also recorded in 1802. The cause of this tree death was discussed in some detail in a newspaper article provided to *The Australasian* (Melbourne, 9 June 1888, p. 51) by a member of the Naturalist's Club:

> *Mention must also be made of the remarkable large tracts here and there of silent dead eucalypt forests, as evinced by the regiments of huge, blanched, gaunt tree-barrels left standing. They convey a weird and melancholy aspect to the scene. It has been reported that bush-fires have been the destroying agency, but some people are of opinion that insect life, or, more probably, local movement of earth and consequent root decay, are at the bottom of the evil.*

There is no mention here of drought. It is possible that the trees had been burnt previously: King Island forests dominated by *E. globulus* and *E. brookeriana* are highly susceptible to fire, and many stands of *E. brookeriana* there today are of a similar age, having sprung up after major wildfires occurred there in the 1930s.[19] Matthew Flinders made no note of past fires while visiting King Island in April 1802,[20] despite personally witnessing the aftermath of a 'general conflagration' on plant life on Kangaroo Island earlier in the same year, but even lower intensity fires can weaken older trees to the point where disease and other environmental stresses set in, ultimately resulting in death. From this perspective it is also interesting that Robbins in 1802 noted that the dead timber occurred mainly on the hills, where he said it was exposed to the salt air – which suggests that he also thought other factors might have been involved.

Ultimately, it is at least possible that the dead and dying trees observed by Flemming had established after fire or some other disaster years earlier and that older, dying or dead trees were either killed by these disturbances or were simply coming to the end of their natural life, worn down by disease, storms and old age. A close reading of Grimes's 1802 map shows that dead stands of timber in various stages of health and decay occurred in a broad range of habitats, including creeks, hills and swamps, and that other species including 'tea trees'[21] were affected. To me, this would be an unusual pattern of drought-induced dieback, for it is more typical for species to suffer high mortality in specific habitats and for this to occur over a matter of months to a couple of years. Conclusions beyond this must remain speculative, but the reports of a great

1700s drought certainly stretch the evidence provided in these historical accounts far beyond the breaking point. Having come up dry, we must turn to other sources, foremost among which is the Aboriginal tale of a great catastrophe that took place high in the mountains above the famous Hunter Valley, long before the arrival of the first Europeans.

Endnotes

[1] *The Australian Star* (Sydney), 14 June 1902, p. 9; *Richmond River Herald and Northern Districts Advertiser* (NSW), 22 August 1902, p. 5.

[2] *Mudgee Guardian and North-Western Representative*, 5 December 1940, p. 9.

[3] Foley JC (1957) *Droughts in Australia: Review of records from earliest years of settlement to 1955.* Commonwealth of Australia Bureau of Meteorology Bulletin No. 43, reproduced by Department of Defence Production, Central Drawing Office, Maribyrnong, WA, p. 53.

[4] *The Gundagai Independent and Pastoral, Agricultural and Mining Advocate*, 20 April 1915, p. 4; cf. *The Age* (Melbourne), 13 March 1923, p. 8.

[5] *The Daily Telegraph* (Sydney), 21 October 1907, p. 3.

[6] *Empire* (Sydney), 11 May 1871, p. 2.

[7] Russell HC, *The Sydney Morning Herald*, 24 December 1875, p. 6.

[8] Russell HC (1877) *Climate of New South Wales: Descriptive, historical, and tabular.* C. Potter, Sydney, NSW.

[9] Russell (1877), p. 181.

[10] This probably refers to the expedition described in Mr Bass's journal in the whaleboat, between the 3rd of December, 1797, and the 25th of February, 1798. *Historical Records of New South Wales* (henceforth *HRNSW*), vol. 3, p. 312.

[11] Full account in Flemming's *The voyage of His Majesty's colonial schooner 'Cumberland,' from Sydney to King Island and Port Phillip in 1802–3. A journal of the exploration of Charles Grimes, Acting Surveyor-General of New South Wales.* In: Shillinglaw JJ (1879) *Historical records of Port Phillip: first annals of the Colony of Victoria.* John Ferres, Melbourne.

[12] Tree death was noted in Letter from Acting Lieutenant Robbins to Governor King, 18 January 1803, *HRNSW*, vol. 5, pp. 5–6, and Letter from Charles Grimes to Governor King, 18 January 1803, *HRNSW*, vol. 5, pp. 6–8.

[13] Willis JH, Skewes CI (1955) Robert Brown's Bass Strait Journal of April/May 1802 (a transcription). *Muelleria* **1**, 46–50.

[14] Charles Grime's Map of King's Island. *HRNSW*, vol. 5.

[15] No other tree species on King Island fits the description provided by Flemming.

[16] *The Daily Telegraph* (Sydney), 29 December 1888, p. 9.

[17] Mr Bass's journal in the whaleboat, between the 3rd of December, 1797, and the 25th of February, 1798. *HRNSW*, vol. 3, p. 312.

[18] Orme RK (n.d.) *Eucalyptus globulus* provenances, http://www.fao.org/3/l1807e/L1807E04.htm.

[19] Donaghey R (Ed.) (2003) *The fauna of King Island.* King Island Natural Resource Management Group Inc., Currie, King Island.

[20] Flinders M (1989) *A voyage to Terra Australis.* South Australian Government Printer, Netley, SA.

[21] Probably *Leptospermum* or *Melaleuca* species.

2
Calamity on the *Coquun*

Borillna yabba caloma, wakka goong, wakka gorroy, wakka chinee!wa, boorahm, coopee, bunggil, gurróman, boorabee, la la talgai talgai, la la gooyoom, doonggee-la dillol-lol!
The old men say a long time ago no water, no rain, no plain turkey, no emus, no 'possums, no grass: only dead trees, dead trees, fire, fire everywhere, and all the creeks full of dry stones!
— *Oral tradition of an elderly Wakka Wakka man, recorded and translated by Archibald Meston, 1890*[1]

WHEN THE FIRST EUROPEAN SETTLERS ARRIVED ON THE SHORES OF Australia, they were completely unaware that volatility in rainfall and temperatures are perhaps the defining climatic features of the new continent. Aboriginal people, of course, had been present for millennia, and possessed a wealth of cultural knowledge about climatic conditions that had been passed down across the generations. Tragically, the rapid cultural disorientation and demographic collapse of Aboriginal communities that closely followed early contact with Europeans and their diseases,[2] along with formidable language barriers – there were more than 290 Aboriginal languages spoken across Australia at the time of European settlement – meant that much of this knowledge was lost, particularly in the south-east of the continent. Luckily, a small number of early explorers and colonists came to realise that both coastal and inland Aboriginal cultures had strong oral histories of great past droughts and took the time to write these traditions down.

One of the first places this occurred was along the Hunter River, known to the Wonnarua[3] people as Coquun, which extends inland for over 200 km from modern-day Newcastle on the NSW coast. The Hunter Valley is of special biogeographical interest because the climate and vegetation of the central and upper catchment, where rainfall averages just 600–650 mm per year and temperatures can reach 45°C or more, resemble those typically found further inland. In evolutionary time the valley has formed a barrier to the movement of plants and animals between the wetter Barrington Tops to the north and the Wollemi region to the south, especially during the arid conditions experienced during the Last Glacial Maximum (around 18 000 years ago), and even today one can see semi-arid trees and shrubs growing here that typically occur further west.[4] Not surprisingly, it was here that Europeans first got a taste of the truly severe conditions typical of the Australian interior when they pushed into the region in the 1820s.

The first account that I am aware of dating to the early period of European settlement in the Hunter Valley was again provided by HC Russell in support of his theory that a great drought occurred in eastern Australia during the mid-1700s. According to Russell:

> *When Singleton [in the Hunter Valley] was first settled, in 1821, the aborigines told the settlers that long before, there was a fearful drought, in which all the lower part of the Hunter River dried up, and the only place that they could obtain water was at the head of the river, amongst the mountain springs; that here all the tribes – even those who bore each other the greatest enmity – collected, and for sake of dear life, lived peaceably for the time. Still the drought dragged on. All the great gum-trees died, and vast numbers of the blacks, who were buried by their friends in a great field. In proof of these statements, the graves and the dead trees still standing in 1822 were shown to the whites.*[5]

In a little-known second account, published in the newspaper media in response to the short but sharp 1888 drought, Russell elaborates further on parts of the story, emphasising that the Aboriginal deaths were due to 'fearful privations caused by the lack of rain', and that while no one could definitively state when the drought occurred, it 'probably took place 60–70 years earlier'.[6] He also noted that the graves were 'about the spring'. Unlike most early oral traditions, these accounts contain enough information for us to test their veracity on historical, ecological and ethnographic grounds, and are therefore some of the most important to emerge from the early colonial era.

Unfortunately, however, we again run into problems of historicity. Russell does not mention a source for the tradition and I have been unable to find any other reference to it either in early explorer or settler accounts from the Hunter Valley, including those written in the 1820s–30s by John Howe, Henry Dangar, Allan Cunningham, Thomas Mitchell and Felton Mathew, or in subsequent historical research that document the Aboriginal history of the region.[7] Perhaps it was passed to Russell directly from people living in the Hunter Valley, since he was born in West Maitland in 1836 and later completed his early years of schooling there.

The dates provided are consistent with the establishment of the town of Singleton in the middle Hunter Valley in 1821[8], although it seems that that the source of the Hunter River was not discovered by Europeans until the late 1820s or early 1830s,[9] rather than in 1822 as we might infer from the text. It may be that the settlers were shown graves along one of the tributaries of the Hunter Valley in the vicinity of Singleton, rather than near its source. I am not aware of any archaeological evidence of a mass burial site in the Barrington Tops or Mount Royal area that might be related to a drought in that region, although it is possible that knowledge of the location was lost during the mid-1800s as Aboriginal people became increasingly alienated from their country.

Nevertheless, geographical details of the account are consistent with the hydrology and geomorphology of the Hunter Valley region. Permanent springs do lie at the headwaters of the Hunter River above Ellerston on the western side of the Mount Royal Range and the Barrington Tops, a massive, isolated plateau that rises to over 1500 m above sea level (Plate 3). Parts of Barrington Tops receive 1000–2000 mm of rain each year, which is cool and wet enough to support forests of ancient Gondwanan species like the Antarctic beech (*Nothofagus moorei*), as well as snow gum (*Eucalyptus pauciflora*) and sphagnum bogs. Like a great sponge, these ecosystems absorb rainfall and snowmelt, which are then slowly released, providing a reliable flow of fresh water to the spectacular Barrington, Chichester, Allyn and William rivers and numerous other freshwater brooks that drain the ranges. The western margins of the Mount Royal Range are drier, and here the permanent springs on the Upper Hunter are still highly valued by local landholders today.[10]

Hydrological evidence from droughts experienced from the 1800s to present also confirm that the lower and middle Hunter River and its tributaries have always been prone to drying up during severe drought lasting for around 3 years or more. In early 1830, the Hunter River in the vicinity of Singleton (probably near Howe's farm) was described as 'a bed of sand and gravel with here and there a dirty pool of stagnant water'.[11] The same situation was observed later by John Henderson, apparently during the 1835–39 drought,[12] and in 1906 the river was so low near Muswellbrook that it could be crossed 'without even damping one's feet'.[13] Between 1936 and 1945 during the prolonged World War II drought, it was virtually dry on multiple occasions, a spectacle that the oldest residents had never seen, and severe enough to apparently kill most fish in the river (Fig. 2.1).[14] Modification of vegetation and watercourses throughout the region, especially completion of the Glenbawn Dam in 1957, means that great care must be taken when comparing the impact of recent and historic droughts on water levels, but during the severe 2017–19 drought the Hunter River above Belltrees was again dry, as were major tributaries including the Allyn, Goulburn, Isis, and Pages rivers and Carrow, Dart, Moonan, Rouchel and Wollombi brooks.[15]

Fig. 2.1: The Hunter River near Muswellbrook reduced to pools during the severe World War II drought, when water levels were among the lowest seen in the past two centuries. The bank erosion, siltation and land clearing visible have drastically altered the hydrology of the catchment. *The Newcastle Sun*, 26 January 1939.

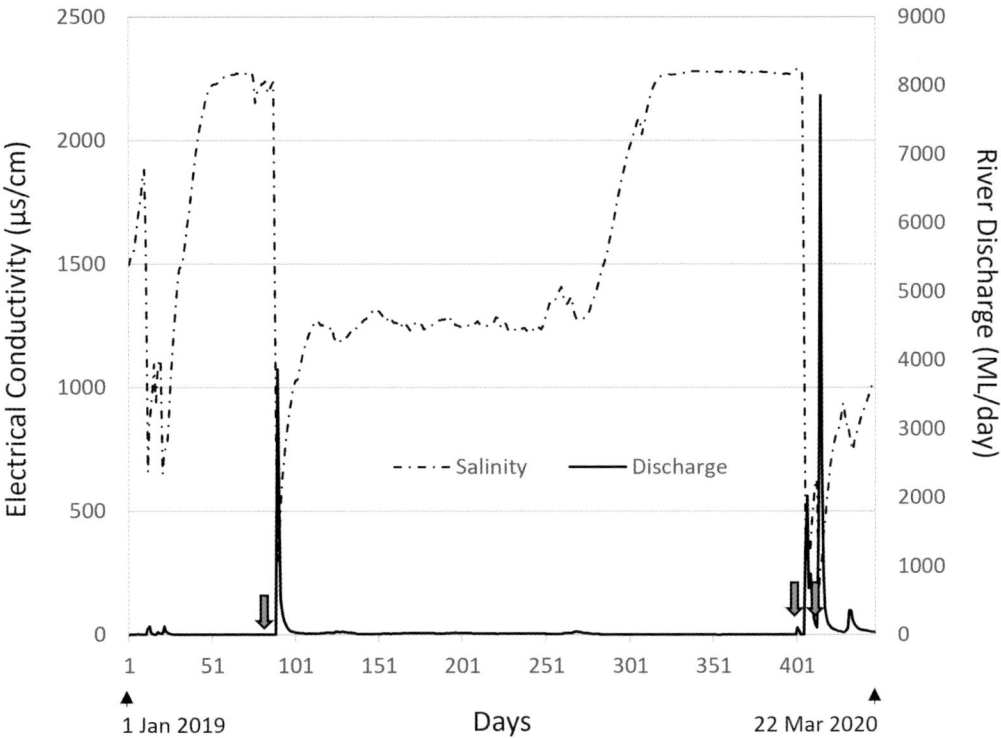

Fig. 2.2: Flow (discharge in megalitres per day; ML/day) and salinity (electrical conductivity in microsiemens per centimetre; μS/cm) of the Goulburn River at Coggan, NSW January 2019–March 2020. During periods of very low flow salinity rises to over 2000 μS/cm and then falls after major rain events (arrows) cause river discharge to increase. Drinking water becomes increasingly unpalatable above 2000 μS/cm.

Based on this it is safe to conclude that water scarcity has always been a problem for people living along the Hunter River during times of drought, a situation exacerbated by the fact that many of its tributaries flow through salt-bearing Permian sedimentary rocks, and so can suffer from high salinity. Land use changes and rising groundwater linked to land clearing have increased this problem, but brackish water in Wollombi Brook and other tributaries was noticed by Thomas Mitchell in 1831 during his expedition in search of the Kindur,[16] and so, as Russell suggests, it is not a new phenomenon. Even in larger rivers salinity can reach quite high levels: data taken from the Goulburn River at Coggan in the western Hunter catchment (Fig. 2.2) show that during the latter stages of the 2017–19 drought salinity reached levels considered marginal for drinking.

There is also historical evidence that Aboriginal people migrated to the mountain in times of water shortage in the lower and middle Hunter. This happened during a severe drought in the 1840s, when local groups living at the property Camyr Allyn near modern-day Gresford departed for the Barrington Tops once the Paterson and Allyn rivers had been reduced to pools.[17] The Aboriginal people involved is not stated, but the Guringay apparently occupied this area, alongside the Wonnarua, Kayawaykal and Gamilaraay of the Upper Hunter and Warrimay and Birrbay of the adjacent coast.[18] The groups that had to cease hostilities during the great drought referred to by Russell are also not known, but one or all could have been involved.

Dating the event, which Russell links to the mid-1700s drought, is also difficult, because to me it seems unlikely, albeit not impossible, that most trees killed 60 to 70 years earlier would still be standing when the European settlers arrived in the 1820s. One account from the lower Hunter suggests that Aboriginal people told settlers there, during the 1827–29 drought, that a large, 18-feet (5.5 m) deep lagoon near the junction of the Paterson and Hunter rivers had gone dry some 15 years earlier (i.e. ca. 1813–15), but that it then did not go dry between 1830 and 1915.[19] Another local Aboriginal tradition from the Upper Hunter indicates that a serious drought occurred around 1800, which 'dried up the brooks of the upper district' to an extent similar to the severe 1827–29 drought.[20] As we will see, both droughts severely affected the Sydney Basin and the adjacent coast. However, the possible dating of the drought closer to 1800 also brings us suspiciously close to two other momentous events: the Settlement Drought, which ravaged the early colony at Port Jackson in 1790–93 (Chapter 4) and the smallpox epidemic of 1789.[21]

The origin of the 1789 (and later 1829) smallpox epidemics is a vexed issue that remains hotly contested among modern scholars,[22] but the disease was almost certainly absent in south-eastern Australia before the arrival of the First Fleet. In 1840, 'oldish' Aboriginal men from the Dungog area who showed smallpox scarring reported that an outbreak when they were young had carried off 'great numbers' of the population, which never recovered. This probably referred to the 1789 epidemic.[23] In addition to direct mortality smallpox could also exacerbate the impact of drought by reducing the ability of people to hunt and gather food, possibly explaining the disaster attributed to Russell's drought in the 1700s. That water shortages could have been solely responsible seems unlikely because springs in the Upper Hunter are not known to have failed even during the severe droughts of the 1930s–40s or 2017–19. Permanent water also exists a short distance to the east in the upper reaches of the Barrington River, which seldom, if ever, stops flowing,[24] and probably also in the Upper Williams, Allyn and Chichester rivers.

Ultimately, if there is any truth to this tale, it would be of immense anthropological and climatic significance. As we will see, most if not all records of similar catastrophic events come from much drier regions, and a drought this severe would exceed any in that region that has occurred since European settlement. What we can say for certain is that great droughts have always had a profound effect on life in this ancient valley, and the devastation seen during the recent 2017–19 drought, within its scenes of dry rivers, dying trees and misery, are nothing new.

Endnotes

[1] Text and translation provided verbatim from Meston A (1890) *The Brisbane Courier*, 21 March, p. 6.

[2] A detailed discussion of the impacts of disease on Aboriginal populations is provided in Butlin NG (1983) *Our original aggression: Aboriginal populations of southeastern Australia, 1788–1850*. G. Allen & Unwin, Sydney, Boston.

[3] Spelling as used in the Wonnarua Language and Culture Archive, https://wonnaruanation.mukurtu-nsw.org.au/.

[4] E.g. *Acacia salicina*, *Geijera parviflora*, *Acacia melvillei-homalophylla* and *Acacia pendula*. The status of the latter in the Hunter Valley is contentious; see Bell SAJ, Driscoll C (2023) *Telopea* **26**, 37–47; Bell S, Driscoll C (2014) *Cunninghamia* **14**, 179–200.

5. Russell HC (1877) *Climate of New South Wales: Descriptive, historical, and tabular*. C. Potter, Sydney, NSW, p. 181.
6. *The Daily Telegraph* (Sydney), 29 December 1888, p. 9.
7. The account is also not mentioned in any source on the Aboriginal history of the Hunter Valley region that I am aware of.
8. Ben Singleton established his 200-acre (81 ha) property on the banks of the Hunter River in 1821; Simpson IM (1978) *Pioneers of a great valley*. Newcastle, NSW.
9. In 1827 no European had apparently yet been to the source of the Hunter; see *The Australian* (Sydney), 14 February 1827, p. 2.
10. Ashley-Brown P (2011) 'Source of the Hunter River'. *ABC Local*, 16 December.
11. Mathew F, Mathew SL (1829–34) Diaries of Felton and Sarah Mathew, 7 volumes. National Library of Australia, NLA MS 15, Libraries Australia ID 8446727.
12. Henderson J (1851) *Excursions and adventures in New South Wales; with pictures of squatting and of life in the bush: An account of the climate, productions, and natural history of the colony, and of the manners and customs of the natives, with advice to immigrants, &c.*. W. Shoberl, London, pp. 180–199.
13. *Mullumbimby Star*, 24 February 1906, p. 2.
14. *The Cessnock Eagle and South Maitland Recorder*, 13 September 1940, p. 7; *Newcastle Morning Herald and Miners' Advocate*, 3 Dec 1936, p. 11; *Warialda Standard and Northern Districts' Advertiser*, 29 April 1940, p. 4.
15. Historical river flow data are available from WaterNSW, https://www.waternsw.com.au/.
16. Mitchell TL (1996) *Three expeditions into the interior of eastern Australia: With descriptions of the recently explored region of Australia Felix, and of the present colony of New South Wales*. Eagle Press, Maryborough, Victoria. Reprint. Originally published: 2nd ed., rev., T. & W. Boone, London, 1839.
17. Archer AC (2019) *The magic valley: The Paterson Valley - then and now*. ACA Books, Lorn, NSW.
18. Spelling follows https://muurrbay.org.au/, https://yuwaalaraay.com/ and https://wonnaruanation.mukurtu-nsw.org.au/.
19. *Observer* (Adelaide), 24 April 1915, p. 12.
20. This drought probably occurred around 1800. See *The Sydney Gazette and New South Wales Advertiser*, 30 June 1829, p. 3.
21. Several authors suggest that the 1789 epidemic may have been chickenpox, but most historical and modern sources support the smallpox theory.
22. European and Indonesian sources have both been suggested; cf. Campbell J (2002) *Invisible invaders*. Melbourne University Press, Carlton South; Willis HA (2010) *Quadrant* **54**; Butlin NG (1985) *Historical Studies* **21**, 315–335; Mear C (2008) *Journal of the Royal Australian Historical Society* **94**, 1–22.
23. Fraser J (1892) *The Aborigines of New South Wales*. Charles Potter, Sydney, p. 62.
24. See hydrological data for the Barrington River at Bobs Crossing, WaterNSW site no. 208001.

3
Vanishing waters

Australia is a land of such paradoxes and illusions.
— The Sun *(Sydney), 10 November 1934*

OUR JOURNEY INTO THE PAST HAS THUS FAR LEFT US WITH MUCH THAT IS obscure or open to interpretation. Fortunately, another avenue remains open to us: the field of *paleohydrology*, literally the study of ancient water. Scientists like HC Russell were early advocates of investigation into the history of floods in Australian rivers and lakes as having the potential to 'throw much light upon the laws which control the changes in seasons that have such prominent effects upon a country like this',[1] but even before this many explorers were well aware that changes in lake and river water levels preserve a record of past climate and rainfall and that these can be seen in contemporary patterns of vegetation and geology. Incongruities between past and current climatic regimes were often particularly striking in ephemeral lakes, billabongs and riverbeds (Fig. 3.1), where

Fig. 3.1: The study of lakes and floodplains can tell us a great deal about historical changes in environmental conditions. These ancient stumps on the bottom of Hut Lake in Barmah Forest, Vic are usually covered with water but were exposed during the 2001–09 Millennium Drought. The modern treeline can be seen on the right. Photo: Keith Ward.

Fig. 3.2: Over the past 20 years it has become common to see river red gum saplings encroaching into rivers, creeks, and lakes across inland New South Wales, due to drought and water extraction. This photograph, taken during the Millennium Drought at Hattah Lakes, shows old river red gum on the left and younger saplings that have established on the lakebed on the right. Photo: Robert Godfree.

small shifts in water availability, especially when above or below critical thresholds, can lead to dramatic and non-linear changes in ecological processes. Unsurprisingly, much evidence of severe, multi-year droughts in the pre-European era comes from these habitats.

In 1836, Thomas Mitchell observed *fully grown* dead trees 'a good way' within the former boundary of Regent's Lake (now Lake Cargelligo) in western New South Wales. Almost certainly river red gum (*Eucalyptus camaldulensis*), these had apparently been killed by too much water and showed 'to what long periods the extremes of drought and moisture have extended, and may again extend, in this singular country'.[2] The floods likely to be responsible were observed by John Oxley's party in July 1817, when Lake Cargelligo had been filled by drought-breaking rains.[3] Mitchell's observation of a forest of dead river red gum in a swamp on the Lachlan River near the junction with the Murrumbidgee, which led him to question whether frequent burning, climatic change or river deposition might be responsible, might also date to this era.[4] Whatever the exact timing, a very long period of lower water levels seems to have occurred in the decades before, allowing the trees to attain this size. Both Mitchell and Charles Sturt (in 1829) also record saplings growing in areas prone to inundation, but these probably date to Sturt's Drought (1826–30, Chapter 9) and maybe the 1813–15 drought (Chapter 8) respectively (Fig. 3.2).[5]

Similar observations followed the spread of European settlers across central Victoria and the drier Wimmera and Mallee country to north, consistent with Aboriginal oral histories telling of periodic droughts lasting for many years in these areas.[6] For example, an 1886 account from Lake Albacutya, at the terminus of the Wimmera River, states that:

> *Lake Albacutya was dry in 1851-2, and had been so for some years. It is believed that it must have been dry for very many years at some period prior to 1852, with the exception of some waterholes in it, because there are large gum trees to be seen in its bed which must have taken very many years to grow, and which would not have grown had the ground over the roots been covered with water.*[7]

Care must be taken, since mature river red gums, unlike submerged seedlings and saplings, are capable of withstanding prolonged flooding. Nevertheless, Lake Albacutya was full in 1830–34,[8] and probably in 1816–24, and so the proposed long period of low lake levels must have occurred earlier. If the trees were 50 years old[9] the evidence again points to a protracted drought in the late 1700s. This would be consistent with the modern hydrological regime of the lake, which has been continuously dry for one to two decades on at least five occasions since 1890, although note that upstream water regulation and extraction has reduced the frequency and depth of flooding in recent decades.[10]

BAD WATER

Lake George, known to Aboriginal people who lived around it[11] as Weereewa, is a mysterious place. When I first saw it in 1985, on the way to the Snowy Mountains for a ski trip, I did not even realise it *was* a lake for at the time it was a great, flat waterless plain fringed by hazy forested hills. Indeed, the complete lack of trees suggested to me that it had been cleared, like much of the southern highlands and tablelands in eastern New South Wales. But Lake George is different. It is what is known by hydrologists as a closed or 'endorheic' basin, a low-lying part of the landscape into which water flows from the surrounding terrain, but then does not flow out again as it does from typical lakes. Australia has many endorheic basins, especially in central Australia, perhaps the most famous of which is Lake Eyre. These lakes are particularly interesting to paleoclimatologists because their water levels are almost solely determined by the delicate balance between evaporation and rainfall over their catchments and they can therefore undergo tremendous swings in volume and depth if this changes.

Lake George is no exception: one year it can be Australia's largest freshwater lake, 23 km long and 10 km wide, a white-capped expanse that sends mist and spray downwind during strong westerly winds, its choppy and treacherous waves lifting 'their crested heads in wild fury'.[12] Yet just a few years later it can be a desiccated barren plain sending up billowing clouds of dust, a formidable barrier to toiling teamsters and drovers during the summer.[13] Local Aboriginal people, who have lived continuously in the region for at least 9000–10 000 years and probably much longer,[14] knew well the volatility of the lake, and, interestingly, the word 'Weereewa' means 'bad water', and has also been linked to 'danger' in the neighbouring Wiradjuri language.[15] In a practical sense, however, the lake presents a more immediate problem: when full its water is fresh, but as it dries it becomes increasingly salty.[16]

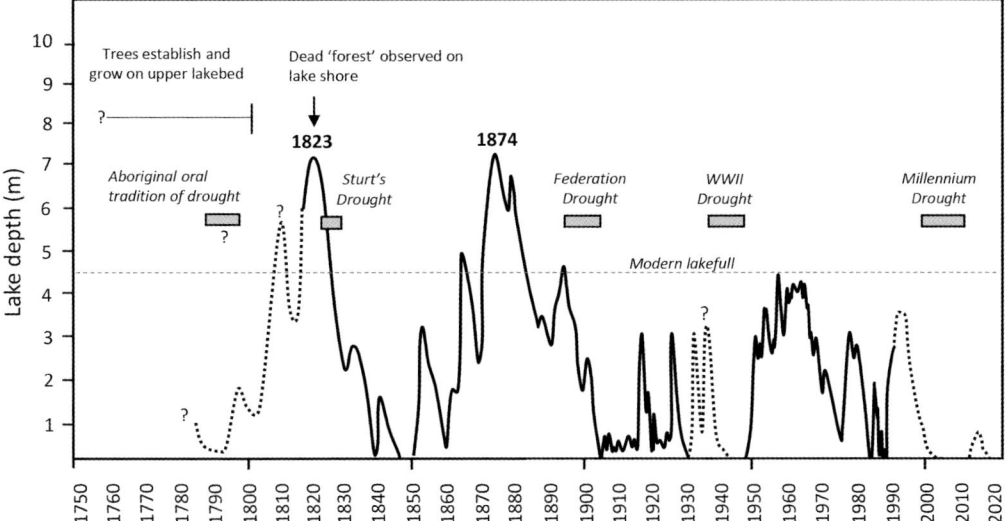

Fig. 3.3: Hydrograph for Lake George, ca. 1788–2020. Depths for 1818–86 redrawn from Russell (1886) and later publications (solid black line). Depths for 1930s and 1940s speculative and based on newspaper reports; data after 1990 estimated from satellite imagery. Lake levels before 1818 are highly speculative and based on Aboriginal oral traditions, rainfall patterns during the time, and observations of tree death at Lake Bathurst. Prior to the major fill event in the early 1800s trees established on the upper lakebed, probably above the typical lakefill level of ca. 4.5 m.

Of specific interest here are the multiple accounts that deep flooding killed extensive stands of large trees around the lake in the early 1820s. If true, a long period of lower lake levels must have preceded the flood; this would also add support to the idea that a long dry period afflicted south-eastern Australia during the 1700s. The first of these accounts was provided by John Thomas Bigge, who visited Weereewa in 1820 with Lachlan Macquarie when compiling information for an inquiry into the state of the settlements in New South Wales. He noted that: 'On approaching the north-east shore of lake George ... dead trees were observed in it to a considerable distance from its present shores'.[17] At that time the lake had reached, or exceeded, it's typical maximum depth of around 4.6 m (15 feet),[18] but following yet more wet years, the lake filled further, reaching a maximum depth of ca. 24 feet (7.3 m) in mid-1823 (Fig. 3.3).

These observations were later confirmed by Thomas Mitchell, who noted in 1828 that the lake was slightly brackish and 'surrounded by dead trees of the eucalyptus measuring about two feet in diameter, which also extended into it until wholly covered by the water'. He also noted that 'an old native female said she remembered when the whole was a forest'. Just as today, however, the lake proved to be ephemeral: by October 1836 it had dried out to become 'a grassy meadow not unlike the plains of Bredalbane'.[19] Interestingly, dead trees were also reported by Deputy Surveyor-General James Meehan on the north-west side of Lake Bathurst, a smaller closed lake that lies 25 km to the east, in 1818,[20] and in the beginning of 1824 this dieback had apparently progressed significantly.[21] Evidence from both lakes therefore points to significant tree death occurring in 1818–23 (Fig. 3.4).

Today, the vast expanse of Lake George, which is dry for years on end, seems an unlikely place to find a forest of trees, and none can be seen on the main lakebed today. Early Europeans saw no sign of dead roots or the trunks of fallen trees, and deep excavations made subsequently across the lake for dams have unearthed no submerged roots or other evidence of dead trees.[22] How then can we account for the early reports of dead forests of trees on the lake by Mitchell and others? The solution lies in the unusual topography and flood history of the Lake George basin itself. For such a vast lake, Weereewa is unusually shallow, and at its northern and southern margins its shores are so gentle as to be barely discerned by eye. In these areas slight differences in water level in the order of a metre or two can result in very large changes in the amount of submerged land along the lake edge. Indeed, at a depth of around 7 m, as it was when visited by Bigge and Macquarie, the lake extended hundreds of metres further inland than it does during normal 3–4 m deep filling events. The dead stands of trees reported by Bigge and Mitchell in 1818–23 could therefore easily have been large enough to be described as 'a forest', depending on the depth of prior floods and the time elapsed since they covered the same terrain.

Once again, we are extremely fortunate that HC Russell turned his attention to this problem in his paper 'Notes upon floods in Lake George', which was read before the Royal Society of New South Wales in December 1886.[23] In it he documents changes in the depth of Lake George between 1828 and 1886 and provides a map that shows the location of dead trees on the lake shore (Fig. 3.4 top) which were apparently mapped in 1832 by Deputy Surveyor-General Robert Hoddle (Fig. 3.5). This map shows areas of dead trees below the 1823 flood depth of 7.3 m (680.3 m asl) but above the level at which it was in 1835, which Russell states was around 2.4 m (675.4 m asl). Most of the trees killed by the 1823 flood therefore grew in stands ca. 200–500 m wide between 4 and 7 m above the floor of the lake. The main species that live in this habitat are *Eucalyptus viminalis* (manna gum) and *E. bridgesiana* (apple box) and, given the shape and size of the trees in Hoddle's watercolour, manna gum was the most likely tree species killed. Manna gum cannot tolerate long-term waterlogging and will die if its roots are submerged for more than a few weeks; this is known to have occurred in 1874 (Fig. 3.4 bottom).[24]

Tree growth is affected by crowding and many other factors, and so is an imperfect measure of age, but at the risk of moving into the realm of what ecologists call 'bucket science'– that is, ballpark figures – it might take *E. viminalis*, which is a very fast-growing tree, perhaps 40–70 years to reach 2 feet (0.6 m) in diameter, which is the size of the trees described by Mitchell. This idea is supported by the fact that trees planted on the lake shore following the fill events of the 1960s–70s are still alive today and now more than 1 foot (0.3 m) across. Based on these data, Lake George must have experienced a dry epoch of at least a half-century in duration before 1818 (i.e. beginning in the mid- to late 1700s) in which the water level never exceeded 4–5 m deep. It might have been longer, because new stands usually develop slowly by gradually spreading outwards from existing forests, although they may have formed from seed washed down creeks and deposited on the shore, perhaps over different lakefill events.[25] The only other direct local evidence of this protracted dry period comes from nearby Lake Bathurst, where the roots of dead 'honeysuckle' trees were observed on the lakebed in 1838. Large specimens of this *Banksia* species would also take decades to grow.[26]

Fig. 3.4: Two early views of Lake George, NSW. Top: Hoddle, R. (1830). *South end of Lake George, New South Wales* [watercolour on card mount]. In this image the steep western escarpment to the left and the flatter Lake George basin in the centre and right are clearly visible, and the tall dead trees in the far right appear to be in roughly the same position as the ones later mapped on the lake shore below the 1823 flood level. State Library of New South Wales, FL3270989. Bottom: *Lake George* [wood engraving]. (ca. 1886). From Garran A (1886) *Picturesque atlas of Australasia*, vol. 1, p. 136. Picturesque Atlas Publishing Co., Sydney. National Library of Australia, nla.obj-1605485403. The dead trees on the lake shore in the foreground were possibly killed by the 1874 flood event.

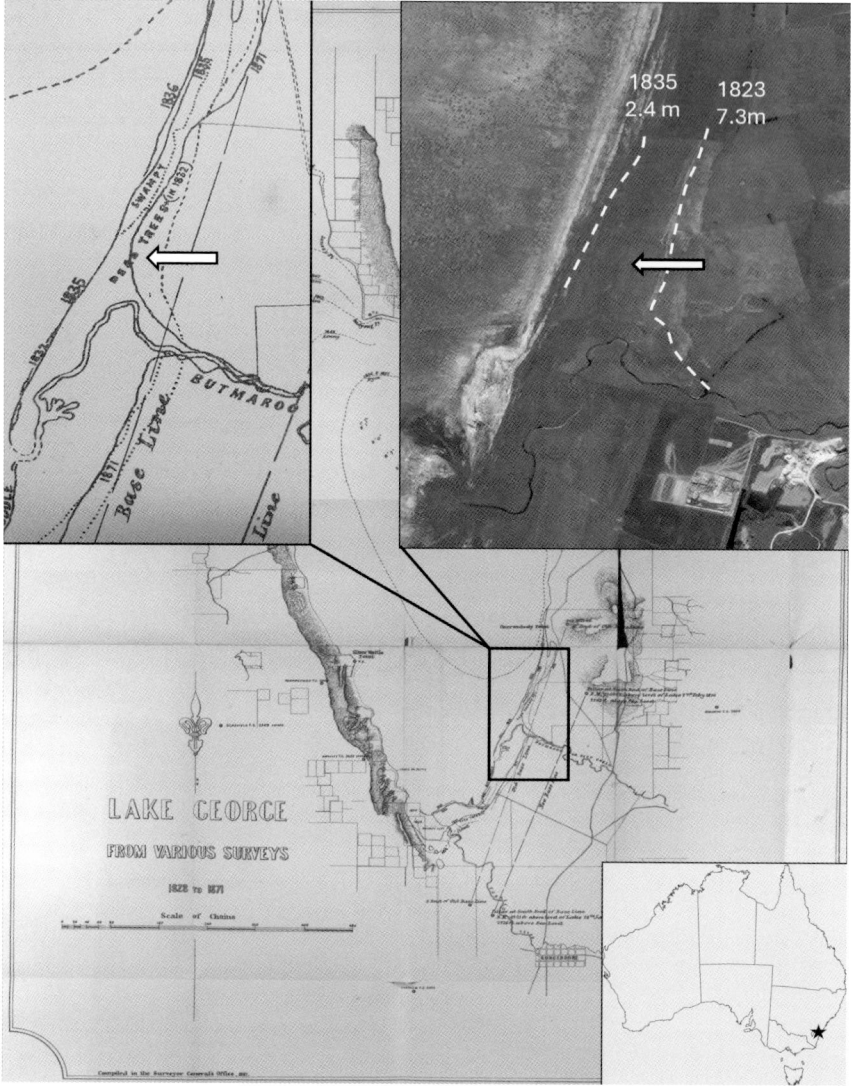

Fig. 3.5: Detail of the southern shore of Lake George based on survey map reported in Russell (1886). Top left: detail of shoreline showing lake depths and location of dead trees (arrow) killed by flooding in 1816–23. Top right: modern satellite image (Google Earth, 4 August 2019) of lake shore with approximate depth of water in 1835 and 1823 and location of dead trees. Multiple stranded shorelines formed during periods of stable lake levels are visible running parallel to the edge of the lakebed.

How would this period of low lake levels in the 1700s compare with those of more recent centuries? Data from ancient shoreline sediments show that as recently as 300–700 years ago (roughly 1300–1700 CE) Lake George reached a staggering depth of 15–18 m, around 10 m deeper than at any time since.[27] The floods that occurred during the 1800s were therefore unexceptional. Since 1874, when the lake filled to a depth of 7.3 m, it has never been more than 4.6 m deep (Fig. 3.3) – ample time for large stands of trees like those seen in the 1820s to have regrown above this level, had the shores of the lake not been cleared and grazed. Thus,

I think it is safe to conclude that a dry period in which Lake George was shallow or dry for many decades did occur during the 1700s, but that this event was probably within the range of hydrological conditions that have been observed over the past 150 years.

Endnotes

[1] Russell HC (1886) *Journal and Proceedings of the Royal Society of New South Wales* **20**, 241–260; *Queanbeyan Age*, 2 November 1887, p. 2.

[2] Mitchell TL (1839) *Three expeditions into the interior of eastern Australia ; with descriptions of the recently explored region of Australia Felix, and the present colony of New South Wales*, vol. 2, second edition. T. and W. Boone, London, p. 34.

[3] Oxley J (1820) *Journals of two expeditions into the interior of New South Wales, undertaken by order of the British government in the years 1817-18*. John Murray, London.

[4] Mitchell (1839), vol. 2, pp. 71–72.

[5] Reported in Sturt C (1833) *Two expeditions into the interior of southern Australia, during the years 1828, 1829, 1830, and 1831: With observations on the soil, climate and general resources of the colony of New South Wales*. Smith, Elder and Co., London, vol. 1, p. 56; Mitchell (1839), vol. 2, p. 118.

[6] *Border Watch* (Mount Gambier), 6 December 1876, p. 4.

[7] *The Australasian* (Melbourne), 14 August 1886, p. 1.

[8] Bren L, Sandell P (2004) *Australian Geographic Studies* **42**, 307–324.

[9] *The Argus* (Melbourne), 26 June 1880, p. 9.

[10] Bren L, Sandell P (2004), p. 318.

[11] The identity of the Aboriginal groups that lived around Lake George is contested; Ngun(n)awal/Gundungurra, Ngarigo and Walgalu speakers may all have been present.

[12] *Queanbeyan Age*, 26 October 1887, p. 2.

[13] *Wagga Wagga Express*, 18 October 1902, p. 4.

[14] C.f., Theden-Ringl F (2016) *Australian Archaeology* **82**, 25–42; Flood J et al. (1987) *Archaeology in Oceania* **22**, 9–26.

[15] Conroy R (2016) *Fusion Journal* **10**, 67–106.

[16] Jacobson G, Schuett AW (1979) *BMR Journal of Australian Geology and Geophysics* **4**, 25–32.

[17] Bigge JT (1823) 'Report of the Commissioner of Inquiry, on the state of agriculture and trade in the Colony of New South Wales'. Ordered, by The House of Commons, to be printed, 13 March 1823.

[18] The typical modern maximum flood depth of the lake is 4.6 m or 15.1 feet; Coventry RJ (1976) *Journal of the Geological Society of Australia* **23**, 249–273.

[19] Mitchell (1839) vol. 2, p. 317; approximately 30 km to the northeast of the north end of Lake George.

[20] Cited in Cambage RH (1921) *Journal of the Royal Australian Historical Society* **7**, p. 234.

[21] Russell HC (1877) *Climate of New South Wales: Descriptive, historical, and tabular*. C. Potter, Sydney, NSW. There may be confusion about the location of this account.

[22] A Mr John King spent time at Lake George in 1834–41 and states that even then trees, stumps and roots were entirely absent from the bed of the lake. Reported in Russell HC (1886) *Journal and Proceedings of the Royal Society of New South Wales* **20**, 241–260. Local people who have undertaken excavations on the lake bed confirm the lack of tree remains deeper in lake sediment.

[23] Subsequently published in Russell (1886).

[24] *Queanbeyan Age*, 16 November 1887, p. 2.

[25] Russell's 1886 map of Lake George shows patches of drowned trees at the termini of Butmaroo and Turallo Creeks on the south shore of the lake.

[26] *The Sydney Monitor*, 18 March 1838, p. 2. Probably *Banksia marginata*.

[27] Fitzsimmons KE, Barrows TT (2010) *The Holocene* **20**, 585–597.

Part 2
An ingratious country

————

4
Unsettling times

I do not think it probable that so dry a season often occurs.
— Governor Arthur Phillip, 4 March, 1791[1]

TODAY, THE STREETSCAPE OF DOWNTOWN SYDNEY IS A LAND OF PLENTY, THE food requirements of its people satisfied by a stream of supplies that pours in continuously from farms and market gardens across the country and from ocean-going transports bringing abundance from overseas. Even when surrounded by barren, drought-ravaged landscapes and inconvenienced by water restrictions and passing dust storms that obscure its famous landmarks, famine still seems far away. Yet entombed beneath the asphalt, steel and glass, the earth tells a different story, one of desperation, starvation and impoverishment, a time over two centuries ago where survival depended on the most meagre supply of water and withered crops. Only glimpses remain of this first European experience of a true Australian drought – a sign, a name or a monument – wished away, perhaps, by a populace more eager to celebrate abundance than scarcity.

This desire is not a modern phenomenon. Just 13 years after James Cook landed at Botany Bay on 29 April 1770, the American sailor James Matra (1746–1806), who accompanied Cook on his famous voyage aboard the *Endeavour*, wrote in glowing terms of the agricultural potential of New South Wales, a settlement 'happily adapted to produce every various and valuable production of Europe, and of the Indies'.[2] Given the perceived salubrious climate and fertile soil of the new land,[3] this idea found considerable support, including from Joseph Banks, who as early as 1779 had advanced the idea that Botany Bay would be a favourable location to establish a convict colony. As president of the Royal Society and undisputed authority on the plants, animals and geology of eastern Australia, Banks's views were highly influential, and by 1786 plans were in place to establish a fleet to begin the transportation of convicts to the new penal colony. Here it was thought that the land was most suitable for allowing the convicts, through industriousness, to obtain a means of subsistence.[4]

A close reading of Cook's journal, however, reveals that this was mostly wishful thinking. Upon landing at a small beach just inside the Kurnell Peninsula headland, on the southern side of Botany Bay, Cook and his men searched for fresh water without success. They later found a meagre supply on the north side, trickling down into pools among the rocks, but this proved difficult to access and so they resorted to digging holes in the sand at the original landing site to resupply the *Endeavour*.[5] This is not surprising, because the peninsula is essentially an extensive sand barrier complex, with large sand dunes connecting low sandstone plateaus to the east and west, completely lacking major waterways. Further up the Georges River the soil was more

> *The climate is a very fine one, and the country will, I make no doubt, when the woods are cleared away, be as healthy as any in the world*
>
> – Arthur Phillip to Lord Sydney, 9 July 1788

favourable, producing timber and 'as fine [a] Meadow as ever was seen',[6] but such areas were very patchy, and along the sea coast to the north of Botany Bay Cook found 'mostly a barren heath, diversified with Marshes and Morasses'.[7] Even Banks himself acknowledged that the proportion of rich soil there was 'small in comparison to the barren'.[8]

The arrival of the First Fleet under command of Arthur Phillip on 18 January 1788 revealed that, if anything, Cook and Banks had underestimated the poorness of the chosen site. A general lack of fresh water was immediately found to afflict Botany Bay,[9] and later the entire coastal strip north to Broken Bay, including Port Jackson.[10]

There was no sign of Cook's meadows, but instead estuarine wetlands dominated by saltmarshes, mangroves, mudflats and sea grass beds,[11] and areas of deep sandy soil with long and coarse grasses and rushes.[12] These provided habitat for a diverse array of migratory birds and numerous reptiles, frogs and mammal species,[13] which along with fish, underpinned the diet of the local Gweagal (= Gwiyagal) people. But the swampy and spongy terrain posed a significant threat to the health of the arriving European population[14] and, to add insult to injury, the sun was brutally intense, the large trees were hollow, and black and red ants 'of an enormous size' were plentiful.[15] As areas of suitable land were too small to support the planned colony (Fig. 4.1), the fleet was ultimately forced to move to Port Jackson on 26 January 1788, finally anchoring in what is now Sydney Cove.

In what became a pattern repeated countless times over the ensuing century, water immediately became central to the selection of land on which Europeans decided to settle. Sydney Cove was identified by Phillip[16] as the only bay within Port Jackson with a reliable source capable of supporting the settlement: a small stream that flowed northwards from marshy ground near the western side of modern-day Hyde Park, between Market and Park streets (Fig. 4.2). The stillness of the ancient wood that bordered the rivulet later known as the Tank Stream soon gave way to 'the busy hum of its new possessors',[17] and in the months that followed rainfall was more a curse than a scarcity. It teemed down for days on end, interrupting labour and exploratory surveys and filling the hospital tents with sick patients.[18] Oppressive heat was followed by violent storms of thunder and lightning which killed livestock and made living conditions extremely uncomfortable in the makeshift tents and huts that lined the cove.[19]

Nevertheless, temperatures in the summer of 1788 were cooler than normal,[20] and heavy rain, gales and cold weather continued throughout mid-1788, resulting in damage to key infrastructure, huts and roads, which were rendered impassable.[21] Hot temperatures, sometimes in excess of 40°C, were reported in January 1789[22] and during December 1789–January 1790,[23] but the winter of 1789 was particularly cold. Conditions remained wet, with torrents of rain

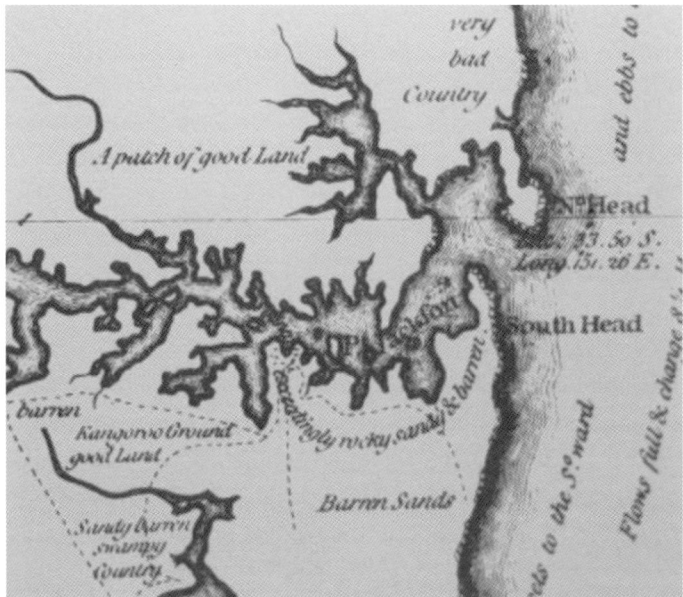

Fig. 4.1: Detail of map from the journals of Watkin Tench, showing the extent of poor agricultural land around Botany Bay and north towards Port Jackson.

falling in early to mid-1790, causing great hardship and despair among the 'miserable mud tenements' of the convicts.[24]

Thanks to the work of climatologists who have reconstructed the state of the El Niño Southern Oscillation (ENSO) back for five centuries, we now know that these first years of European settlement coincided with a prolonged La Niña event that extended from 1785 until 1790.[25] The relationship between ENSO and the Sydney climate is not always especially strong,[26] but the accounts above certainly indicate that abnormally cold and wet weather accompanied the peak of the La Niña during 1788. Whatever the cause, these conditions, combined with the interruption of plant and seafood collection by Aboriginal people, and even the taking of these by force,[27] loss of livestock to predation by dingos, unreliability of the fish supply,[28] the robbery of food stores by rats,[29] the wrecks of the supply ships *Sirius* and *Guardian*, and the arrival of the hundreds of malnourished convicts on the Second Fleet inevitably led to very serious food shortages.

The deciding factor, however, was the 'everlasting and unconquerable sterility'[30] of the soil, especially at the 'sandy desert'[31] of Sydney Cove, which resisted all efforts of the settlers to establish crops and reduce their dependence on imported goods. This problem was by no means lost on Phillip and others, who had watched crop seeds rot in the ground and wheat crops fail in this 'vile country' during the spring of 1788.[32] In response, Phillip made plans for a settlement on the more fertile clayey soils at Rose Hill (now Parramatta) and by March 1789 clearing and cultivation was underway.[33] Soils with an abundance of clay particles have a physical structure that allows them to hold more nutrients than sandy soils, and to retain a much higher volume

4 – Unsettling times

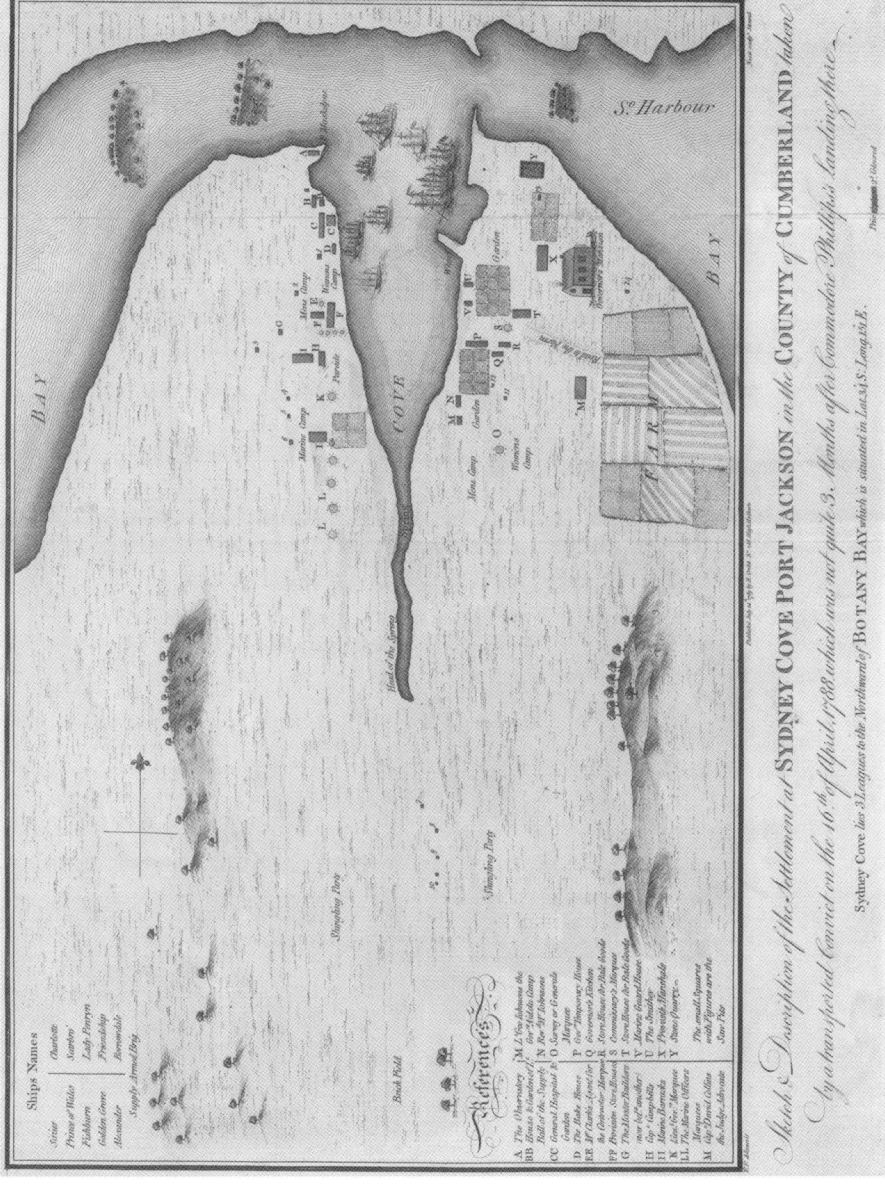

Fig. 4.2: Fowkes, F. (ca. 1788–89). *Sketch & description of the settlement at Sydney Cove Port Jackson in the County of Cumberland taken by a transported convict on the 16th of April, 1788, which was not quite 3 months after Commodore Phillip's landing there* (detail). The infertility of the soil quickly led to the establishment of farms on more productive land at Parramatta. Note the location of the spring on the Tank Stream. National Library of Australia, nla.obj-230578175.

Fig. 4.3: The *Neptune*, captained by the brutal Donald Traill, arrived in Sydney on 22 June 1790. The hundreds of sick and dying convicts aboard the ships of the Second 'Death' Fleet placed an immense burden on the colony at Sydney just as the Settlement Drought was about to begin. Wikimedia Commons.

of plant-available water during periods of low rainfall. The shift in agricultural effort to Rose Hill was soon rewarded with the growth of more vigorous crops, and in December 1789 a good harvest of corn was taken in.[34]

THE DROUGHT BEGINS

Some droughts creep up stealthily, marked only by a gradual diminishment of rainfall and slowly falling water levels in dams, creeks and rivers that takes place almost imperceptibly over many months. But the Settlement Drought did not begin in this way: all eyewitness accounts indicate that the steady, reliable rainfall of the past two-and-a-half years ended suddenly in June 1790, and within 6 months the colony was suffering under full-fledged drought conditions. The last rainfall of consequence seemed to occur in the first week of June 1790,[35] just as the ships of the notorious Second Fleet began to arrive in Port Jackson (Fig. 4.3). Little fell through July and August, by which time the extremely dry weather was beginning to affect crops.[36] Some rain occurred at the end of September, but this was inadequate to recharge the soil, which by October had become parched again following the onset of strong north-west winds.[37] This situation, in which only a very small fraction of rainfall is available for plant use after accounting for losses due to evaporation and other factors (known as *effective precipitation*), is related to the small and sporadic nature of rainfall events and the higher temperatures typically experienced during drought. The fact that a 4-month drought could have such a significant impact on agriculture might seem strange but are in fact typical of sandy soils.

It is difficult to know exactly how dry this period was, but we have several sources that shed light on this question. The most detailed of these are the meteorological records of the scientist Lieutenant William Dawes, which span the period 14 September 1788–6 December 1791.[39] This important dataset, which was only discovered in 1977, contains data on wind speed and direction, barometric pressure, temperature and precipitation collected at a small observatory that was located on the western side of Sydney Cove near a southern pylon of the Sydney Harbour Bridge. Unfortunately, Dawes included only brief notes on the timing and intensity of precipitation without providing rainfall amounts, but these data do support the conclusion that rainfall between July and October 1790 was unusually light. Dawes mentions only drizzle or brief showers during each of these months, apart from 21 August, where a moderate rain fell from 8 am to sunset, and a longer period of sparse showers, drizzle and light rain in October. July was particularly dry, with light precipitation noted on only three separate days. Watkin Tench (Fig. 4.4), who also kept some meteorological records, thought that all the showers of the last four months (ca. July–October) would not 'make 24 h rain',[40] broadly consistent with Dawes's observations.

Fig. 4.4: Portrait of Watkin Tench (detail) [photographic copy of painting]. (Post-1821; possibly by R.T. Penreath.) Tench's *A complete account of the settlement at Port Jackson* is a rich, compassionate, and often humorous account on life in the early settlement. His knowledge of agriculture and science provide us with key insights today into the severity of the Settlement Drought. Mitchell Library, State Library of NSW, FL14298195, and Courtesy of the owner of the original work.

Assuming that these falls were light or moderate, it seems very unlikely that more than 100 mm fell during this period, and it may well have been half of this or less. The long-term rainfall dataset recorded at Sydney Observatory Hill, which lies just east of Sydney Cove, helps place this in context: between 1858 and 2016 fewer than 4% of years received less than 100 mm

> *all the grain of every kind that we have been able to raise in two years and three months would not support us three weeks, which is a very strong instance of the ingratitude and extreme poverty of the soil and country at large*
>
> – *Chief Surgeon White to Mr Skill, 17 April 1790*[38]

of rain from July through October, and less than 50 mm has occurred just once, in 1907. These data support the view, held by Tench and Phillip, that mid- to late 1790 was extremely dry indeed. November was probably wetter, with periods of rain noted on 18, 20, 21, 24 and 29–30 November, but these did little to relieve the drought. The frustration of the hungry settlers at Sydney Cove can only be imagined as their small vegetable gardens and cereal crops were either dug up prematurely or failed altogether.[41]

At Parramatta the situation was little better: Tench, on a tour of the area in November 1790, observed that wheat sown in June and July (probably too late) were already turning yellow, the oats were in ear at just 6 inches (15 cm) high, the barley was little better and only patches of maize had reasonable prospects of returning a decent yield. On hearing of an expected yield of 400 bushels from 55 acres (22 ha) of drought-stricken wheat and barley, he drily wrote, 'Appearances hitherto hardly indicate so much.'[42] The first water shortages also began to affect the community, as smaller sources dried up across the harbour. The Parramatta River became so brackish that it was suggested that a dam should be constructed to prevent the tides moving upriver,[43] and the Tank Stream, now the lifeblood of the main settlement at Sydney Cove, was much reduced in volume but still flowing.[44] Circumstances deteriorated further during the summer, by which time the formation of a strong El Niño was well underway. Major heatwaves swept in from the dry interior in December 1790 and again in February 1791, the latter so tremendously hot that local wildlife was killed (Chapter 5).

The evaporation rate must also have been very high, because by March the Tank Stream had been reduced to little more than a trickle. Indeed, concerns over the potable water supply led to what might have been the first instance of environmental protection undertaken by Europeans in Australia in response to drought: the construction of a ditch on either side of the stream and palings on the bank to reduce the destruction of vegetation by livestock.[45] Recent investigation of buried plant fossils indicate that the vegetation that once stabilised this ground consisted of riparian vegetation dominated by swamp oak (*Casuarina glauca*) and rainbow fern (*Calochlaena dubia*), and drier vegetation further up the catchment that contained a rich variety of shrubs and trees such as hop bush (*Dodonaea triquetra*), black she-oak (*Allocasuarina littoralis*) and red bloodwood (*Corymbia gummifera*).[46] It seems that trampling and disturbance were already in the process of destroying the understorey, ultimately leaving silt to run into the unprotected stream.

STARVATION AND CONFLICT

By April 1791 the settlement's crops had largely failed, and with provisions running short, rations were cut to 3 pounds (1.4 kg) each of flour, weevil-infested rice and ill-flavoured rusty pork (or 4.5 pounds (2 kg) of unpalatable beef) for each man, woman and child older than 10 years.[47] Although the nutritive value of the food is uncertain, this one-quarter cut to rations almost certainly reduced the caloric intake to 1500 calories or less per day, which for people engaged in hard labour results in slow starvation and terrible hunger. Furthermore, rations for some convicts might have been significantly lower. Watkin Tench, whose compassion and humanity are unmistakeable in much of his writing, noted at the time that 'I every day see wretches pale with disease and wasted with famine, struggle against the horrors of their situation', a plight that inevitably led to the theft of food supplies and the plundering of gardens and crops and even instances of assault.[48]

Fishing tackle and other equipment was also stolen from local Aboriginal people, including Daringha (Dar-ìn-ga; Fig. 4.5), the wife of Gadigal (Cadigal) warrior Colbee (Cole-bee, Coleby or Colebee), a man who had been captured in November 1789 and held at Government House before escaping. Despite the severe flogging of the convict responsible in front of those affected, further such transgressions, including the destruction of a canoe owned by a Burramattagal man called Bal-loo-der-ry, led to a breakdown in relations with Europeans, and in particular the cessation of critical trade in fish for bread or salted meat.[49]

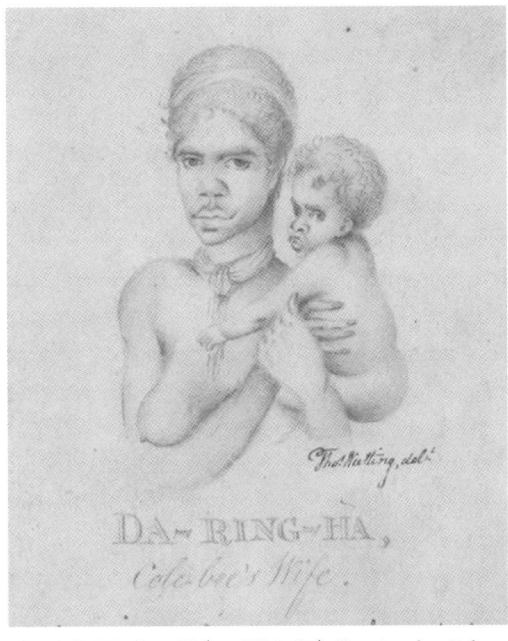

Fig. 4.5: Watling, T. (ca. 1792–97). *Da-ring-ha, Colebee's wife*. Pressed by hunger and drought, convicts resorted to stealing fishing supplies from local Aboriginal people including Dar-ing-ha. Evidently a compassionate person, she reportedly wept during the flogging of the convict responsible. Watling Drawing No. 31, courtesy The Trustees of the Natural History Museum, London.

After an exceptionally dry period during which the ground became so hard that it could barely be broken with a hoe, rain fell and extensive crops were again sown at Parramatta, including around 350 acres (142 ha) of maize and 44 acres (18 ha) of wheat.[50] Much of this seed sat in the ground for an extended period before germinating, but the ensuing growth looked promising. However, by October it was clear that the 'great drought[s]',[51] now 18 months long, had returned, and the wheat and barley again became stressed. By early December, Watkin Tench, while surveying the cultivated land in the Parramatta region, noted that the wheat, maize and garden vegetables looked very poor, and estimated the production of many wheat crops to be in the order of 8–9 bushels per acre. It was later reported that the crops overall averaged 15 bushels per acre,[52] and assuming 57.5 pounds of wheat per bushel,[53] this still equates to an optimistic estimate of yield of around 862.5 pounds per acre or 967 kg per hectare. This compares with the modern Australian rain-fed wheat yield which

The throes of hunger will ever prove too powerful for integrity to withstand
– *Watkin Tench, April 1791*

> *I cannot acquiesce with you in thinking that the ration served from the public stores is unwholesome; I see it daily at my own table*
>
> – Arthur Phillip to Major Gross, October 1791

averages 1.74 tonnes per hectare and 3 or more in favourable areas. Of course, modern values largely reflect decades of scientific progress in the development of drought-tolerant wheat varieties and improved cultivation techniques, but the comparison certainly demonstrates the extent to which the early colonists were constrained by low crop yields. Indeed, the 1791 wheat harvest still only provided little more than enough seed for the next season's planting.[54]

Hot weather resumed in December 1791, by which time flow in the Tank Stream had become so low that Governor Phillip was forced to forbid all ocean transports from resupplying their water at Sydney. The drought also clearly demonstrated the need for water storage during periods of low rainfall, a problem that still plagues Australia today, and stonemasons were employed to cut three tanks into the sandstone bedrock for this purpose. Reportedly 10 feet (3 m) in depth,[55] these were located near what is now the intersection of Pitt and Hunter Streets, around 600 m north of the stream's source. A further tank with a well 15 feet (4.5 m) deep in the centre and capable of holding 7996 gallons (more than 35 000 L) of water was constructed in May 1792 to ensure a clean supply.[56]

By February 1792, convict labourers, sick from the mistreatment and malnourishment that they had endured on the voyage of the *Queen* from Ireland, which arrived in September 1791, were daily dropping into graves dug in foreign soil. The terrible mortality continued through May, and starvation prompted theft of perhaps one-sixth of the season's corn crop.[57] Of the more than 120 male convicts who arrived on the *Queen*, only 50 were still alive at the beginning of May 1792[58] – not surprising given their skeletal state upon arrival at the colony, a harrowing account of which is provided by Mary Ann Parker, wife of Captain John Parker, commander of the *Gorgon*.[59] Even David Collins, not a man known to be overly emotional in his writing, noted that, as bad as it was to witness the actual deaths of the convicts, 'it was far more afflicting to observe the countenances and emaciated persons of many that remained soon to follow their miserable companions'(Fig. 4.6).[60] The arrival of these dead and dying people in a colony that had been plagued by serious food shortages for over a year, and which was unable to ease their suffering, marks one of the more tragic events in the history of Australian drought.

Just as circumstances were at their worst, however, the fortunes of the settlement began to improve. The corn harvest was better than that of the previous year, and the fresh meat provided by shooters to the hospital, combined with an abundance of turnips apparently grown at Parramatta, led to an improvement in the patients' health. On 20 June 1792 the storeship *Atlantic* arrived at Sydney with a cargo of rice, soujee and dholl, its arrival casting 'a gleam of sunshine which penetrated everyone capable of reflection'.[61] The meat supply was still limited, but notwithstanding some theft of grain and stock, disaster had been averted. Shortly after,

Fig. 4.6: Rowlandson, T. (ca. 1790). *Convicts embarking for Botany Bay, ca. 1790* (detail) [watercolour on paper]. Large numbers of convicts arrived at Port Jackson in a malnourished and sickly state only to subsequently die while working the drought-stricken fields. Mitchell Library, State Library of New South Wales, FL8862380.

the storeship *Britannia* arrived, and weekly rations returned to something close to normal in quantity, although certainly not in quality. Prior efforts made clearing the land for cultivation had also borne fruit, and more than 200 acres (81 ha) of wheat and 1100 acres (445 ha) of maize had been planted by October 1792.[62] Set up by several days of heavy rain in September, these crops yielded well, and for the first time the colony was approaching an adequate supply of grain. So difficult had conditions been over the previous 4 years that when Phillip departed for England in December 1792, he singled out his efforts to achieve agricultural self-sufficiency as one of his major gubernatorial achievements.[63]

But in a pattern that we will see borne out in future decades, just when optimism returned, the long drought again tightened its grip on the land, before finally relenting. By the beginning of summer in late 1792 conditions were again very hot and dry, and in December fires, perhaps lit by Aboriginal people, sprang up on the west side of Sydney Cove, fuelled by grass that had grown during the wetter months. Whipped up by north-westerly winds and temperatures reaching the century mark (100°F or 37.8°C), the furious fires destroyed a house and several garden fences before being brought under control.[64] The dry conditions must have continued through early autumn, because heavy rains in May 1793 came too late to save the summer corn crop, and by August 1793 the wheat crop was also looking yellow and parched.[65] However,

3 days and nights of rain fell in late August, growing conditions remained excellent through October, and heavy rains fell again in November.

In January 1794 the crops yielded nearly 7000 bushels of wheat, with the more productive farms yielding 30 bushels or more to the acre, more than double that of the 1791 harvest.[66] While severe shortages of rations continued, these cannot be attributed to low rainfall, since there is little or no mention of drought during 1794, and in January 1795 extremely heavy rainfall caused major flooding along the Hawkesbury. With that, the long drought had been consigned to memory.

THE LOSS AND REDISCOVERY OF THE SETTLEMENT DROUGHT

Recently, scholars with a renewed interest in colonial-era climatic reconstruction have noted that before the late 20th century, the Settlement Drought was seldom mentioned by climate historians, and that it was generally considered to have been neither particularly long nor severe.[67] Of course, early writers had very limited access to many of the sources available to us today, some of which, like Dawes's meteorological journal, have only recently been discovered. We must also keep in mind that descriptions of other early droughts (e.g. 1797–99, 1813–15) suffered from a similar lack of detail, and it was not until the 1820s that researchers had a broader range of eyewitness accounts written by early European explorers and settlers on which to draw. Even so, early writers such as William Jevons and Samuel Bennett do mention the event.[68] In 1877 HC Russell detailed evidence of drought and heat from 1790–93 despite lacking knowledge of the ENSO phenomenon that would have helped him construct a more complete understanding of Australian drought and flood cycles,[69] and a later writer, W Allen, noted in 1895: 'The pioneers of settlement in Sydney Cove in 1788 had to counter all the disadvantages of a serious and prolonged drought' which was 'very severe in 1791–2'.[70]

We now know that the Settlement Drought lasted for 3 years (July 1790–August 1793), was particularly severe in 1791–92, at the height of the El Niño, and featured the pattern of extremely low rainfall and high summer temperatures typical of the most severe Australian droughts. Above all, however, its human impacts clearly rank among the most severe of any drought that has occurred since European settlement. It is, I think, often overlooked that the human impact of drought depends not only on the severity of rainfall deficits but also on the state of agriculture, governance, resource availability, land degradation, and the size, technological capability and underlying health of the affected population. For a newly arrived convict, emaciated and ravaged by disease, forced to grow varieties of wheat susceptible to drought and disease on poor soils, lacking any agency over the storage or dissemination of the harvest and facing corporal or capital punishment for food theft, the Settlement Drought could hardly have been a more serious threat to life.

It would be interesting to contrast this experience with that of local Aboriginal people, who had been living with drought in the Sydney region for tens of thousands of years. Unfortunately, early European writers left us with little insight into how changing environmental conditions affected their food supply beyond mentioning the vague belief that a lack of fish perhaps caused starvation or hardship among Indigenous people in June 1788. David Collins also mentioned the occasional theft of grain and provisions from the settlement.[71] It is a mistake to dismiss these observations out of hand, because, as we will see, severe droughts elsewhere in Australia certainly placed the food and water supplies of Aboriginal people under immense pressure.

Nevertheless, there can be no doubt that in the Sydney region Aboriginal society was almost immediately thrown into disarray following European contact, and groups in the vicinity of Sydney Cove and Parramatta frequently had food and equipment stolen by convicts and settlers. Even more devastating was the 1789 smallpox epidemic, which decimated populations around Sydney and across the south-east of the continent, reducing some to just a handful of members. By the end of the drought conflict with the European population was escalating, and the traditional lifestyle had already been significantly disrupted or displaced over a large part of the Sydney Basin. Even if we knew how Aboriginal people responded to the harsh environmental conditions of the Settlement Drought, it occurred so soon after the smallpox disaster, and in such a new socio-ecological setting, that we would probably be able to infer little about how these compared to the impact of previous droughts.

Endnotes

[1] *Historical Records of New South Wales* (hereafter *HRNSW*) vol. 1, part 2, p. 470 (4 March 1791).

[2] Martra JM (1783) *A proposal for establishing a settlement in New South Wales. HRNSW* vol. 1, part 2, p. 2.

[3] *HRNSW*, vol. 1, part 2, p. 17 (18 August 1786).

[4] *HRNSW*, vol. 1, part 2, p. 14 (18 August 1786).

[5] Cook J (1893) *Captain Cook's journal during his first voyage round the world in H.M. bark 'Endeavour' 1768-71.* (Ed. WJL Warton). Elliot Stock, London (hereafter Cook 1893); journal entry Sunday, 29 April 1770.

[6] Cook (1893); journal entry Thursday, March 3, 1770

[7] Cook (1893); journal entry Saturday, March 5, 1770.

[8] *Journals of the House of Commons*, 19 Geo. III, 1779, vol. 37, p. 311.

[9] Collins D (1798) *An account of the English colony in New South Wales with remarks on the dispositions, customs, manners, etc, of the native inhabitants of that country. Volume 1.* AH & AW Reed in association with The Royal Australian Historical Society, Sydney, p. 2.

[10] Tench W (1979) *Sydney's first four years; being a reprint of A narrative of the expedition to Botany Bay and a complete account of the settlement at Port Jackson.* Library of Australian History and Royal Australian Historical Society, Sydney, pp. 65–66.

[11] Evans MJ, Williams RJ (2001) *Wetlands Australia Journal* **19**, 61–71.

[12] Smyth AB (1979) *The Journal of Arthur Bowes Smyth: Surgeon, Lady Penrhyn 1787-1789.* Australian Documents Library, Sydney, p. 57; Tench (1979), p. 215.

[13] Kermode SJ *et al.* (2016) *Marine and Freshwater Research* **67**, 771–781.

[14] *HRNSW*, vol. 1, part 2, pp. 121–122 (15 May 1788).

[15] Smyth (1979), p. 57.

[16] Phillip 'fixed on the one [cove] that had the best spring of water'. *HRNSW* vol. 1, part 2, p. 122 (15 May 1788).

[17] Collins (1798), p. 4.

[18] Collins (1798), pp. 14–15; *HRNSW* vol. 1, part 2, p. 125 (15 May 1788).

[19] Described in Collins (1798), p. 14 and Smyth (1979), pp. 66–67.

[20] Gergis J, Karoly DJ, Allan RJ (2009) *Australian Meteorological and Oceanographic Journal* **58**, 83–98; Gergis J, Garden D, Fenby C (2010) *Environmental History* **15**, 485–507.

[21] Collins (1798), p. 30; also p. 73 in Hunter J (1793) *An historical journal of the transactions at Port Jackson and Norfolk Island with the discoveries which have been made in New South Wales and in the southern ocean since the publication of Phillip's voyage, compiled from the official papers; including the journals of Governors Phillip and King, and of Lieut. Ball; and the voyages from the first sailing of the Sirius in 1787, to the return of that ship's company to England in 1792.* John Stockdale, London.

[22] Collins (1798), p. 43.

[23] Gergis *et al.* (2010), p. 494; Gergis *et al.* (2009), p. 89.

[24] Collins (1798) pp. 79, 82–83, 93.

[25] Gergis JL, Fowler AM (2009) *Climatic Change* **92**, 343–387.

[26] Gergis *et al.* (2009), p. 96.

[27] Cf., *HRNSW*, vol. 1, part 2, p. 178 (10 July 1788); *HRNSW*, vol. 1, part 2, p. 208 (30 October 1788); *HRNSW*, vol. 1, part 2, p. 150 (9 July 1788).

[28] *HRNSW*, vol. 1, part 2, pp. 126–127 (15 May 1788).

[29] *HRNSW*, vol. 1, part 2, p. 297 (12 February 1790).

[30] Tench (1979), p. 262.

[31] *HRNSW*, vol. 1, part 2, p. 369 (26 July 1790).

[32] *HRNSW*, vol. 1, part 2, p. 212 (16 November 1788); Collins (1798), p. 33.

[33] Collins (1798), p. 46.

[34] *HRNSW*, vol. 1, part 2, p. 299 (12 February 1790).

[35] Collins (1798), p. 93, perhaps 96; Tench (1979), p. 167.

[36] Collins (1798), p. 113.

[37] Hunter (1793), p. 472; Collins (1798), p. 113.

[38] *HRNSW*, vol. 1, part 2, pp. 332–333 (17 April 1790).

[39] McAfee RJ (1981) *Dawes's meteorological journal*. Australian Government Publishing Service, Canberra.

[40] Tench (1979), p. 192.

[41] Tench (1979), p. 192.

[42] Tench (1979), pp. 194–195.

[43] Hunter (1793), pp. 488–489.

[44] Hunter (1793), pp. 488–489.

[45] Collins (1798), p. 130.

[46] Macphail M, Owen T (2018) *Australasian Historical Archaeology* **36**, 16–28.

[47] Collins (1798), p. 131; Tench (1979), p. 220.

[48] Tench (1979), pp. 220–221.

[49] Collins (1798), pp. 137–139; Tench (1979), pp. 221–222.

[50] Collins (1798), p. 157.

[51] *HRNSW*, vol. 1, part 2, p. 529 (24 October 1791).

[52] Tench (1979), p. 255.

[53] Tench (1979), p. 248.

[54] *HRNSW*, vol. 1, part 2, p. 598 (19 March 1792).
[55] Campbell JS (1924) *Journal and Proceedings of the Royal Australian Historical Society* **10**, 63–103.
[56] Collins (1798), p. 178.
[57] Collins (1798), p. 176.
[58] Collins (1798), p. 175. According to Collins (p. 149) there were 126 male convicts, 23 female convicts and three children aboard the *Queen*.
[59] Parker MA (2010) *A voyage round the world, in the Gorgon Man of War*. Cambridge University Press, Cambridge.
[60] Collins (1798), p. 175.
[61] Collins (1798), pp. 183–184.
[62] Collins (1798), p. 209.
[63] *HRNSW*, vol. 1, part 2, pp. 645–646 (2 October 1792).
[64] Collins (1798), p. 216.
[65] Collins (1798), pp. 239, 259.
[66] Collins (1798), p. 284.
[67] This argument appears in Gergis J *et al*. (2010).
[68] Bennett S (1867) *The history of Australian discovery and colonisation*. Hanson and Bennett, Sydney; Jevons WS (1859) Some data concerning the climate of Australia and New Zealand. In *Waugh's Australian Almanac, For The Year 1859*, pp. 47–98. James William Waugh, Sydney.
[69] See Jevons (1859), Bennett (1867) and Russell HC (1877) *Climate of New South Wales: Descriptive, historical, and tabular*. C. Potter, Sydney.
[70] Russell (1877); Allen W (1895) *The Capricornian* (Rockhampton), 23 March 1895, p. 27.
[71] E.g. Collins (1798), p. 239.

5

The summer of discontent

Were I asked the cause of this intolerable heat, I should not hesitate to pronounce, that it was occasioned by the wind blowing over immense desarts [sic], which, I doubt not, exist in a north-west direction from Port Jackson

– Watkin Tench[1]

IN THE SYDNEY REGION, THE HOTTEST SUMMER WEATHER TYPICALLY occurs ahead of the passage of cold fronts over the southern part of the continent, when hot, dry air is drawn from the interior in a south-easterly direction and out over the Tasman Sea (Fig. 5.1). During the most extreme heatwaves, temperatures can soar to well above 40°C, even near the coast, with relief from the stifling heat coming only with the arrival of cool southerly winds behind the cold front. In the first 2 years after the arrival of the First Fleet, the colonists had experienced a few days of such weather, particularly in January 1789 and

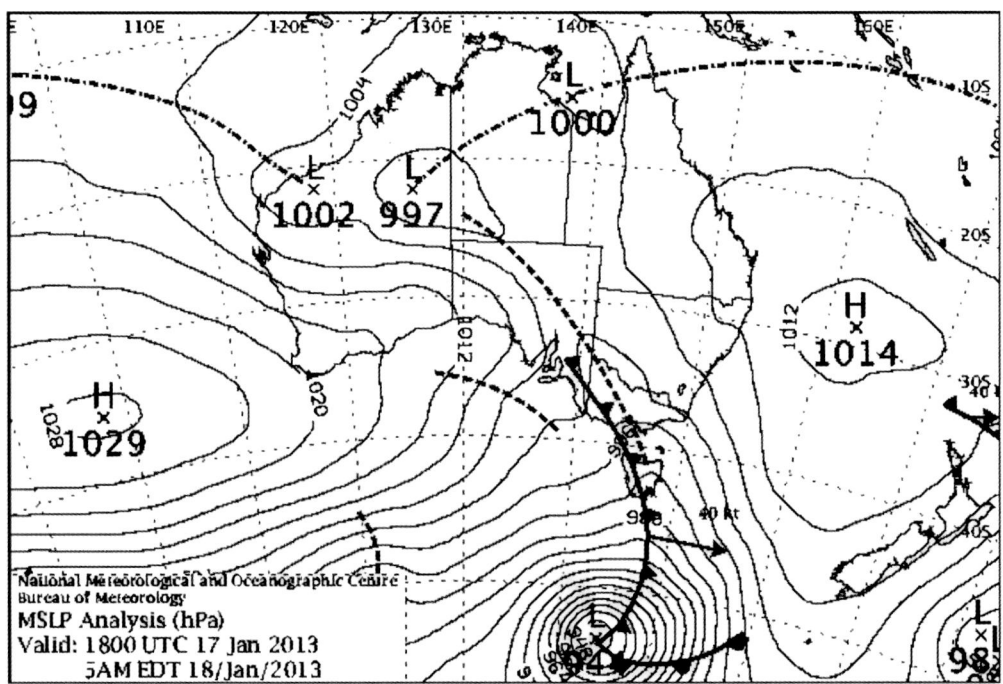

Fig. 5.1: Weather map for 5 am, 18 January 2013, showing a cold front approaching New South Wales from the south-west. The temperature reached 45.8°C at Sydney Observatory later in the day, the hottest day on record. Courtesy Australian Government Bureau of Meteorology (CC BY 3.0 AU DEED).

December 1789–January 1790, when temperatures rose to over 40°C in the shade. Most had never experienced such heat, which they described as 'intolerable' and 'suffocating'. Yet this was just a taste of what was to come when the Settlement Drought had finally stripped the moisture from the land, leaving nothing to cool the savage skies.

> ### Box 5.1: The Stevenson screen
>
> The establishment of a standardised method for measuring air temperature stands as one of the most significant developments to have occurred in meteorology and climatology in the last 150 years. In recent decades it has taken on added importance as concerns over anthropogenic (human-caused) climate change have made comparisons of historical and modern temperatures extremely important. Prior to the late-1800s thermometers were placed in a range of settings or shelters with differing exposure to radiation and air movement, including under open shelters or on south-facing walls or fences. Unfortunately, air temperatures are extremely sensitive to even small differences in these factors: even the beloved thermometer 'under the verandah' which lacks any direct exposure to sunlight can give readings that differ by several degrees from those made inside a properly shielded, ventilated and positioned instrument box. Furthermore, even small design differences across purpose-built shelters can generate significant biases in temperature data.
>
> A breakthrough came in 1864, when Thomas Stevenson, father of Robert Louis Stevenson, invented the Stevenson screen. His inexpensive, box-shaped shelter, which was to be positioned 4 feet (1.2 m) above the ground, featured a double layer of louvre-boards that prevented direct sunlight from any direction entering the screen while still allowing adequate air flow. White paint reduced absorption of solar radiation by the structure itself. This design, with several additional modifications, quickly became popular, and from 1887 they began to be used by Australian meteorologists. By the late 1880s many meteorological stations across Queensland had installed Stevenson screens, and this continued in other states through the early 1900s. Since then they have provided critical data that can be used in climatological research. Technological advances never stop, however, and in recent years new, cheaper and less bulky screens that provide similar or even higher quality data have been developed to accommodate automated weather stations and other modern instruments.
>
>
>
> A Stevenson screen located outside the Sydney Observatory, 1903, around the time that they were coming into widespread use in New South Wales. Courtesy National Archives of Australia.

On 27 December 1790, the colony was greeted with a hazy sky and strengthening north-westerly winds that felt like 'the blast of a heated oven'.[2] We are fortunate that regular meteorological observations were being made at the time by the astronomer and scientist William Dawes, and by Watkin Tench, who had a keen interest in unusual weather phenomena. Together, their data and written accounts provide us with not only the first scientific description of a great Australian heatwave, but also the opportunity to compare it with more modern episodes of extreme heat. Since standards for thermometer exposure have changed over the years, such efforts have become highly controversial, especially when they extend to exploring the links between heatwave severity and rising global temperatures. However, they can also provide us with tremendous insights into past Australian climates and an opportunity to consider both the potential pitfalls and benefits associated with such comparisons.

Dawes's measurements, which were taken inside his observatory (Fig. 5.2) at Sydney Cove, form the basis of recent reconstructions of the early settlement-era climate.[3] Although his thermometer was positioned in a shaded, ventilated place that lacked direct exposure to solar radiation, the temperatures he recorded are not strictly comparable to those made under modern, standardised conditions inside a Stevenson screen (Box 5.1), which were not in general use until more than a century later. As such, it is possible that his high temperature observations have a slight (perhaps up to 1–2°C) warm bias compared with modern measurements. The maximum temperature that he recorded during the December 1790 heatwave was 39.7°C (or 103.5°F) at noon on 27 December (see Table 5.1). However, Watkin Tench provided more detailed measurements, and his notes clearly show that he was tracking the temperature throughout the day.[4]

The location of Tench's thermometer is not known, but it was certainly also in Sydney Cove, perhaps not too distant from Dawes's observatory. His noon temperature (40.0°C) was essentially the same as Dawes's, but critically he recorded a steep rise over the next hour, and a maximum temperature of 42.8°C at 2:20 pm. Dawes, lacking a minimum–maximum recording thermometer, did not record this high temperature, and as such his data probably failed to capture the full intensity of the heatwave. According to Tench (and Dawes) the next day started out even hotter but peaked at 40.3°C at 12:15 pm before rapidly falling following a cool southerly change that arrived just after 1 pm. Dawes made no measurements between 8 am and 4 pm, and so missed this second consecutive day of high temperatures altogether (Table 5.1).

Tench stated that these days were the 'hottest we ever suffered at Sydney'. But on 11 February the hot north-westerly wind returned, pushing temperatures at Dawes's observatory up to 38.6°C (101.4°F) at noon, and following a very hot night (28.9°C at sunrise), to 38.3°C at noon on the 12th. According to David Collins, the temperature reached 105°F (40.6°C) on one of these days,[5] and Tench suggests a peak temperature of 42.2°C. This is hot even by the standard of modern heatwaves, but since they fall well short of the record temperatures observed in January 2013 (45.8°C) and January 1939 (45.3°C), they appear quite plausible.

Tench specifically mentions, however, that on these days the most intense heat occurred not at Sydney Cove but to the west at Parramatta where 'it [the heat] was allowed, by every

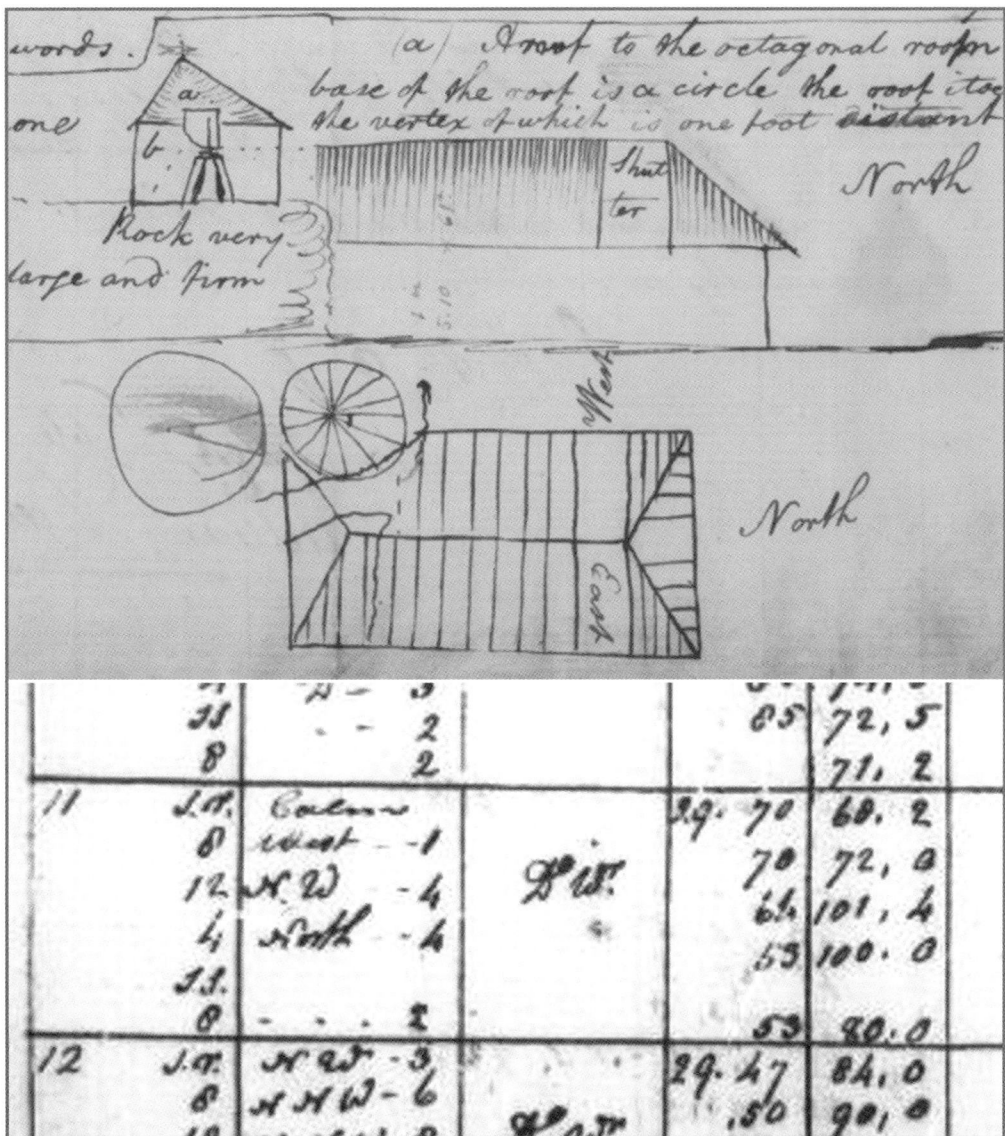

Fig. 5.2: Top: sketch of Dawes's observatory at Port Jackson, April 1788. From William M. Dawes's correspondence during the establishment of an observatory at Port Jackson, New South Wales. Reproduced by kind permission of the Syndics of Cambridge University Library, RGO 14/48: 281-282. Bottom: extract from the meteorological journal of William Dawes documenting the high temperatures and strong north-westerly winds that occurred during a heatwave on 11–12 February 1791 at Sydney Cove.

person, to surpass all that they had before felt, either there, or in any other part of the world'.[6] The impact on local wildlife, especially bats, was devastating: an immense flight of flying foxes was seen that 'covered all the trees around the settlement, whence they every moment dropped dead, or in a dying state, unable to endure the burning heat of the atmosphere'.[7] Some fell dead while in flight, others dropped from trees into waterways, creating a terrible stench and fouling the water. The numbers were incredible: 20 000 were seen dead within the space of a mile.

Table 5.1: Temperatures (in °C) recorded by William Dawes and Watkin Tench during the 27–28 December 1790 heatwave.

		27 Dec		28 Dec	
	Time	Tench	Dawes	Tench	Dawes
am	Sunrise		18.9		23.3
	8		23.9	30.0	26.7
	9	29.4			
	10			33.9	
	11			38.3	
pm	12	40.0	39.7	39.7	
	12:30	41.9		40.3	
	1	42.5		38.9	
	1:15			31.7	
	2:20	42.8			
	4		36.8		23.9
	5			22.8	
	Sunset	31.7	33.4	20.8	
	8		28.6		22.2
	11	25.8			

Birds fared little better, with the ground strewn with dead 'perroquettes', and in several parts of the harbour different sorts of small birds, some dead, and others gasping for water, were found.[8] The impact was clearly widespread, because dead birds were also observed in other locations around Port Jackson.

These observations tell us much about the severity of the heatwave because fruit bats are very sensitive to high temperatures. Indeed, they are now used by scientists as bioindicators of extreme heat events and, being important seed dispersers and pollinators, broader ecological change under global warming. The fruit bat referred to in these accounts is the grey-headed flying-fox (*Pteropus poliocephalus*), which occupies a large but shrinking range from coastal central-eastern Queensland to eastern South Australia. Recent research has shown heat-related mortality in this species usually happens when the air temperature exceeds a critical threshold of 42°C, and major die-offs in flying-fox colonies have occurred in the nearby suburbs of Cabramatta and Gordon when temperatures reached 43.3–44.5°C (e.g. December 1994, January 2003).[9]

Over the past decade similar heatwaves have killed tens of thousands of fruit bats in many locations across eastern Australia. In south-eastern Queensland an estimated 45 000 bats, mainly the black flying-fox (*P. alecto*) were killed in temperatures of 43–45°C on 4 January 2014, while tens of thousands of the spectacled flying-fox (*P. conspicillatus*) were killed during a heatwave in northern Queensland in late 2018. While clearly not a new phenomenon, the demographic impact of these die-offs for the grey-headed flying-fox population is increasing, since numbers of this species have declined from many millions in the 1930s to fewer than 400 000 today. Habitat loss and fragmentation and persecution are also major factors in the decline of these species.[10]

Allowing for some heat bias compared with modern measurements, these accounts all suggest that temperatures at Parramatta during the February 1791 heatwave reached somewhere between 42 and 45°C, compared with 39–41°C at Sydney Cove. This is a reasonable scenario, because during hot north-westerlies temperatures at the modern Sydney Observatory Hill site are often, though not always, a couple of degrees cooler than they are at Parramatta, which lies 20 km further inland. Perhaps a similar situation occurred on 18 January 2003, when the temperature reached 44.5°C at Parramatta North and 39°C at Sydney Observatory Hill. Therefore, I think we can safely conclude that while the February 1791 heatwave was obviously abnormally hot, even at Parramatta the highest maximums were within the range of modern extremes. Indeed, temperatures of 44–47°C have been observed here eight times since the year 2000.[11]

While I see no reason to doubt these accounts, there are two final aspects of it that remain perplexing. The first is that David Collins mentions that the bats killed during the heatwave were chiefly male, which is inconsistent with modern research that has shown that females and the young are far more likely to die during such events.[12] One possibility is that both sexes were affected, but by chance only males were observed. This is plausible because flying foxes show strong seasonal sexual segregation, and while this usually occurs in autumn and winter camps,[13] perhaps on this occasion male and female bats responded with a different pattern of movement during the hot, windy weather (since they were 'driven before the wind') or had some degree of geographical segregation before onset of the heatwave.

The second interesting aspect of the heatwave is that there appears to be some suggestion that the higher temperatures might have been influenced by fires lit nearby by Aboriginal people. However, Watkin Tench specifically mentions that methods used by some people to estimate the intensity of the heat were 'equally unfair and unphilosophical' and emphasises the fact that his thermometer was 'hung in the open air, in a southerly aspect, never reached by the rays of the sun, at a distance of several feet above the ground'.[14] I am sure that many readers who have heard legendary tales in local outback pubs and on radio talk shows of temperatures reaching over 50°C for days at a time[15] will appreciate Tench's concerns, and to me it is fascinating that the same difficulties that bedevil attempts to obtain a scientific understanding of Australian climatic extremes today were present as far back as the earliest days of European settlement.

Endnotes

[1] Tench W (1979) *Sydney's first four years; being a reprint of a narrative of the expedition to Botany Bay and a complete account of the settlement at Port Jackson.* Library of Australian History and Royal Australian Historical Society, Sydney, p. 266.

[2] Tench (1979), p. 265.

[3] Gergis J, Karoly DJ, Allan RJ (2009) *Australian Meteorological and Oceanographic Journal* **58**, 83–98.

[4] Tench (1979), p. 265.

[5] Collins D (1798) *An account of the English colony in New South Wales with remarks on the dispositions, customs, manners, etc, of the native inhabitants of that country. Volume 1.* AH & AW Reed in association with The Royal Australian Historical Society, Sydney, p. 127.

[6] Tench (1979), p. 266.

[7] Tench (1979), p. 266.

[8] Tench (1979), p. 266; Collins (1798), p. 127.

[9] Welbergen JA *et al.* (2008) *Proceedings of the Royal Society B* **275**, 419–425.

[10] Woinarski JCZ, Burbidge AA, Harrison PL (2014) *The action plan for Australian mammals 2012.* CSIRO Publishing, Collingwood, Victoria.

[11] Modern measurements in these areas may also have been affected by the urban heat island effect.

[12] Welbergen *et al.* (2008).

[13] Nelson JE (1965) *Animal Behaviour* **13**, 544–557.

[14] Tench (1979), p. 266. This was common practise until the introduction of the Stevenson screen.

[15] Temperatures of 50°C or more occur exceptionally rarely in Australia, and there has never been an instance of 51°C recorded under modern instrumental conditions.

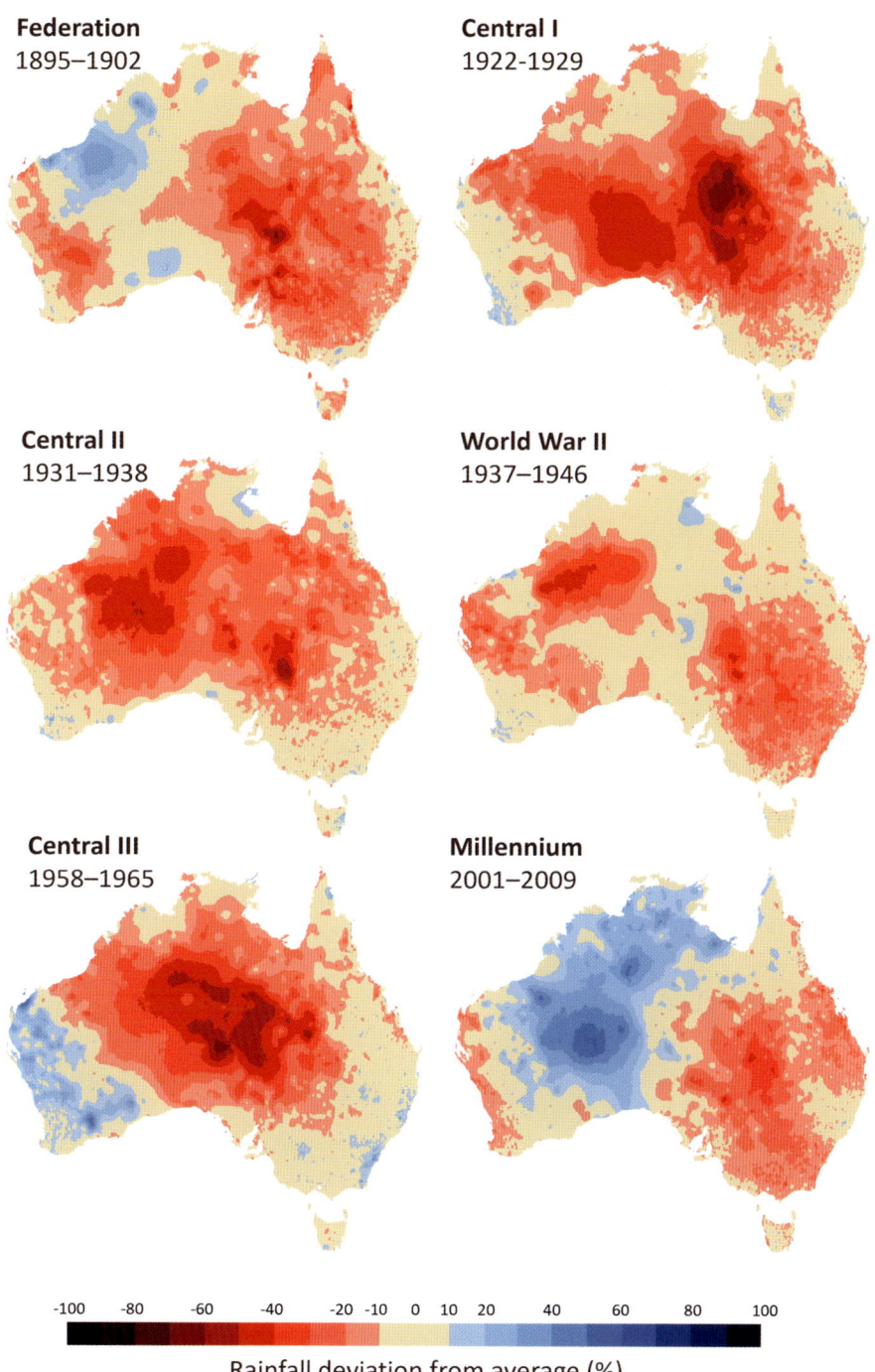

Plate 1: Six continental-scale, quasi-decadal droughts, 1895–2009, based on rainfall data from the SILO climate database.

DROUGHT COUNTRY

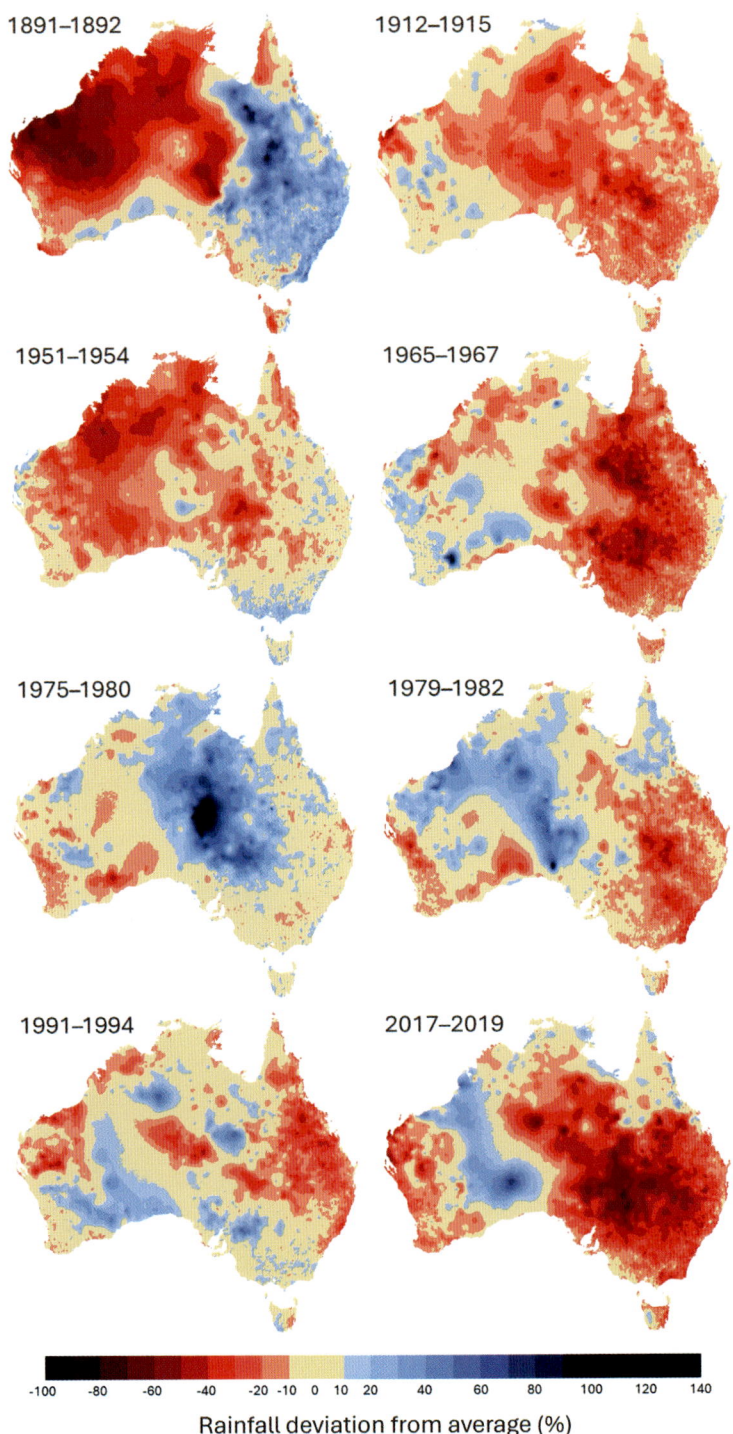

Plate 2: Major biennial to semi-decadal droughts, 1891–2019, based on rainfall data from the SILO climate database.

Plate 3: *View on the Upper Hunter, New South Wales* [watercolour] (1864). Rex Nan Kivell Collection, National Library of Australia, nla.obj-133057179. The banks of the Hunter River are shown covered in thick vegetation, as they would have in the pre-European era. The mountains depicted in the background are probably the Mount Royal Range.

Plate 4: Sedimentation of a creek crossing on the author's property following drought breaking rains in early 2020. The amount of silt is remarkable given that virtually the entire catchment was protected in a National Park with no livestock. Photo: Robert Godfree.

Plate 5: Lagoon behind Potato Point Beach, north of Potato Point. When George Bass travelled between Potato Point and Lake Tuross in December 1797, the area was affected by drought, and waterholes dug in the centre of the largest of the swamps by Yuin people were completely dry. Photo: Robert Godfree.

Plate 6: In addition to his exploits as an explorer, Gregory Blaxland was an astute observer of the land and among the first to describe the decline in native perennial grasses that follows heavy grazing. Shown here is a memorial that lies near to the site of Blaxland's farm 'Leeholme' on South Creek, the point from which his party departed to cross the Blue Mountains on 11 May 1813. The original native understorey of the woodland in the background has mostly been replaced by weeds. Photo: Robert Godfree. Inset: Gregory Blaxland Esq. formerly of Brush Farm. Mitchell Library, State Library of New South Wales, ML 143.

Colour plates

Plate 7: While accompanying Governor Macquarie on his 1815 western tour to Bathurst, John Lewin produced beautiful watercolours that now provide key insights into the nature of landscapes at the time. Top: Lewin, J. W. (ca. 1815–16) *Sydmouth Valley* [watercolour]. Here, pools are shown in the bottom of the Sydmouth Valley, which is green and lush. Many watercourses on the tablelands and ranges of New South Wales once contained similar 'chains of pools' which were a reliable water source, even during drought. Bottom: Lewin, J. W. (ca. 1815–16). *Campbell River* [watercolour]. This image shows Campbells River upstream of its junction with the Fish River. The lack of an understorey shrub layer characterises the open woodlands adjacent to the river, consistent with traditional Aboriginal burning practises. While the landscape appears dry, the painting suggests that pools in Campbells River were full at the time. Mitchell Library, State Library of New South Wales, PXE 888.

Plate 8: A drought-blasted riverine plain south of Tilpa on the Darling River in March 2019. The landscape that Charles Sturt encountered north of the Barwon River in March 1829 must have looked much like this, for he travelled all day without seeing a drop of water or blade of grass, nor virtually any living creature. Photo: Robert Godfree.

Plate 9: Shrublands like this once dominated vast tracts of the Riverina, Darling and Murray plains. The large species is oldman saltbush (*Atriplex nummularia*), a drought-hardy species that was one of the first to disappear due to overgrazing in the 1840s–80s. In recent decades it has recovered in some areas due to careful management or removal of grazing and is now planted as a source of feed during drought. Photograph: Robert Godfree.

Colour plates

Plate 10: The introduction of livestock and later the spread of cats, rabbits and foxes devastated native mammal populations in western New South Wales and South Australia. Top: the greater stick-nest rat (*Leporillus conditor*), a key source of food to local Aboriginal people, was widespread along the Murray River prior to European settlement. Blandowski Collection, Museum für Naturkunde, Humboldt Universität, Berlin. Bottom: apparent remains of a burrowing bettong (*Bettongia lesueur*) warren on Langawirra Station, Broken Hill. This warren, which now contains rabbits, was constructed on calcareous clay–limestone flats, and measures 20 m across. Photo: Robert Godfree.

Plate 11: Minchin R. E. (1867). *Far North* [watercolour]. This image is one of the very first to show the tragedy of drought in arid and semi-arid Australia. Here, the breakdown in the socio-ecological system is complete: livestock lie dead, buildings sit abandoned, and the once rich landscape has been reduced to a barren wasteland. State Library of South Australia, PRG 119/28/22.

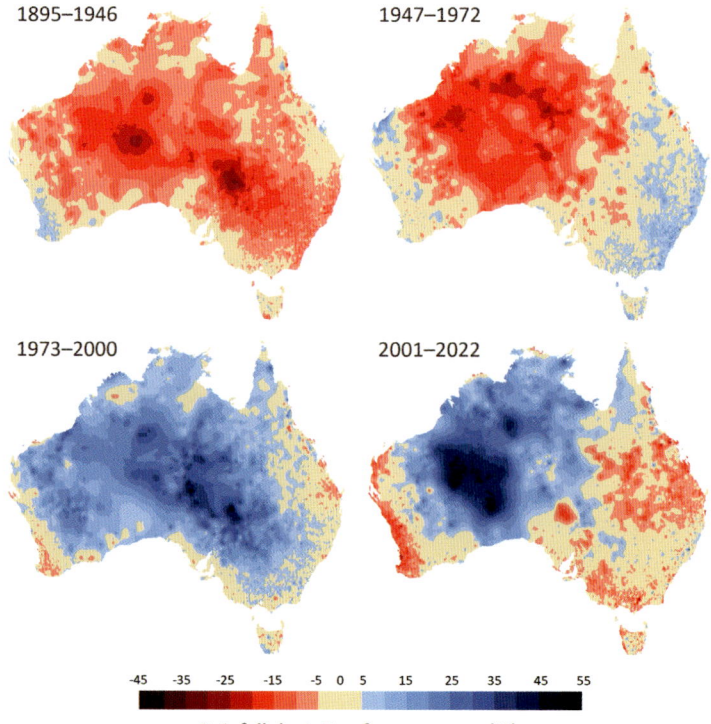

Plate 12: Four major rainfall epochs, 1895–2022.

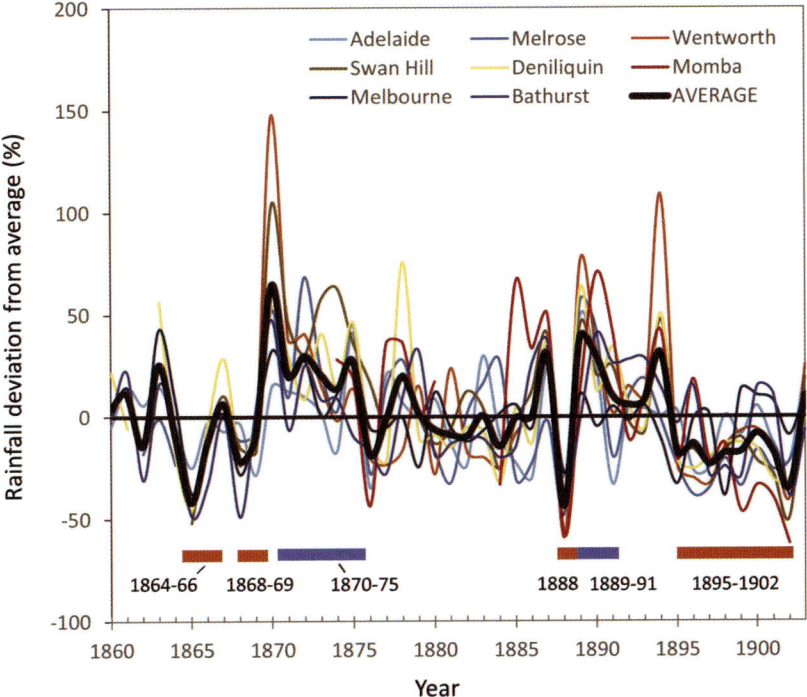

Plate 13: Rainfall, as a deviation from the long-term average, at eight locations in south-eastern Australia, 1860–1903. Stations are arranged from wettest (Melbourne) in dark blue to driest (Momba Station) in red. The tendency for rainfall in drier inland locations like Momba, Wentworth and Deniliquin to be more variable than wetter locations is obvious. The data show that the 1864–1866 and 1888 droughts were synchronous and severe across all stations.

Plate 14: Drysdale, R. (1946). *Crucifixion* [oil on plywood]. Generously provided by the Art Gallery of NSW and the Drysdale Estate.

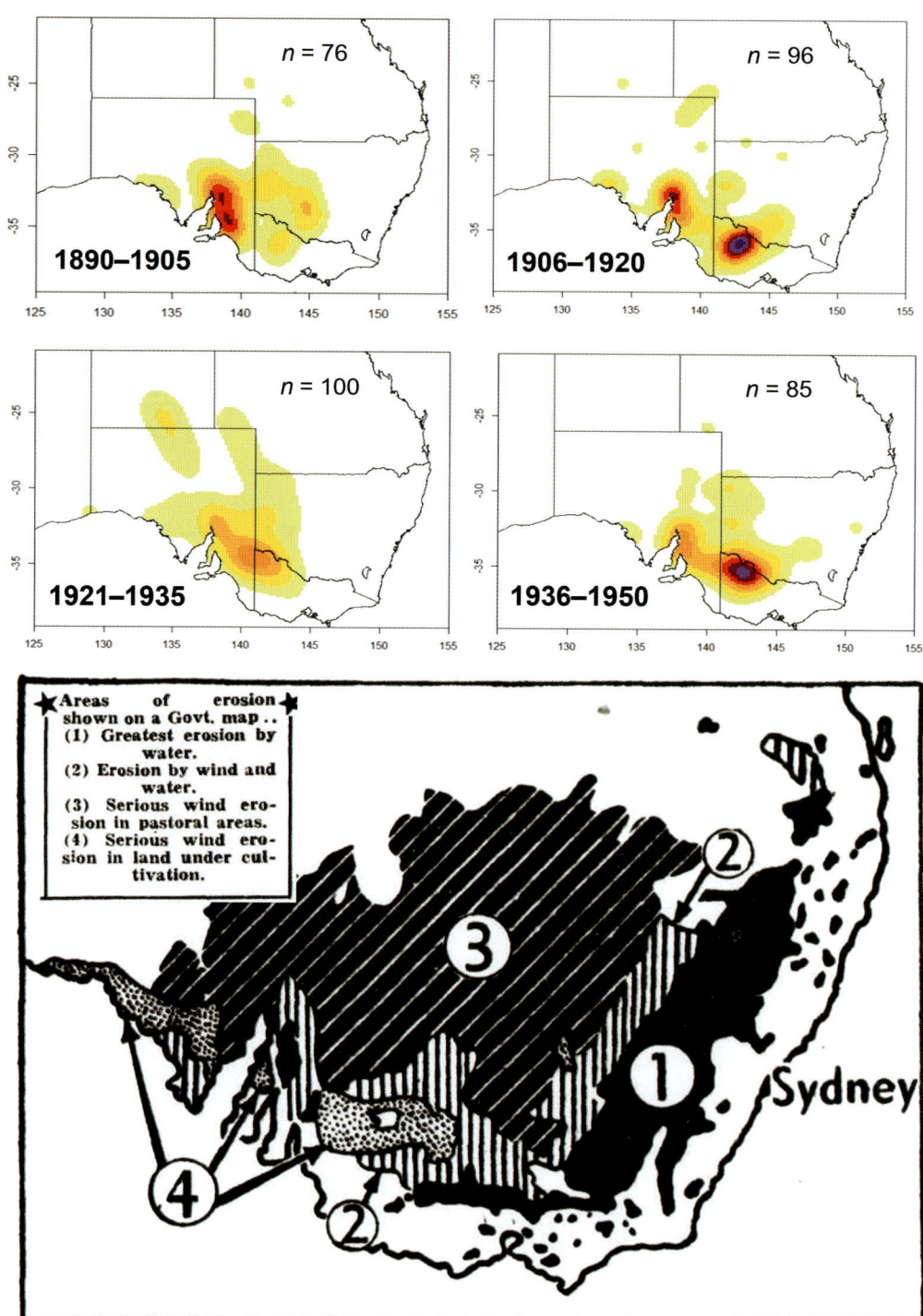

Plate 15: Maps of soil erosion in south-east Australia during the 'dust bowl' era. The top four panels contain density maps of sand and soil drift in four key time intervals. The number of newspaper and other accounts use to generate each map are shown in the top right of each panel. The bottom map of soil erosion appeared in *The Sun* (Sydney) on 19 June 1950.

Colour plates

Plate 16: Top: vast areas of scalded country that developed during the 1880s–90s and then increased in scale over the next 50 years can still be seen in western New South Wales today. This scald is located north of Menindee. Bottom: mulga country west of Cobar, NSW in 2018, showing the ravages of decades of overgrazing and drought. Note the near total absence of ground cover and the lack of young mulga plants. Photos: Robert Godfree.

Plate 17: Top: a great dust storm about to arrive at Broken Hill, 15 December 1907. National Library of Australia, nla.obj-138224063. Middle: famous photograph of an immense dust storm captured on Monday, 2 February 1903 at Narrandera, NSW, by the photographer Carl Dugdale. A description is found in the *Narandera Argus and Riverina Advertiser*, 6 February 1903. This recoloured image beautifully captures the fiery appearance of the approaching wall of dust. National Library of Australia, nla.obj-136754952. Bottom: dust storm – Woolworths building, Sydney Town Hall and George Street, September 2009. This dust storm, known as the 'Red Dawn' storm, was more than 3000 km long, and deposited more than 2 million tonnes of soil in the Pacific Ocean. Photographer: Helen Grant, courtesy City of Sydney Archives, A-00029932.

Plate 18: True colour satellite image of South Australia taken on 11 September 2019, showing a belt of green vegetation below Goyder's Line (in yellow). Image courtesy of NASA Worldview.

Plate 19: The Macquarie Marshes stand out like emeralds in this satellite image. The major watercourse on the far left is the Bogan River. Image courtesy Google Earth, compiled from multiple sources and dated December 2017.

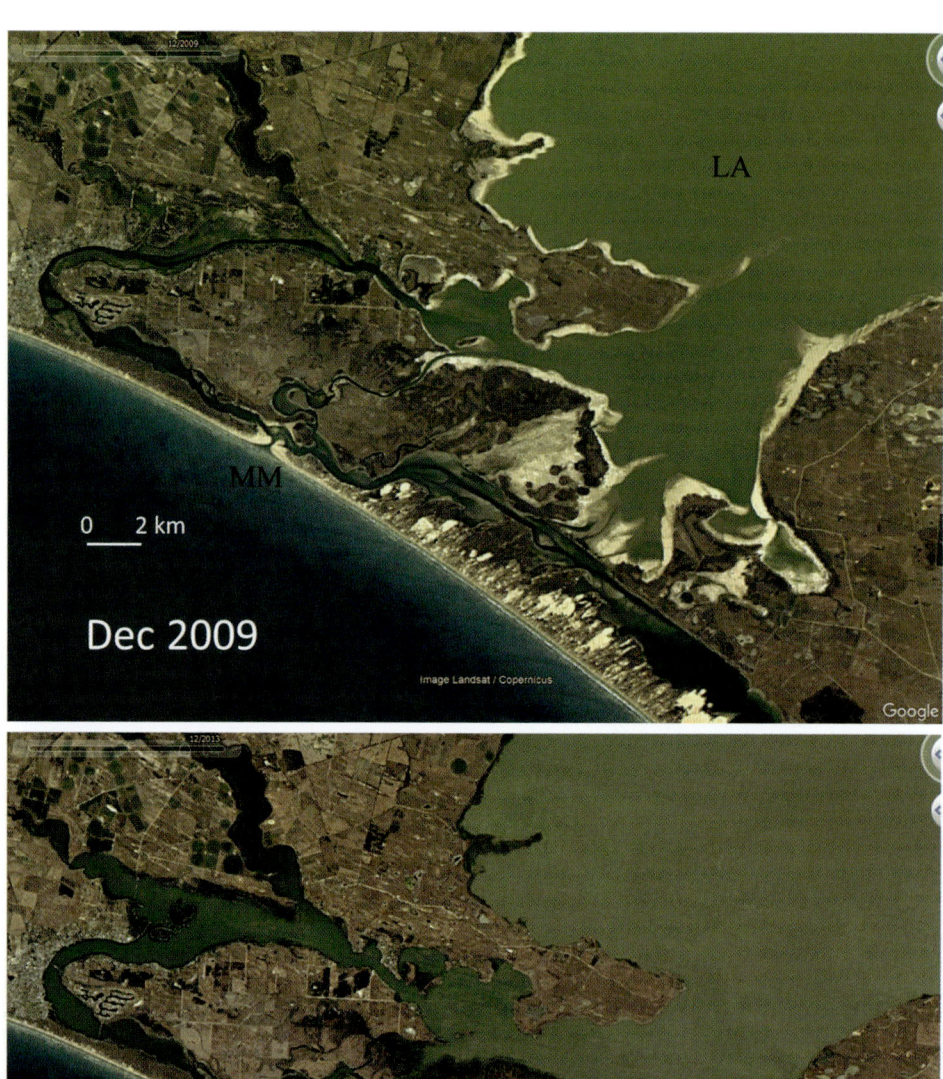

Plate 20: The Coorong (C), Lake Alexandrina (LA) and Murray Mouth (MM) during and after the Millennium Drought. Top: by late 2009 Lake Alexandrina had fallen below sea level, causing the shoreline to retreat. This exposed large areas of the lakebed. Bottom: the same area in 2013 after the break of the drought. Images courtesy Google Earth, compiled from multiple sources and dated December 2009 and December 2013.

Colour plates

Plate 21: Top: dust rises from a paddock near Mullaley, north-west NSW, in late 2018. Bottom: smoke from megafires in the Blue Mountains settles on drought-stricken farmland near Wellington, 150 km away, in December 2019. Photos: Robert Godfree.

Plate 22: Good times, bad times. Top: satellite image (Terra Modis, True Colour) of northern New South Wales and southern Queensland taken on 13 August 2016. Bottom: satellite image (SumoiNPP/VIIRS True Colour) of the same region on 19 November 2019. Fires can be seen on the Northern Ranges, Mid-North and Blue Mountains. Both images courtesy of NASA Worldview (https://worldview.earthdata.nasa.gov/).

6

An uncommon and tedious drought

Quis talia fando, temperet a lachrymis?
Who can relate such woes without a tear?
— *David Collins, from Virgil's* 'The Aeneid'

FOLLOWING THE BREAK-UP OF THE SETTLEMENT DROUGHT, THE COLONY experienced several years of higher rainfall and more favourable agricultural conditions. The 1795–96 harvest was spectacular, yielding 35 000–40 000 bushels of grain, enough to supply bread to the colony for 12 months.[1] This was welcome news to John Hunter, who had taken up the office of Governor in September 1795, reporting the harvest to be 'in point of quantity as well as quality, very superior to anything which this country has before experienced'.[2] Indeed, apart from the destruction of several wheat fields by a violent storm with massive hailstones 'three inches square' (7 cm), climatic conditions during this time proved the least of his problems. Upon arrival, the 60-year-old Hunter, who had been seeking a comfortable and rewarding post, instead found that the colony established by Phillip had been run down and corrupted by years of what amounted to military control under Lieutenant Governor Francis Grose. Apart from a burgeoning traffic in spirits established by military officers, and the poverty and disorder that this apparently caused among the 'lower orders', the colony also suffered from almost nightly robberies, a shortage of important store provisions, increasing conflict with the local Aboriginal population and a lack of convict labour for public works.

Nevertheless, while the security of the food supply was hindered by a shortage of agricultural tools needed to cultivate the land, and the lack of any building in which to store the grain harvest or mills to produce flour, prospects for food production remained good. For the first time, the colony was nearing self-sufficiency in flour, and by late 1796 the area of cultivated land had expanded to more than 5400 acres (2185 ha), almost half belonging to settlers. Numbers of livestock owned by the government and officers had also risen to more than 200 cattle, 50 horses and nearly 5000 sheep, hogs and goats.[3] The wet weather continued through December 1796, when heavy rain prevented the harvesting of wheat crops. Abruptly, however, the prosperous times ended with the onset of a sustained period of high summer temperatures and low rainfall between January 1797 and February 1799. Although it has received relatively little attention by climate historians, who have focused instead on the Settlement Drought, at its height this 'great drought' was considered by David Collins to be the worst since settlement.[4] Furthermore, it occurred during a period of increased sea and land exploration across the south-east and marks the beginning of written eyewitness accounts of Australian drought from outside of the Sydney Basin.

> *The night was dark, the wind high, and the fire, from its extent, and the noise it made thro' lofty blazing woods, was truly terrible*
>
> – Governor Hunter to the Duke of Portland, 10 June 1797

The suffocating heat and dry weather that commenced in January 1797 had immediate consequences for the colony. Temperatures were so hot that the leaves of the culinary plants, presumably garden vegetables, were 'reduced to a powder'.[5] Stripped of its moisture by hot winds, the thick and luxuriant vegetation that had accumulated during the preceding wet years became tinder dry, and fires, which might have been intentionally lit, sprang up around the colony, threatening grain stores and livestock and burning down houses and barns.[6] The parched conditions and dryness of the fuel load must have been extreme, for Hunter wrote, 'Trains of gunpowder cou'd scarcely have been more rapid in communicating destruction, such was the dry'd and very combustable state of every kind of vegetation, whether grass or tree [sic]'. These fires were evidently very intense since they burnt to the top of 'lofty woods'.[7] During these conflagrations 800 bushels of wheat burnt, a significant loss to the colony which was compounded by the destruction of large areas of pasture and ensuing grain feeding of cattle.

There is some evidence that conditions ameliorated following rainfall in March and April 1797, and then alternated between being very dry in May to July to periodically wet from August to November.[8] While we lack detailed weather records for this period, rainfall was probably substantial, since an expedition to the Cow Pastures (or Cowpastures) region – roughly extending between Camden, Menangle, Douglas Park and Picton on the Nepean River floodplain – in September 1797 reportedly found 'a luxuriant and well watered pasturage'.[9] In October the wheat crop was also performing well, and the weather remained unsettled during November.[10] However, by January 1798 dry weather had returned, and the maize crop was showing signs of stress.[11] Sporadic but heavy rains fell during June and July, placing construction projects on hold and damaging the road network, but the final and most severe phase of the drought began in August 1798, when fires again sprung up around the colony. At least some, including the one that destroyed the home of the Surveyor-General, Augustus Alt, were probably deliberately lit, but in general the conditions point to rapid drying of the landscape. I suspect that winter rainfall was insufficient to alleviate soil water deficits that had gradually accumulated over the previous summer and autumn, and it is not unusual to see

> *the very flattering prospect of ample crops, which I saw with so much satisfaction, and had every reason to expect wou'd have furnish'd a supply of wheat for at least twenty months to come, exclusive of considerable crops of maize, is at present in a very precarious state from an uncommon and tedious drought, attended with very sultry weather*
>
> – Governor Hunter to the Duke of Portland, 25 September 1798

green vegetation rapidly dry off after just a week or two of high spring temperatures and drying winds under these circumstances.

By September the gardens and pastures were suffering from want of rain and Hunter was voicing concerns over the now precarious state of the wheat crop.[12] These fears proved well founded, because virtually no meaningful rain fell over the next 2 months, and the wheat harvest was only a third of what might be expected in favourable years. Temperatures near the Hawkesbury River climbed to 107°F (41.7°C) in December, with parching northerly winds and widespread fires, the few showers immediately evaporating in the excessive heat. By early January David Collins, who had witnessed conditions in 1790–93, reported that the drought had become the worst since establishment of the settlement, where 'so much continued drought and suffocating heat had not been experienced'.[13] The drought broke suddenly in March 1799 with several days of heavy rain, which refreshed the vegetation but also resulted in a damaging flood along the Hawkesbury River, where houses, infrastructure, livestock, harvests and one human life were lost.[14] Flooding and severe weather occurred again in June 1797 and wet conditions prevailed for the rest of the year before major flooding occurring again in March 1800.

The 1797–99 drought in the Sydney Basin might be described as a relatively short and intermittent period of low rainfall. Only during the last 6–7 months did continuous dry weather prevail at Sydney and judging from the fact that no reports were made of drought and heat stress among native wildlife, maximum summer temperatures were almost certainly lower than those experienced during the Settlement Drought. While grain shortages occasioned by the dry weather were significant, the depressed state of the colony in the latter stages of the drought clearly owed much to corruption and violent behaviour among the soldiery and the exhaustion of stores of clothing and bedding. This left many people, including convict labourers, nearly naked and, as Collins puts it, susceptible to 'the annihilation of morality, honesty and industry'.[15] These problems plagued Hunter until his return to England in 1800, and he bitterly observed that it would have been far easier to have planted a new colony than to have attempted the recovery of one so 'shamefully plungd in profligacy & licentiousness'.[16]

Nevertheless, from a historical perspective the 1797–99 drought is of considerable interest because it marked the first time that several important hydrogeological phenomena were recorded by Europeans. The first of these was that salinity of water in ponds around the colony tended to increase as they dried up, even in areas not prone to marine inundation. The implications of this were understood by Collins, who in February 1799 noted, 'From this circumstance, it was conjectured, that the earth contained a large portion of salt, for the ponds even on the high grounds were not fresh'.[17] Recent research conducted in the western Sydney Basin has revealed that this salt is derived from a combination of ocean spray, dust, weathering of rock and sedimentary sources, and salt load is particularly high in areas associated with shale formations (e.g. Wianamatta Shales). As Europeans explored more of the continent it became clear that salt is prevalent in many Australian soils, and rising concentrations due to land clearing and associated changes in groundwater dynamics, known as *dryland salinity*, now threaten millions of hectares of agricultural land and river systems across Australia, including in the Sydney Basin.[18]

A second geophysical property of landscapes that contributed to the disastrous Hawkesbury flood of May 1799 was also reported by Collins, probably based on data or despatches provided by Hunter. In his account, he proposed that following the long period of drought the earth lacked the ability to absorb water and so ran down the hillsides as if from 'mountains of solid rock'.[19] Interestingly, it is only in recent years that hydrologists and ecologists have come to fully appreciate the importance of this phenomenon, and studies are now being undertaken worldwide to determine why soil run-off often increases dramatically during drought. These have revealed that this pattern can largely be explained by an increase in the fraction of bare soil, reduced water storage capacity of vegetation and leaf litter, and an increase in the *hydrophobicity* (water repellence) of the soil surface as it dries.[20] These changes tend to be self-reinforcing – known as *positive feedbacks* – in which further drying of the soil results in further loss of vegetative cover and increased hydrophobicity, and so the drought history of soil becomes increasingly crucial in determining its ability to absorb rainfall.

The most important consequence of these changes is that during prolonged dry periods soils are often extremely prone to erosion during occasional heavy rainstorms and thunderstorms.[21] Paradoxically, sedimentation of dams, creeks and rivers can be much higher during drought than in wetter periods when soils have greater vegetation cover and lower hydrophobicity. Indeed, on my own property in northern New South Wales, far more erosion damage occurred during the first rains that broke up the severe 2017–19 drought than in the previous three years combined, with waterways, even in ungrazed catchments, becoming rivers of mud and debris (Plate 4). Overgrazing compounds these problems by further reducing water infiltration into the soil, lending credence to the adage 'first into drought, last out'.

The accounts of Collins and Hunter show that by the late 1790s the European settlers were certainly aware that severe, recurring rainfall deficiencies are a natural feature of the Australian climate and that during such periods hydrogeological changes that have the potential to cause land degradation and disrupt agricultural development occur. It is therefore interesting that most of these lessons were quickly forgotten or ignored, and seldom discussed in scientific or government discourse at the time – even eminent scholars such as Joseph Banks, President of the Royal Society, did not appear to see drought as a factor that would significantly influence the fortunes of the settlement. I suspect that this can be attributed to the domination of both scientific writing and government despatches by individuals who were present in the colony only briefly (e.g. Tench, Hunter, Bass; the obvious exception was Collins) or, like Banks, personally invested in the success of the colony. These limitations were compounded by the reluctance or inability of early writers to document the rich knowledge of climate history and landscape processes held by Aboriginal people, despite considerable interest in their culture and customs. This recalcitrance proved detrimental to the early colonists on many occasions, such as during the Hawkesbury River flood of March 1799, when settlers on the river flats suffered devastating losses of infrastructure and livestock. Aboriginal people (probably Darug or Dharug) knew of the danger posed by heavy rainfall in the upper Blue Mountains catchment, and reportedly foresaw the event.[22]

Finally, the 1797–99 drought marks the first time in which shifts in occupation of the landscape by Aboriginal people during periods of extended drought, a strategy that proved central to survival and sustainable water resource management in Indigenous societies across the continent, were recorded. It is at this point that we must turn to the writings of the surgeon, naturalist and explorer George Bass, whose short but productive career marked the beginning of serious Western scientific investigation of natural Australian ecosystems.[23]

Endnotes

[1] *Historical Records of New South Wales* (henceforth *HRNSW*), vol. 3, p. 38 (28 April 1796).

[2] *HRNSW*, vol. 3, p. 30 (3 March 1796).

[3] *HRNSW*, vol. 3, p. 99 (1 September 1796).

[4] Collins D (1802) *An account of the English colony in New South Wales with remarks on the dispositions, customs, manners, etc, of the native inhabitants of that country. Volume 2*. AH & AW Reed in association with The Royal Australian Historical Society, Sydney, p. 103.

[5] Collins (1802), p. 12.

[6] Collins (1802), pp. 12, 15, 17; *HRNSW*, vol. 3, p. 219 (10 June 1797).

[7] *HRNSW*, vol. 3, p. 219 (10 June 1797).

[8] Collins (1802), pp. 20–46.

[9] Collins (1802), p. 37.

[10] Collins (1802), pp. 41, 46.

[11] Collins (1802), p. 60.

[12] Collins (1802), p. 92; *HRNSW*, vol. 3, p. 493 (25 September 1798).

[13] Collins (1802), pp. 98, 100, 103.

[14] I use March 1799 as the termination date of the 1797–99 drought, although sources vary as to the precise timing.

[15] Collins (1802), pp. 7, 154; *HRNSW*, vol. 3, p. 493 (25 September 1798).

[16] Letter from John Hunter to Joseph Banks, 30 March 1797, Banks Papers, Series 38: Correspondence, being mainly letters received by Banks from John Hunter, with related papers, 1795–1802, 1807 (Mitchell Library).

[17] Collins (1802), p. 143.

[18] E.g. NSW Department of Environment, Climate Change and Water (2011) Hydrogeological landscapes for the Hawkesbury-Nepean Catchment Management Authority, Western Sydney; Pannell DJ (2002) *The Australian Journal of Agricultural and Resource Economics* **45**, 517–546.

[19] Collins (1802), p. 143; *HRNSW*, vol. 3, p. 668 (1 May 1799).

[20] E.g. Gazol A *et al.* (2018) *Forest Ecology and Management* **422**, 294–302; Gimbel KF (2016) *Hydrology and Earth Science Systems* **20**, 1301–1317.

[21] Shakesby RA *et al.* (2000) *Journal of Hydrology* **231**, 178–191; Allen PM (2011) *Journal of Hydrology* **407**, 1–11.

[22] Collins (1802), p. 143.

[23] Bowden H (1952) *George Bass, 1771–1803: His discoveries, romantic life and tragic disappearance*. Oxford University Press, Melbourne.

7

Bass and the South Coast drought

... we had every reason to believe that the country at this time to be everywhere unusually dry

– *George Bass,* Journal in the Whaleboat, *21 December 1797*

BETWEEN 3 DECEMBER 1797–25 FEBRUARY 1798, GEORGE BASS CONDUCTED HIS now famous voyage from Sydney to Western Port, Victoria, in an open 28′7″ (8.7 m) whaleboat.[1] During this expedition, Bass, who brought a keen scientific mind to the study of natural phenomena, reported that he had encountered increasingly dry conditions as he proceeded from Port Jackson along the coast, especially south from the Jervis Bay area. Indeed, the crew were forced to resort to tapping trickles of fresh water that drained from beach-edge sandhills for their needs, and at Wingan Inlet in far eastern Victoria, Bass feared that the expedition would need to return to the north to procure fresh water.[2] At Ninety Mile Beach he found dried up swamps, and at Western Port had great difficulty in finding fresh water, there being 'every appearance of an unusual drought in the country'.[3]

Bass's observations at Tuross, some 265 km south of Sydney, were particularly interesting. After landing just south of Lake Tuross,[4] Bass and his crew spent the day investigating the rocky headlands, forests, swamps, sand bars and estuaries that comprise the coastal landscape. Here, he writes: 'The country seems to be at all times but sparingly watered, but it is now in a state of drought. In the course of our round of not less than 12 or 14 miles we could not find a drop of fresh water, altho' the heat of the day made us search for it with extreme eagerness.' During this expedition, he also 'met with numbers of native huts deserted, the cause of which appeared when we traced down their paths to the dried up waterholes they had dug in the very heart of the largest of the swamps'.[5]

Clearly, Bass thought that the apparent absence of Aboriginal activity in the area was due to lack of fresh water – similar to what he proposed at Western Port. These observations are ethnographically significant because they suggest that Aboriginal societies were forced to reorganise during drought not just in arid and semi-arid areas, as is discussed in later chapters, but even in wetter temperate coastal environments where dry periods are much less severe. As such, we must further scrutinise them on climatic, hydrological and anthropological grounds.

First, Bass's observations of drought along the south coast of New South Wales in December 1797 appear to be inconsistent with conditions in the Sydney Basin, where average to wet conditions prevailed between July and November 1797. This does not mean that Bass was mistaken, because key drivers of climate and rainfall (e.g. ENSO, Indian Ocean Dipole, position of the subtropical ridge) exhibit strong regional variability across south-eastern Australia,[6] and severe droughts restricted to southerly coastal areas of New South Wales and Victoria have occurred on many occasions (e.g. in March 2008–February 2009 during the Millennium Drought).[7] Bass's report of increasing dryness as he progressed south is therefore plausible, but I think it surprising that neither he nor Matthew Flinders mentioned this during their 7 October 1798 to 12 January 1799 voyage aboard the *Norfolk* at a time when drought at Sydney had grown very intense.[8] However, during his visit to Preservation Island, which lies roughly 30 km north of Tasmania near Cape Barren Island, Bass did state that unusual falls of rain had recently filled small creeks and swamps on the island, but that water had been very scarce when the survivors of the shipwrecked *Sydney Cove* had been stranded there in February to June 1797, just before he made his whaleboat voyage along the southern mainland coast.

From a hydrological perspective, Bass's explorations in 1797 were mainly restricted to coastal and near-coastal tidal estuaries and sandy areas between Potato Point and Lake Tuross where, as Phillip discovered at Botany Bay, fresh water can be scarce even during wet years. For example, Lake Tuross and the Tuross River estuary, which Bass specifically mentioned, are usually open to the ocean and thus salty, and tidal influence in the Tuross River extends more than 10 km from the coastline. Even in March 2021 during a very wet year I found that water in the river at the bridge crossing north of Bodalla, which is far beyond the distance inland that Bass is likely to have ventured, was too brackish to drink.[9] Most other waterbodies that lie close to the coast comprise either shallow lagoons or creeks that are intermittently open to the ocean and are therefore slightly to strongly brackish and unfit for drinking. During his journey on 17 December 1797 Bass evidently covered much terrain, but it is possible that his inability to find fresh water was due more to this regional preponderance of saline waterbodies than to drought.

Nevertheless, I find this explanation unsatisfying. Bass was by this time a seasoned explorer who was aware of at least the basic hydrological features of estuaries along coastal New South Wales. Furthermore, he directly observed swamps that Aboriginal people had been using as a water source but had since gone dry. Where might these swamps have been? The most likely candidates are the small chain of lagoons and swamps that lie behind dunes at Jemison's, Potato Point and Blackfellows Beaches between Jemison's Point and Tuross Lake (Plate 5). These lie along the apparent path that Bass took in 1797 and are known to dry up during severe drought – the shallowest, near Jemison's Beach, took only around 6 months to do so during extremely dry weather in 2002 and more recently in 2018–19.[10] Others persist longer, but even much larger swamps near Moruya Heads dried up during the height of the Millennium Drought (ca. 2000–09). It is thus plausible that Bass may have seen some of these lagoons dry, or low, in 1797, after perhaps 6–12 months of drought.

The quality of water in these swamps presents a more complex picture. When I tested their salt content in March 2021, a relatively simple procedure that is based on measuring electrical conductivity, I found that water from these swamps had an electrical conductivity of around 4700 microSiemens per centimetre (µS/cm). This is classified as moderately brackish: very much below that of seawater (55 000 µS/cm) but significantly above the maximum acceptable standard for human drinking water (ca. 2000 µS/cm). I also tested water at a swamp behind

Box 7.1: The monster bird of Tuross

On 17 December 1797 George Bass and six seamen landed in a whaleboat about a mile (1.6 km) to the south of Tuross Head while on his voyage to Bass Strait and the southern coastline of Victoria. While investigating the swamps and lagoons of the Tuross River estuary, he observed that many huts of the local Yuin people appeared to be deserted, which he attributed to the fact that the area was in a state of drought. Nearby the huts, waterholes had been dug in the middle of the largest swamps, but these too were dry. He encountered no Aboriginal people despite walking more than 10 miles (16 km) during his exploration of the area.

However, one local oral tradition suggests that far from being deserted, local Yuin people were in the area and witnessed Bass's arrival. In an article published in the *Canberra Times* (28 October 1989, p. 25), the Eurobodalla historian Noel Warry retells the story of a young boy named Cooral, who woke while sleeping with his people among cliffs near Potato Point to see the sails of Bass's whaleboat out to sea. Thinking they were a great white bird of prey, they fled to near Coila Lake, leaving all their possessions behind. Later a small number returned to their camp, which they found unmolested, and were terrified to find the tracks of 'toeless creatures' in the sand. After again fleeing to spend another uncomfortable night among the trees, the group returned, and never saw the 'white bird' again. Later, similar stories trickled in from elsewhere along the coast, and the 'coming of the spirits of men, turned white'.

This is perhaps the first instance in which strong counter-narratives developed based on differing European and Aboriginal experiences of drought in Australia. While Bass saw a drought-affected landscape deserted by its inhabitants, Cooral's account suggest that that Yuin people were simply living in a different location at the time, perhaps closer to permanent springs – a strategy frequently adopted during times of contracting water supply.

Left: The Tuross area, home to the Yuin people, looking north from near Blackfellows Point, with Tuross Head and Tuross Lake in the far background. Photo: Robert Godfree. Right: Stamp commemorating Bass's navigational achievements, with whaleboat in the background. Designer: Walter Jardine.

Blackfellows Beach and while this was fresh and drinkable, it appears that earthworks have altered connectivity with the ocean. One explanation for this discrepancy is that fresh springs at these swamps might become accessible during drier periods – according to the oral tradition of local families of the Yuin Nation, marshland near Blackfellows Point was fed by a freshwater spring the water of which rose with the incoming tide.[11] Changes over the last 100–200 years in these areas due to cessation of Aboriginal burning have also probably been very significant, as they have across much of Australia, for today the landscape around these marshes contains a very dense understorey with thickets of species like the aptly named tall sawsedge (*Gahnia clarkei*), along with healthy numbers of paralysis ticks!

Does this mean that drought had caused Aboriginal people to abandon the region, as implied by Bass? No. A fascinating oral tradition said to be based on the recollections of young Yuin man named Cooral, an eyewitness to Bass's arrival (Box 7.1), shows that Aboriginal people were still present nearby on the cliffs at Potato Point. In this version of events, local Yuin people simply fled northwards towards Coila Lake after seeing Bass's whaleboat out to sea, returning only after his departure. This account specifically mentions springs, which unbeknownst to Bass are scattered throughout the area – including the south side of Tuross Lake. The impact of drought in coastal areas has for too long been overlooked, and historical and traditional narratives of this kind add much to our understanding of Aboriginal water use in these environments.

THE END OF THE EARLY COLONIAL PERIOD

Bass's voyage and the 1797–99 drought mark the close of the early colonial period in New South Wales, a time when the Sydney settlement had only a tenuous grip on survival. Here, the colonists had experienced a taste of the climate that Australian was to become famous for: severe summer heatwaves, bushfires fanned by furnace-like winds and recurring multi-year droughts. On two separate occasions these conditions had pushed the settlement to the brink of despair, and from the perspective of human suffering the impacts of these droughts, and particularly the famine endured during the Settlement Drought, were undoubtedly among the worst ever faced by Europeans in Australia. However, even by the onset of the 1797–99 drought, several agricultural improvements, such as expansion of the cultivated land base, improvements in grain storage and rapid growth in livestock numbers had significantly improved the ability of the colonies to withstand periods of low rainfall, and by 1800 self-sustainability was firmly in sight. Yet success brought new pressures, as the land and resource requirements of the colony soon began outgrowing those available in the Sydney Basin. Plans were soon afoot to expand across the Blue Mountains and northwards into the Hunter Valley, and here settlers were to encounter droughts of a new and more formidable kind. It is to this era of expansion that we now turn.

Endnotes

[1] Cole H, Cole V (1997) *Mr Bass's Western Port: The whaleboat voyage.* Hastings-Western Port Historical Society in conjunction with South Eastern Historical Association, Hastings, Vic.

[2] Mr. Bass's Journal in the Whaleboat, between the 3rd of December, 1797, and the 25th of February, 1798. *Historical Records of New South Wales* (hereafter *HRNSW*), vol. 3, p. 319.

[3] *HRNSW*, vol. 3, p. 324.

[4] Bass landed near Potato Point, also known as Marka Point, to the south of the outlet of Lake Tuross; *HRNSW* vol. 3, p. 317.

[5] *HRNSW* vol. 3, p. 318.

[6] Duc *et al.* (2017) *Theoretical and Applied Climatology* **127**, 169–185; Timbal B, Drosdowsky W (2013) *International Journal of Climatology* **33**, 1021–1034.

[7] Other recent annual to 18-month periods of drought in south coastal New South Wales and southern Victoria occurred in 2002–03, 1982–83 and 1972–73.

[8] The narrative of the voyage is provided in *HRNSW*, vol. 3, Appendix B, pp. 769–818. David Collins provides an account of Bass's notes in Collins D (1802) *An account of the English colony in New South Wales with remarks on the dispositions, customs, manners, etc, of the native inhabitants of that country. Volume 2*. AH & AW Reed in association with the Royal Australian Historical Society, Sydney, pp. 103–139.

[9] Goodrick GN (1970) 'A survey of wetlands of coastal New South Wales'. Division of Wildlife Research Technical Memorandum No. 5, CSIRO, Canberra.

[10] Wet–dry cycles in these lagoons can be observed in Google Earth historical imagery.

[11] As reported by Lionel Mongta (2006); cited in Yuin Moruya Community storylines from Blackfellows point, Potato Point and Jamison Point at https://moruya.storylines.com.au/2016/10/12/blackfellows-point-potato-point-and-jamison-point/.

Part 3
A sky untroubled by clouds
———

8

The poverty of confinement

if, without any preparation whatever against the withered herbage or the rivulet's suspended flow, to varying chance some still confide their care, they cannot well complain should chance at some eventful period deceive them.
– Sydney Gazette and New South Wales Advertiser, *19 October 1811*

WHEN THE 1797–99 DROUGHT FAILED TO REPEAT THE TERRIBLE PRIVATIONS experienced 5 years earlier, it had become clear that climatic extremes no longer posed an existential threat to the food security of the colony. Furthermore, new discoveries, such as the presence of exportable quantities of coal at the mouth of the Hunter River near Newcastle[1] and areas of fertile land along the southern Australian coast and Tasmania, showed that the prospects for expansion were good. True, agriculture remained almost solely restricted to the Cumberland Plain, but the wet conditions that prevailed in late 1799–1801 favoured grain production, particularly in late 1799, when the crops were reportedly the richest ever beheld in this or indeed any country.[2] While occasional setbacks still occurred – severe storms and grubs later damaged much of the grain and fruit and garden crops, and floods in 1800 and 1801 destroyed tens of thousands of pounds worth of grain and property on land adjacent to the Hawkesbury and Georges rivers[3] – by 30 June 1801 the area of crops planted had expanded to 5333 acres (2158 ha) of wheat and 3864 acres (1564 ha) of maize, an 86% increase in just 5 years.[4]

Livestock numbers in the colony were also beginning to rise rapidly. Prior to the mid-1790s growth in the national sheep and cattle supply had been painfully slow, with the tiny herds so unsuited to conditions at Sydney Cove (Fig. 8.1)[5] increasing to just 40 cattle and 526 sheep by June 1794 and to 176 cattle and 832 sheep by June 1795.[6] Most of the increase in cattle numbers during this time can be attributed to the importation of new animals from overseas, particularly 19 from the Cape Colony by Lieutenant-Governor King aboard the *Gorgon* in 1791 and 132 from Surat, India, in May 1795 aboard the *Endeavour*.[7] Other key early importations of livestock, especially sheep, came aboard the *Atlantic* (1792), *Daedalus* (1793), *Shah Hormuzear* (1793), and the *Reliance* and *Supply* (1797).[8]

However, once the open woodland and grasslands of the western Sydney Basin were opened to graziers, raising livestock proved increasingly attractive, especially compared with the arduous labour that was required to clear land for grain production. Sensing opportunity, wealthy and influential landholders like John and Gregory Blaxland began to seriously develop the embryonic beef and wool industries, but as it often came at the exclusion of raising grain, the increased focus on beef production drew the ire of Governor Lachlan Macquarie. Inevitably,

Fig. 8.1: Bolger, J. (1803). *Walloomoolloo The Seat of Jno Palmer Esqre Port Jackson* [watercolour]. Palmer's 100-acre Woolloomooloo Farm at Garden Island Cove, now Farm Cove. The horned cattle depicted did not thrive in swampy and sandy areas close to the coast, but livestock enterprises expanded rapidly on the more fertile pastures in the central and western Sydney Basin. By 1803 there were more than 2000 cattle and 9000 sheep in the colony, with thousands being added annually. Mitchell Library, State Library of NSW, SV1A/Wllo/3.

however, control of the national herd gradually shifted away from government and into the hands of private interests.

Less than 5 years later, in June 1801, the colony boasted 1242 cattle (excluding the hundreds of wild cattle living in the Cowpastures region) and 7046 sheep, 29% and 89% of which respectively were now privately owned. So great was the rate of growth that Governor Phillip Gidley King predicted that within 2 years there would be no further need for imported stock.[9] This was put to the test when rainfall tapered off in mid-1802, ushering in a phase of intermittent dryness that was to last until February 1805. By November 1802 nearly 4 months had passed without meaningful falls,[10] but since soil water had built up during the previous wet years, crop production remained 'tolerably good'.[11] Further intensification of rainfall deficiencies throughout February 1803 resulted in the destruction of gardens and the retardation of maize and potato crops planted in the spring and summer,[12] but wet weather in autumn saw a break in the conditions. Rainfall was patchy in the spring and summer of 1803–04, and dry conditions continued throughout January 1805, when there were complaints of drought on the Hawkesbury with stubble corn being affected and peaches and other fruit falling off trees.[13]

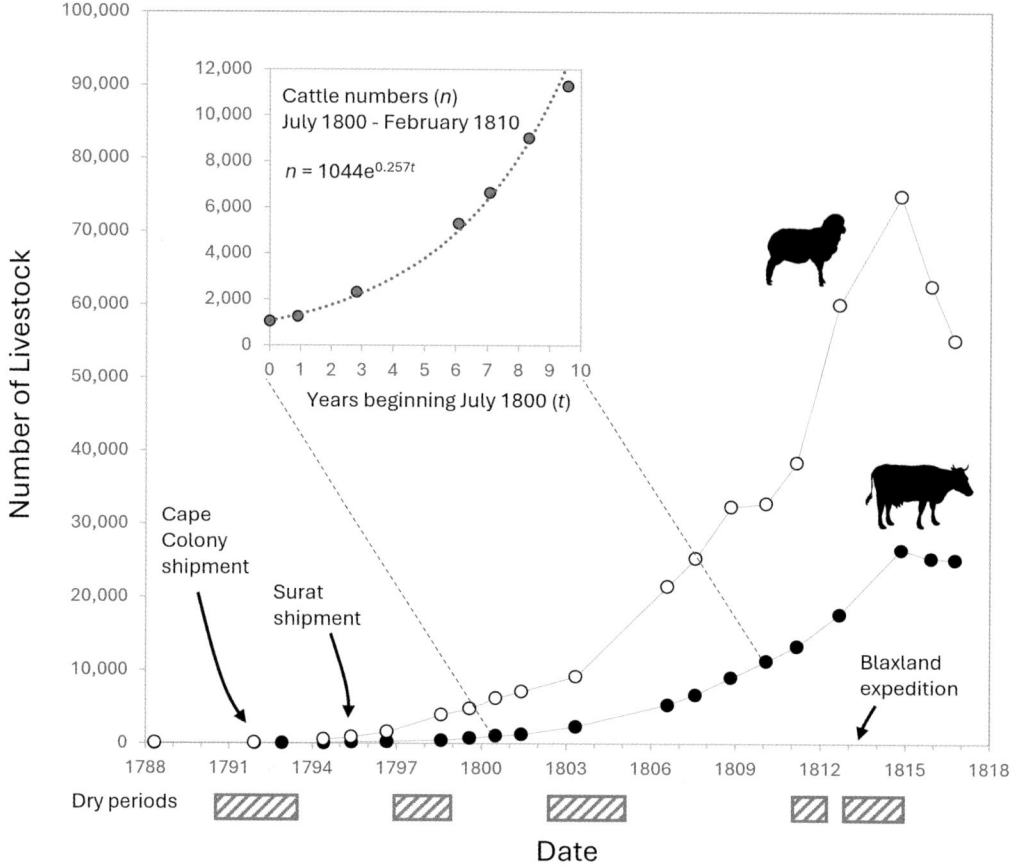

Fig. 8.2: Approximate numbers of cattle and sheep in the Sydney Basin based on official livestock returns, 1788–1816. The inset shows cattle numbers during the period July 1800 to February 1810, which show a very close fit to an exponential growth function. Based on this, the doubling time (t_d) for cattle numbers over this 10-year period can be estimated as $t_d = \ln(2)/0.257 = 2.7$ years or ca. 32.4 months. The rate of growth slowed slightly in the years immediately before the onset of the 1813–15 drought, but numbers were still increasing by thousands of head per year. By 1813 the Sydney Basin was overstocked, leading to mortality among cattle and sheep when dry conditions intensified.

Just like the 1797–99 drought, however, this spell of intermittent dry weather did nothing to slow the growth in stock numbers. Livestock returns (Fig. 8.2) show that by August 1806 the colonial herd had grown to 5286 cattle (a 326% increase from June 1801) and 21 457 sheep (205% increase), an exceptional rate of growth, and not far less than the maximum possible growth rate of managed cattle populations under ideal breeding conditions. Even allowing for some inaccuracies in livestock counts and the possible inclusion of both imports and wild cattle from the Cowpastures (estimated in 1806 at up to 4000 head)[14] into official returns during this period, livestock numbers in the colony were not fundamentally restricted by the food supply, since graziers could support hundreds to thousands of new individuals per year by accessing new country on the southern and western fringes of the colony.

Flooding rains in February emphatically terminated the 1802–05 dry period, and, as often occurs, an unusually long run of rainy seasons and excellent pasturage ensued. Conditions in early 1805 were so damp that a large part of the maize crop rotted, and a great flood occurred on the Hawkesbury River in 20–23 March 1806 after months of wet weather and a deluge in the week prior.[15] Floods now proved a greater threat to security of the grain supply than drought, and again the colony endured a severe food shortage. But apart from setbacks in 1808, when smut (a fungal pathogen) killed much of the grain crop,[16] and in mid-1809 when floods impacted on farmers on the Hawkesbury, Grose, Nepean and Georges rivers,[17] conditions were favourable for graziers, and by February 1810 an estimated 11 276 cattle and 32 818 sheep were depastured across the colony (Fig. 8.2).

DOUBLE TROUBLE

It is time to use simple mathematical models to reveal some interesting features of this population growth and the real-world consequences for land management. First, between 1800 and 1810 an exponential growth curve provides a very close fit to cattle numbers reported across the colony (Fig. 8.2). Populations that are growing exponentially are defined as undergoing a constant percentage increase in population size per unit time, and double within a fixed time interval. In this case, starting with 1044 cattle in July 1800, the population doubled roughly every 2.7 years (32 months 12 days). We can express this as using the equation $n(t) = 1044 \times 2^{(t/2.7)}$ where $n(t)$ is the population size after a given number of years t.

The most important feature of such populations is that they can stay small for an extended period before undergoing rapid, explosive growth that catches land managers by surprise. This dynamic was clearly underway between July 1800 and May 1803, when the population grew more (1253 head in 34 months) than it had during the previous 12 years (1044 head). By 1806 thousands of animals were being added annually. The problem with such systems is that they are inherently unstable, because in the absence of immediate negative feedbacks that limit population growth per capita, resource availability declines drastically as the population rises, provided the total size of the resource is constant. A famous example is that of a Petri dish containing a bacterial population that doubles every day – if the Petri dish is completely full at 100 days, half of the growing space still exists after 99 days. More sophisticated models can incorporate negative feedbacks as populations as their populations approach the maximum size that the environment can sustain (termed the *carrying capacity* or K), but population overshoot and subsequent collapse can still occur in these situations.[18] The critical point is that failure to recognise and plan for declining resource availability for exponentially growing populations is a recipe for disaster.

The second and perhaps more important consequence is that the problem of declining per capita resource availability is compounded when the underlying resource base, and hence K itself, undergoes a sudden reduction. A familiar situation is when a cattle herd that has expanded in size on the back of a run of good seasons is suddenly faced with a 90% or more reduction in pasture growth during drought. If the manager has not been prudent and secured an alternate food source like hay, grain, or offsite pasture agistment, the herd is suddenly faced with starvation. Modern livestock management practises emphasise rigorous planning for food shortages during drought, and I have frequently heard the argument made that the devastating stock losses that occurred during, for example, the 1895–1903 Federation Drought resulted from a lack of appreciation by early graziers of the inherent volatility of the Australian climate and pasture base. However, the crisis that was to unfold during the 1813–15 drought suggest that ignorance is, at best, an incomplete explanation of the drivers of such events.

Among the first to recognise the fundamental limits to the grazing capacity of the Sydney Basin and take action was Gregory Blaxland (Plate 6), one of the very first 'gentleman settlers' who arrived in the colony under an arrangement with the government to bring capital and other effects to the colony in exchange for the promise of land. Soon after arriving in April 1806, he and his brother John purchased two leaseholds and established a dairy at Sydney. Like others before him he found that cattle fared poorly on the sandy coastal soils, and so relocated to more favourable land in the Orchard Hills area between South Creek and the Nepean River. Here he was ultimately granted more than 4000 acres (1619 ha) by Governor Macquarie in 1810, along with convicts to work the land.[19] Macquarie had little regard for gentleman settlers, whom he considered a drain on the colony, later describing the Blaxlands as 'the most discontented unreasonable and troublesome persons in the whole country.'[20] Whatever the truth, the move to the colony proved lucrative, for by November 1812 Gregory Blaxland owned 140 head of horned cattle and 795 sheep; the cattle alone were worth £2100 at a time when the cost of providing provisions, clothing and bedding for a convict was around £25 annually.[21]

Unfortunately, he soon observed that despite the superior soils in this new country 'a great part of the grass when hard fed died away', and that the pasturage when closely grazed and manured quickly became less productive.[22] I suspect that Blaxland was observing the decline of native grasses such as *Themeda triandra* (kangaroo grass), which were once common across south-eastern Australia but quickly replaced by nitrophilous (nitrogen-loving) introduced grasses and forbs once fertilised or grazed heavily (Plate 6). His brother John also noted that these species tended to grow rapidly and become coarse and rank, prompting graziers to regularly burn it off to provide cattle with fresh growth. Unfortunately, this destroyed whatever pasture base was available, in turn requiring graziers to run their mobs at some distance from their properties until the grass recovered.[23] These factors, along with an invasion of army worms, the conversion of surrounding land into cultivation, and the growing shortage of unoccupied pastures east of the Blue Mountains led Gregory Blaxland to realise that within a few years *even in the absence of further increases in numbers* he would need to move his herd to new land.[24]

HUMAN ART AND PROVIDENCE

The first serious shock to graziers arrived with the onset of extremely dry conditions in 1811 after 6 years of relatively high rainfall. It is difficult to determine when this period of drought started, but conditions were apparently good in late November to mid-December 1810 when Macquarie visited western parts of the colony.[25] The following summer must have been essential rainless, because by March 1811 the maize crop was reportedly beyond recovery and the normally reliable tanks on the Tank Stream at Sydney had been dry for several weeks, with the small quantities being available from a spring upstream being sold at 4–6 pence per pail.[26] Conditions were probably drier that at any time since the Settlement Drought, although we cannot rule out the possibility that landscape changes (e.g. clearing) and increased demand also contributed to the diminishment of the water supply. But by October the long succession of dry weather had still not broken, and the *Sydney Gazette and New South Wales Advertiser* stated that so intense a dry spell at this time of year not seen since 1789 – probably referring to the extremely dry period of July to October 1790.

Cattle and sheep were presumably kept alive during this period on the pasture base that had grown during previous wet years, but it was becoming clear that livestock numbers had increased to unsustainable levels, and that the visitation of another drought as severe as 1789 would be 'incalculably severe'.[27] The blame for this lack of foresight was placed squarely at the feet of landholders who, motivated by self-interest, had pursued an aggressive expansion of livestock numbers for more than a decade. Sympathy was expressed for the predicament of drought-affected livestock in newspaper columns at the time, marking the embryonic beginnings of the movement within Australia to hold livestock owners responsible for the welfare of animals in their care, a battle that still rages today.[28]

While they bear ultimate responsibility, many farmers at the time had only a tenuous foothold on the land and so also deserve some sympathy for their predicament. Their concerns, and those of more influential administrators, were understandably focused on the short-term, since the devastation wrought by drought, hail, fire, disease and pests had serious implications for crop production, the food supply, and hence the social stability of the colony itself. People faced with the immediate problem of survival or living 'hand-to-mouth' have little opportunity to plan for future events since time itself becomes a scarce commodity, and much is required to allocate and manage scarce resources that are critical for survival. Thus, in addition to the huge herds run by wealthy landowners, poverty also probably contributed to the general problem of overstocking across the colony.

Providence intervened in March 1812, when the situation was saved by general rains that put several freshes down the Hawkesbury River.[29] Heavy rains and thunderstorms continued through the end of 1812, resulting in flooding and minor damage to infrastructure.[30] Livestock numbers swelled further (Fig. 8.2), and wild cattle residing in the Cowpastures also now numbered in the thousands. Some rain fell, perhaps in local storms, during February 1813, but dry weather returned in March 1813 with the failure of the expected autumn rains.[31] Nevertheless, food shortages were far from everyone's mind, for the wheat and maize crops

> *it is our duty fervently to pray that Providence may defend those creatures that instinctively lay claim to her benign consideration and protection. By every aid that human art can give, those nevertheless that are immediately interested in their increase should to the best of their ability most certainly provide*
>
> – The Sydney Gazette and New South Wales Advertiser, *19 October 1811*

had been so prolific in all parts of the colony and in Tasmania that there was no market for the grain,[32] and settlers began to feed it to horses, cows, pigs and dogs. By August 1813 this ill-advised profligacy had exhausted any remaining surplus grain, much to the chagrin of Governor Macquarie, who feared that the quantity remaining would be insufficient to last the colony until the next harvest.[33] Virtually no rain had fallen for 5 months.

THE ROAD TO SALVATION

Thanks to the detailed (albeit imperfect) record keeping of colonial officials and decades of work on pasture productivity conducted by agricultural scientists, it is possible to go beyond descriptive accounts and measure the squeeze placed on the colony by the combination of reduced land availability and high livestock numbers in 1813. We know that in September 1812 there were 17 678 cattle, 59 949 sheep and 1618 horses (along with 2683 goats and 15 710 hogs) in Sydney, Parramatta, Windsor and the vicinity, all living on 127 136 acres (51 450 ha) of pasture country and perhaps a portion of the 14 860 acres (6013 ha) sown to wheat, maize, barley and oats.[34] The most common approach to compare the food requirements of these different classes of livestock is to use the *dry sheep equivalent*, or DSE (Box 8.1), which is the amount of feed needed by a two-year old, 45–50 kg Merino wether to maintain its weight. Mature cattle and horses typically consume the feed requirements of around 9–10 dry sheep, and females that are pregnant or feeding young have even higher nutritional demands.

Conservative 'back of the envelope' calculations suggest that cattle, sheep, horses, goats and hogs across the colony in 1812 together comprised at least 330 000 DSE, excluding several thousand wild cattle living in the Cowpastures.[35] At the time there was 51 450 ha (127 136 acres) of pasture available for these animals, or around 6.5 DSE per hectare, which is towards the maximum carrying capacity of unimproved coastal pastures, although the decline in productivity observed by Blaxland suggests that the value may have been significantly lower. The inescapable conclusion is that the colonial pastures were at best fully stocked, and more likely heavily overstocked, by the end of 1812. The shortage of pasture had also been exacerbated by landholders who burned off their remaining dry grass in mid-1813 in the hope that rain would stimulate new, more palatable growth. While this practice can be effective in good seasons, it was ill-advised in 1813, because it destroyed a valuable drought resource. It effectively ended when Governor Macquarie recommended in a Public Notice on 18 September 1813 that landholders should strictly avoid burning off until conditions improved, and that anyone intentionally or inadvertently burning the land of others would be rigorously prosecuted and punished.[36]

Lacking new areas for expansion and surplus grain supplies to make up the shortfall, the situation for graziers had become critical. The pasture now exhausted, malnourished ewes lost an estimated 80% of their lambs after weaning, and then began to die from disease. Cattle, enfeebled by the need to travel further afield for the diminishing feed supply, lay down and perished in the paddocks, became bogged in the mud that surrounded the drying waterholes, or succumbed to illness. Overcrowding contributed to mortality, since it was remarked that certain herds were confined to paddocks far smaller than they required, and that splitting them up would have reduced mortality.[37] By late September 1813 reports of

Box 8.1: Dry sheep and stocking rates

As sheep and cattle graziers spread into different Australian environments and began to optimise numbers of livestock and other animals on their properties, it became important to develop a simple method for estimating the feed requirements of their herds and the carrying capacity of their land. By the mid-1850s it was becoming normal to describe carrying capacity in terms of numbers of dry sheep, and graziers were familiar with rough conversions among different classes of livestock based on feed consumption (e.g. one steer is the equivalent to eight dry sheep; *The Australasian* (Melbourne), 21 Nov 1896). Interest in the capacity for feral competitors to displace sheep grew with spread of rabbits across runs in Victoria, South Australia and New South Wales (e.g. six rabbits eat as much as one sheep; *The Capricornian* (Rockhampton), 14 June 1884), and understanding of the comparative nutritional requirements of livestock increased dramatically from 1900 to the 1930s (e.g. *Western Mail* (Perth), 24 September 1925; *Weekly Times* (Melbourne), 16 May 1936).

By the 1940s the concept of the 'dry sheep equivalent' or DSE was in widespread use and in subsequent decades it became the standard means of determining stocking rates and managing the feed supply across Australian mixed agricultural systems. One DSE is defined as the energy required to maintain the weight of a 2-year-old, 45 kg or 50 kg (depending on the source) Merino wether or non-lactating ewe, which is ca. 7.5–8.8 megajoules (MJ) per day. Typically, one DSE consumes 1 kg of green dry matter (pasture) per day. Tables are available that provide typical DSE values of other livestock classes; for example, a 40–50 kg Merino ewe with twin lambs at foot has a DSE rating of around 3, while for a large lactating cow it is around 20. By making some simple assumptions about the demographic structure of livestock and wild animal populations (e.g. rabbit = 0.07–0.12 DSE, kangaroo = 0.7 DSE) it is possible to determine the total feed requirements of grazing animals on a property or in a region.

Left: *Launceston Examiner*, 7 August 1852. Right: Merino ram from Bando Station on the Liverpool Plains. *Illustrated Sydney News*, 16 February 1866.

the magnitude of the losses rolled in: numbers of cattle and sheep had fallen by at least 3000 and 5000 head respectively from those reported a year earlier, indicating that many thousands had recently died.[38] Perhaps three-quarters of the wild cattle on the Cowpastures also succumbed (Fig. 8.3). While these numbers pale in comparison to the slaughter that

Fig. 8.3: The Cowpastures area near Menangle in the western Sydney Basin, where the lost seven cattle from the First Fleet were rediscovered in 1795. By 1810 Governor Macquarie estimated that the Cowpastures contained 4000–5000 cattle, but in 1817 only around a quarter remained, with 'immense numbers' dying during the 1813–15 drought. Photo: Robert Godfree.

occurred during later droughts, the disaster was nevertheless the worst seen in the colony since European settlement, and the first to bring the issue of livestock management into mainstream discussion.

In the meantime, Gregory Blaxland was following through on his conviction that the path to greener pastures lay westward over the Blue Mountains. Recently historians have suggested that John Wilson, a freed convict who lived with Aboriginal people near the Hawkesbury River, may possibly have crossed or skirted the mountains using Aboriginal 'highways' to reach the upper Coxs River valley before 1797, and that this information had been suppressed by colonial powers who were motivated to dissuade free settlers or escaped convicts from leaving the Sydney Basin. Nevertheless, attempts to cross the Blue Mountains immediately to the west of the colony had eluded notable early explorers such as George Bass (1796) and George Caley (1804), and in 1812 there remained no direct route over the mountains known to Europeans, much less an established track along which to transport stock.

The crucial breakthrough, at least from a political perspective, was made by Blaxland in May 1813, when, together with William Wentworth, Lieutenant William Lawson and a small number of men, he descended into the upper reaches of the Coxs River valley in the vicinity of modern-day Hartley Vale 2 weeks after departing from his farm near South Creek (Plate 6). Here they found extensive grassy meadows, clear of trees and well watered by Coxs River and the River Lett, which they thought to be sufficient to provide feed for the colonial livestock for the next 30 years.[39] This assessment proved overly optimistic, since the party had actually not crossed the Great Dividing Range and the more fertile and extensive pastures of the Bathurst Plains lay slightly further west.

Nevertheless, Macquarie found these reports compelling enough to have the route confirmed, and then extended to the Bathurst Plains, by Assistant-Surveyor George William Evans in November 1813 to January 1814. The first European road, little more than a cart track, was constructed by Lieutenant William Cox and a team of convicts, settlers, soldiers and Aboriginal guides between July 1814 and January 1815. Thus, after more than two decades of stagnation, new pastures had been discovered and opened to development within a span of less than 2 years. Interestingly, multiple additional routes were quickly discovered across the Blue Mountains as soon as colonial powers and settlers had become motivated to open the country for settlement.[40]

While obviously a historically important event that ultimately led to rapid settlement of the Bathurst region and a frontier war with the Wiradjuri people in 1824 (known as the 'Bathurst Wars'), the new road across the Blue Mountains actually did little to alleviate the impact of drought in the Sydney Basin. Here, extremely dry conditions continued through the end of 1813, and by January 1814 the full weight of both another harvest failure and ongoing livestock losses was falling heavily on the shoulders of the European lower and middle settler classes.[41] Extensive fires were burning in the Blue Mountains, and once again a shortage of water and feed was causing cattle to sicken and die[42] and, despite occasional rain, by September 1814 pastures and the wheat crop were again looking poor and the condition of stock was falling off.[43]

But Macquarie took a cautious approach to settlement expansion, and even when concerns were again raised about impending stock mortality in May 1815[44] a public order prevented passage over the mountains to the new western pastures without a written pass.[45] Cattle movement over the mountains was therefore very limited: in mid-1815 only a small number of government stock were depasturing near Coxs River, and in October 1815, 100 head were driven back across to the Sydney Basin for slaughter. But by the time larger drives might have alleviated pressure on the Cumberland Plain, the drought there was sputtering to an end. Rain fell early in the spring of 1815, and after a brief return to dry conditions,[46] the decisive break came in January–February 1816 with heavy rains that spoiled much of the wheat crop, saturated the earth, filled lagoons and hollows, and sent freshes of water down the Hawkesbury River.[47] This marked the onset of one of the wettest periods in south-east Australia since European settlement, which, as previously discussed, culminated in the filling of Lake George in 1823.

WEST OF THE RANGES

Surprisingly, the first significant accounts of conditions west of the Cumberland Plain, which occur in the journals of the 1813 Blaxland expedition, reveal that in May wet conditions prevailed in and west of the Blue Mountains. The explorers encountered swamps and runs of water in the heads of gullies in the vicinity of the modern towns of Linden and Hazelwood,[48] and Coxs River and the River Lett near Hartley were both flowing swiftly with an abundance of grass.[49] We must be careful when comparing historical with modern river observations to infer drought severity, since catchments across much of Australia have since been altered by impoundment, vegetation clearing, soil erosion and water extraction. However, the River Lett's rugged catchment remains forested and unregulated, and yet it still stops flowing during severe droughts, most recently in 2019.[50] I take this to mean that any dry conditions in this area prior to May 1813 must have been comparatively mild and brief.

Six months later, when livestock were dying on the drought-stricken Cumberland Plain, George Evans also reported the River Lett[51] to be flowing well, and that water was plentiful in the drought-resilient chains of pools that formed in low-lying parts of the surrounding terrain (Plate 7).[52] On the western side of the divide, Campbells River was flowing above its junction with the Fish River and the grass was so green and thick that it was fatiguing to walk through.[53] A year later William Cox and his crew endured heavy rain, cold and sickness during their construction of the Bathurst road, and on 27 November 1814 he reported the Fish River to be high following days of rain.[54] These records all confirm that severe drought during 1813–14 did not penetrate inland beyond the coastal hinterland.

The first solid evidence of drier conditions in this region were reported by Macquarie himself in May 1815, when he noted that Campbells River had been essentially reduced to a chain of pools several kilometres south of its junction with the Fish River. While on the same tour Henry Antill observed that the Fish River had been reduced to a small stream and that country near the junction of the Campbells and Fish rivers had recently been transformed by grass fires. On his expedition from Bathurst to the Lachlan River in May 1815, Assistant-Surveyor George Evans found the Belabula River to the south of Mt Canobolas reduced to pools, and deeper sections of the large creeks that lie between Canowindra and Cowra dry. When Evans arrived on the banks of the Lachlan River on Saturday 27 May 1815, he found it flowing, although too shallow to be navigable by boats.[55]

Nevertheless, Macquarie's suggestion that drought west of the Great Dividing Range had been similar in duration and severity to that experienced in the Sydney Basin seems questionable. We are lucky to have a painting of Campbells River made by John Lewin in 1815, and even allowing for some artistic licence, the image depicts pools full of water and waterfowl in the river (Plate 7). Winburndale Rivulet, which flows into the Macquarie River north of Bathurst (Fig. 8.4), was still running in 1815, as was Coxs River and streams arising nearby in the Great Dividing Range,[56] but today those that remain unregulated (e.g. Clear Creek) typically stop flowing after 1 or 2 years of drought.

Furthermore, the main tributaries of the upper Macquarie River catchment were always prone to periodic, severe drying even before the construction of impoundments and dams in

> *the extraordinary drought which has apparently prevailed on the western side of the mountains, equally as throughout this colony for the last three years, has reduced this river so much that it may more properly be called a chain of pools than a running stream*
> – *Governor Lachlan Macquarie on his tour over the Blue Mountains, April–May 1815*

Fig. 8.4: The Winburndale Rivulet was flowing when visited by Henry Antill in mid-1815. The streambed and banks then probably looked much as they do today, although the Winburndale Dam now lies upstream, which affects flow. Clear Creek, which is of similar size but unregulated, stops flowing in modern times during severe droughts. Photo: Robert Godfree.

the early to mid-1900s. Campbells River stopped flowing completely in 1833, 1897, 1902, 1913 and 1915,[57] and above the Ben Chifley Dam it did so in 2014 after a dry 7-month spell, and more than once during the 2017–19 drought.[58] The Fish River, still flowing in 1815, almost stopped in 1833 and (probably) in 1888, and was dry for 'miles' in 1897 during the Federation Drought.[59] To the west, the Belabula River stopped flowing during the protracted Federation (1899) and WWI (1915) droughts, and even in the short but severe drought of 1888–89.[60] The Lachlan River was reduced to pools downstream of Forbes in 1836, and stopped flowing in the vicinity of Cowra during the relatively short droughts of 1862 and 1881–82, and on multiple occasions during the Federation Drought (Fig. 8.5).[61]

Collectively, these facts leave little doubt that any drought experienced in the upper catchments of both the Macquarie and Lachlan rivers during 1815 was mild, involving no more than 6 months of low rainfall before Macquarie's tour. It must have broken well before October 1815, when escaped convicts described luxuriant grass along the Macquarie River north-west of Bathurst.[62] The stark difference in drought severity experienced west and east of the Blue Mountains is in itself not surprising, as many droughts restricted to coastal areas have occurred since. Perhaps the best explanation of these events comes from the 1813 observations of William Wentworth, who astutely linked the wet conditions he experienced in the Blue Mountains to the passage of cold, rain-bearing winds from the west. The passage of such cold fronts from the south-west typically lead to the uplift of moist air and orographic rainfall on the western

Fig. 8.5: Even before construction of major impoundments the Lachlan River regularly stopped flowing during more severe droughts. This photograph shows the river returning to Hillston in May 1903, after the breakup of the Federation Drought. When Evans saw the river in mid-1815 it was still flowing at Cowra. Carrathool Shire - W.G. Parker Memorial Library historical photograph collection.

side of the ranges, but dry conditions in the Cumberland Plain, which lies in the rain shadow of the Blue Mountains.

THE LEGACY

Among important droughts of Australian history, the 1813–15 drought is arguably the most enigmatic. Contemporary accounts of long rainless periods, crop losses and pasture shortages in the Sydney Basin all indicate that this period was undoubtedly dry, and as early as 1817 it had become accepted wisdom that these conditions were the main impetus behind the Blaxland Party's 1813 expedition over the Blue Mountains.[63] But compared to the Settlement Drought and many that were to follow, the drought does not appear to have been unusually severe, and even in late August 1813, when cattle and sheep were dying *en masse* in the Sydney Basin, conditions were considered no drier than those experienced during 1803–04.[64] Perhaps a better explanation is that the impact of the drought was greatly exacerbated by near-exponential growth of sheep and cattle numbers, overstocking and

> *Envy, malice, and other base incitements are however capable of inventing excuses for the most ridiculous as well as the most perfidious contrivances.*
> – Sydney Gazette and New South Wales Advertiser, *25 September 1813*

THE DOG IN THE MANGER
A FABLE OF AESOP

A Dog made his bed in a Manger, and lay snarling and growling to keep the horses from their provender. "See," said one of them, "what a miserable cur! who neither can eat corn himself, nor will allow those to eat it who can."

Fig. 8.6: The 'dog in the manger' metaphor has been used since the early 1800s by Australian writers to condemn covetous behaviour of governments or individuals. In the case of the 1813–15 drought it was used to describe landholders who lacked stock but would burn or deny access to the starving herds of their neighbours. The earliest textual references to the Greek fable date to the 1st and 2nd centuries CE but contrary to the caption it does not have a clear Aesopic origin. *Worker* (Brisbane), 29 August 1939.

mismanagement of the pasture base, marking the first time in post-European history that the needs of human populations and their agricultural systems had outstripped the capacity of the land to provide.

The 1813–15 disaster also throws shade on several key narratives about how early Australian societies understood, and responded to, the origin and progression of drought. First, far from being unaware of the volatility of the Australian climate and the potential for drought to impact grazing operations, colonial commentators and landholders alike were well aware of the consequences of overstocking should drought return for more than a year. This is unsurprising, because many had first-hand experience of the 1797–99 drought, and even the horror of the Settlement Drought still lingered on in the public consciousness. Second, despite arguments that the uniquely Australian concept of 'mateship' and associated solidarity in the face of adversity was taking root among convicts and the poorer classes to at least the 1820s, I remain sceptical that such values had extended to broader society. Indeed, at least some jealous landholders either burnt their own grass or that of their neighbours, much like the dog in the manger who would 'neither eat the hay himself, nor suffer the ox to feed upon it', thus taking pleasure in the suffering of the unfortunate (Fig. 8.6).[65] As I will show later, such behaviour emerges when economic fortunes of the wealthiest and poorest landholders diverge.

Ultimately, the 1813–15 drought had few lasting impacts on Australian agriculture or society. True, it stimulated expansion of the livestock industry to the western inland, but the limits imposed by the drought would have been naturally reached no more than a decade later, for by 1823 the Cumberland Plain had been fully occupied by graziers and farmers. With this irresistible requirement for growth, expansion into new terrain was inevitable. The dip in cattle and sheep numbers after 1813 was also only temporary: between 1816 and 1821 these had rebounded to 68 000 and 120 000 respectively.[66] By 1829 there were 600 000 sheep in New South Wales, and by 1851, 7 million.[67] The writings of Blaxland and Macquarie quickly faded into history, along with an understanding of the need for careful forward planning of livestock numbers and pastures in the unpredictable Australian climate. Tragically, this would soon prove to be a lesson more easily forgotten than learnt.

Endnotes

[1] Coal was reported in the area by escaped convicts as early as 1791, and specimens were brought to Sydney by fishermen in June 1796.

[2] Collins D (1802) *An account of the English colony in New South Wales with remarks on the dispositions, customs, manners, etc, of the native inhabitants of that country. Volume 2*. AH & AW Reed in association with The Royal Australian Historical Society, Sydney, p. 194.

[3] Russell HC (1877) *Climate of New South Wales: Descriptive, historical, and tabular*. C. Potter, Sydney, pp. 64–65.

[4] *Historical Records of New South Wales* (henceforth *HRNSW*), vol. 4, p. 465 (21 August 1801).

[5] *HRNSW*, vol. 1, part 2, p. 143 (5 July 1788).

[6] Collins D (1798) *An account of the English colony in New South Wales with remarks on the dispositions, customs, manners, etc, of the native inhabitants of that country. Volume 1*. AH & AW Reed in association with The Royal Australian Historical Society, Sydney, p. 314; *HRNSW*, vol. 2, p. 311, enclosure (15 June 1795). Early livestock counts had known inaccuracies and should be regarded as approximate.

[7] *HRNSW* vol. 1, part 2, p. 529 (24 October 1791); *HRNSW* vol. 2, p. 300 (1 June 1795); *HRNSW*, vol. 2, p. 306 (15 June 1795).

[8] A detailed history of early sheep imports is provided in Garran JC, White L (1985) *Merinos, myths and Macarthurs: Australian graziers and their sheep, 1788-1900*. Australian National University Press, Canberra.

[9] *HRNSW*, vol. 4, pp. 465, 473 (21 August 1801).

[10] *HRNSW*, vol. 4, p. 847 (2 October 1802); *HRNSW*, vol. 4, p. 882 (1 November 1802).

[11] *HRNSW*, vol. 4, p. 929 (31 December 1802).

[12] *The Sydney Gazette and New South Wales Advertiser* (henceforth *Sydney Gazette*), 5 March 1803, p. 3.

[13] *Sydney Gazette*, 13 January 1805, p. 2.

[14] *HRNSW*, vol. 6, p. 23 (2 March 1806).

[15] *HRNSW*, vol. 5, p. 715 (1 November 1805); Howe R (1952) Part 1. 1788–1828. In *A chronology of momentous events in Australian history, 1788-1846*. (Ed. G Mackaness) pp. 3–53. DS Ford, Sydney.

[16] *Historical Records of Australia* (henceforth *HRA*), series 1, vol. 7, p. 1 (20 February 1809).

[17] *HRNSW*, vol. 7, p. 218 (14 October 1809).

[18] For example, when breeding occurs during discrete intervals.

[19] Houison JKS (1936) *Journal and Proceedings of the Royal Australian Historical Society* **22**, pp. 1–41.

[20] *HRA*, series 1, vol. 7, p. 560 (17 November 1812).

[21] *HRA*, series 1, vol. 7, p. 579 (17 November 1812).

[22] Gregory Blaxland's 'Narrative', submitted to Mr. Commissioner John T. Bigge, 1819 (henceforth Blaxland's 'Narrative'). In *Fourteen journeys of the Blue Mountains of New South Wales 1813-1841* (henceforth *Fourteen Journeys*). (Ed. G Mackaness) pp. 13–15. Horwitz-Grahame, Sydney, 1965.

[23] *HRNSW*, vol. 7, p. 235 (27 November 1809).

[24] Blaxland's 'Narrative', pp. 13–15.
[25] *HRNSW*, vol. 7, p. 468 (15 December 1810).
[26] *Sydney Gazette*, 2 March 1811, p. 2.
[27] *Sydney Gazette*, 19 October 1811, p. 3.
[28] The appalling conditions that animals experience in the live export trade are particularly contentious.
[29] *Sydney Gazette*, 21 March 1812, p. 2.
[30] *Sydney Gazette*, 12 December 1812, p. 2 and 28 November 1812, p. 2.
[31] *Sydney Gazette*, 20 February 1813, p. 2 and 27 March 1813, p. 2; Ellis MH (1973) *Lachlan Macquarie: His life, adventures and times*. Fifth edition. Angus and Robertson, Sydney, p. 262.
[32] *HRA*, series 1, vol. 7, pp. 707–708 (28 June 1813).
[33] *HRA*, series 1, vol. 8, p. 58 (7 August 1813).
[34] *HRA*, series 1, vol. 7, p. 639 (17 November 1812).
[35] Assumed dry sheep equivalents as follows: male horse = 13, female horses = 16 with foal (25%) and 10 without foal (75%), bulls = 9 , cows = 20 with calf (50%) and 8 without calf (50%), oxen = 9, male sheep = 1, female sheep = 3 if lactating (25%) and 1 if non-lactating (75%), male meat goat = 1.5, female goat = 3 if lactating (25%) and 1 if non-lactating (75%), hogs = 1 (allowing for some hand feeding).
[36] *Sydney Gazette*, 18 September 1813, p. 1.
[37] *Sydney Gazette*, 28 August 1813, p. 2.
[38] *Sydney Gazette*, 25 September 1813, p. 2.
[39] Blaxland G. A journal of a tour of discovery across the Blue Mountains, New South Wales, in the year 1813 (henceforth Blaxland's Journal), p. 10. In *Fourteen Journeys*, pp. 2–10.
[40] Perry TM (1955) *Historical Studies: Australia and New Zealand* **6**, 376–395.
[41] *Sydney Gazette*, 8 January 1814, p. 1.
[42] Assistant-Surveyor Evans's Journal, 1813–1814. In *Fourteen Journeys*, pp. 18–32; *HRA*, series 1, vol. 8, p. 118–123 (19 January 1814).
[43] *Sydney Gazette*, 1 October 1814, p. 2.
[44] *Sydney Gazette*, 20 May 1815, p. 2 and 3 June 1815, p. 2.
[45] *HRA*, series 1, vol. 8, p. 576 (10 June 1815).
[46] *Sydney Gazette*, 2 December 1815, p. 2.
[47] *Sydney Gazette*, 13 January 1816, p. 2; 20 January 1816, p. 2; 24 February 1816, p. 2.
[48] Blaxland's Journal, pp. 5–6.
[49] Lawson W (1822) *William Lawson Journal of an Expedition from Bathurst to the Liverpool Plains, 9-24 January 1822*. Mitchell Library, call number C 120 vol. 2; Blaxland's Journal, pp. 9–10.
[50] Horton K (2019) 'Hartley Vale residents call for flow to revitalise the dry River Lett'. *Lithgow Mercury*, 25 November.
[51] Evans uses the name 'Riverlett', meaning 'rivulet'.
[52] E.g. Eyles RJ (1977) *Australian Geographer* **13**, 377–386; Mould S, Fryirs K (2017) *Catena* **149**, 349–362; cf., Butzer KW, Helgren DM (2005) *Annals of the Association of American Geographers* **95**, 80–111.
[53] Assistant-Surveyor Evans's Journal, 1813–1814. In *Fourteen Journeys*, pp. 18–32.
[54] Journal kept by Mr W. Cox in making a road across the Blue Mountains from Emu Plains to a new country discovered by Mr. Evans to the westward. In *Fourteen Journeys*, pp. 34–63.
[55] *HRA*, series 1, vol. 8, pp. 611–619 (13 May 1815).
[56] Antill HC (1815) Journal of an excursion over the Blue or Western Mountains of New South Wales to visit a tract of new discovered country, in company with His Excellency Governor and Mrs. Macquarie and a Party of Gentlemen. In *Fourteen Journeys*, pp. 74–75.
[57] *Sydney Gazette*, 14 March 1833, p. 2, *Barrier Miner* (Broken Hill), 21 May 1897, p. 3; *Bathurst Free Press and Mining Journal*, 19 May 1902, p. 2; *The Bathurst Times*, 10 February 1914, p. 2.
[58] During this 7-month period in 2014 only 230 mm of rain fell at Bathurst, slightly more than half of normal. Flow in Campbells River has probably been affected by willow removal, entrenchment and tree clearing.

[59] *Sydney Gazette*, 14 March 1833, p. 2; *Australian Town and Country Journal* (Sydney), 15 December 1888, p. 1189; *Bathurst Free Press and Mining Journal*, 13 April 1897, p. 2.

[60] *Sydney Morning Herald*, 6 February 1889, p. 7; *Evening News* (Sydney), 11 January 1899, p. 5; *Molong Express and Western District Advertiser*, 1 May 1915, p. 15; *Lithgow Mercury*, 19 February 1915, p. 3.

[61] *Empire* (Sydney), 23 December 1862, p. 2; *The Sydney Mail and New South Wales Advertiser*, 18 March 1882, p. 441; *Maitland Mercury and Hunter River General Adviser*, 18 January 1881, p. 6; *Freeman's Journal* (Sydney), 8 July 1899, p. 13; *Evening News* (Sydney), 20 April 1897, p. 6; Mitchell TL (1996) *Three expeditions into the interior of eastern Australia: With descriptions of the recently explored region of Australia Felix, and of the present colony of New South Wales*. Eagle Press, Maryborough, Victoria. Reprint. Originally published: 2nd ed., rev., T. & W. Boone, London, 1839.

[62] *Sydney Gazette*, 2 December 1815, p. 2.

[63] Oxley J (1820) *Journals of two expeditions into the interior of New South Wales, undertaken by order of the British government in the years 1817-18*. John Murray, London.

[64] *Sydney Gazette*, 28 August 1813, p. 2.

[65] Dyrenfurth N (2015) *Mateship: A very Australian history*. Scribe Publications, Brunswick, Victoria.

[66] *HRA*, series 1, vol. 10, p. 577, enclosure no. 3 (30 November 1821).

[67] Garran JC, White L (1985) *Merinos, myths and Macarthurs. Australian graziers and their sheep, 1788-1900*. Australian National University Press, Canberra, p. 218.

9
The stillness of death

It almost appeared as if the Australian sky were never again to be traversed by a cloud.
 – *Charles Sturt,* Two expeditions into the interior of South Australia, vol. 1.

ON 2 FEBRUARY 1829, AN EXPLORATION PARTY LED BY CAPTAIN CHARLES STURT arrived on the banks of a great river that wound its way across the parched and blistered plains of north-west New South Wales. After weeks of arduous travel and battered by thirst, heat and voracious swarms of biting insects (Fig. 9.1), the water that lay at the bottom of the channel seemed like a small of mercy for man and beast alike, a reward for their trials and exertions. But the Australian interior is fickle: no sooner had the men clambered down the 40-feet-high (12 m) riverbank to the water below when cries of amazement and disappointment disrupted the stillness of the drought-stricken landscape, and terrified faces turned to Sturt, who waited above. As he put it, 'The cup of joy was dashed out of our hands before we had time to raise it to our lips', for the water was saline.[1]

Sturt had the misfortune of undertaking his expedition to the Darling River (Fig. 9.2) during the most severe and protracted drought to occur in eastern Australia since the establishment of the Port Jackson colony four decades earlier. Rain had deserted a vast swathe of land from south-eastern Queensland to Tasmania since April 1826, a year that marked 'the commencement of

Fig. 9.1: Charles Sturt's expedition to the Darling River in 1828–29 was undertaken during one of the most severe droughts in the post-settlement era, making his achievements even more impressive. This sketch, which accompanies an article on his 1844 expedition into central Australia, where he also experienced water shortages and tremendous heat, appeared in *The Sydney Mail*, 19 January 1938.

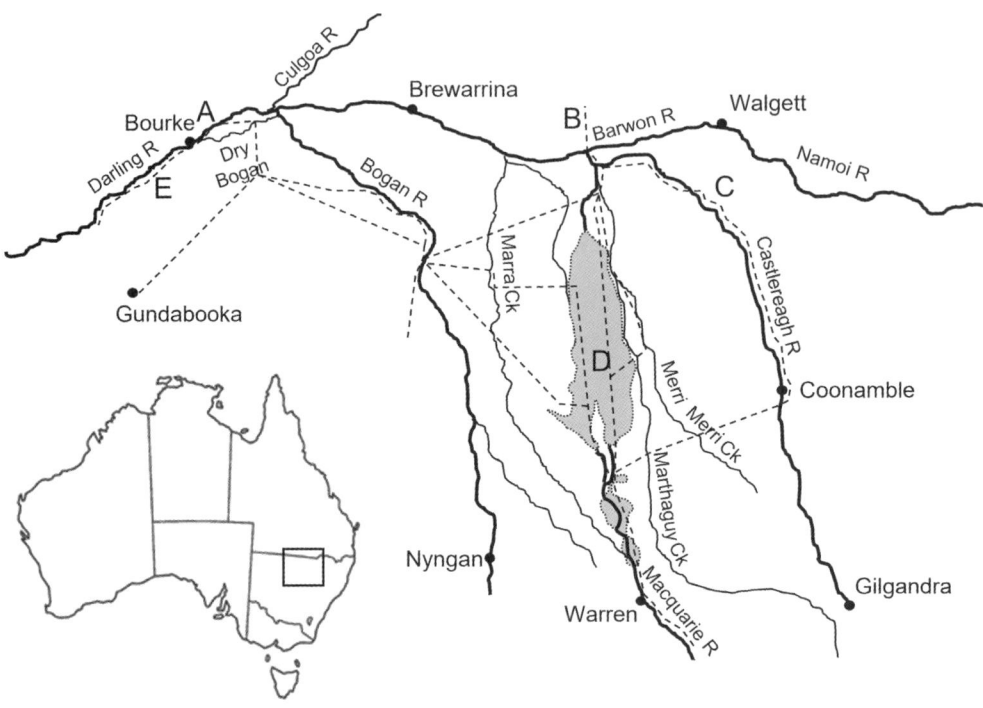

Fig. 9.2: The Charles Sturt expedition to the Macquarie Marshes and Darling River, November 1828–April 1829. The dashed lines indicate the approximate routes followed by Sturt or other members of the party while the grey shading shows the location of the Macquarie Marshes. Significant events occurred at: A) Darling River found to be saline, B) extreme drought conditions north of the Barwon River, C) long stretches of Castlereagh River dry; Aboriginal people digging holes in bed of river, D) fish killed by drying of the Macquarie Marshes, and E) Aboriginal people affected by smallpox. Map redrawn from Cumpston JHL (1951) *Charles Sturt: His life and journeys of exploration.* Georgian House, Melbourne.

one of those fearful droughts to which we have reason to believe the climate of New South Wales is periodically subject'.[2] Ironically, Sturt found himself in this desperate situation because dry conditions were thought necessary to cross the Macquarie Marshes – a wilderness of virtually impenetrable rushes and vast floodplains fed by Macquarie River. John Oxley's inland expedition had prematurely ended there during the extremely wet year of 1818, and the dry weather had rekindled hopes that this barrier could be breached, allowing exploration of the interior to continue.[3] But setting out from Sydney on 10 November 1828, Sturt was soon to find out just how punishing droughts in inland Australia can be.

It is difficult for someone who has never been to the riverine plains of western New South Wales to appreciate the bleakness and monochromaticity of the landscape when all moisture has been stripped away by years of relentless heat and sunshine. The grasses and forbs first wither and turn yellow, but then transform to darker shades of brown, and finally take on a silvery-greyish hue that marks the terminal point of senescence. Starving livestock, feral animals and kangaroos soon strip the ground bare, revealing vast expanses of unprotected alluvial soil that

Box 9.1: Allan Cunningham: Botanist in a 'poor and hungry' land

From the earliest days of colonial Australia botanists featured heavily among those drawn to the possibility of discovery in the unexplored expanses of the great southern continent. Among these was Allan Cunningham (1791–1839), who was sent to Australia to collect seeds, bulbs and living plants for the Royal Gardens at Kew by Sir Joseph Banks and William T Aiton. He immediately became familiar with inland Australia after being attached by Governor Macquarie to the Lachlan and Macquarie River expedition with John Oxley in 1817, where he collected specimens of more than 400 plant species. Later expeditions with Phillip Parker King proved equally productive.

Unfortunately, Cunningham found that subsequent governors were less willing to fund scientific work with scarce colonial resources. He was therefore forced to map new agricultural land in New South Wales, much of which was botanically boring, to fund his later expeditions. In 1827, however, his services were secured by Governor Darling to explore northwards from the Liverpool Plains in search of land for settlement. His (European) discoveries on this expedition, which culminated in the sighting of Cunningham's Gap, marked the apogee of his career.

The experience was soured, however, by a severe drought that raged at the time. On the Liverpool Plains in northern New South Wales he found most watercourses, including Warrah and Quirindi creeks, reduced to chains of stagnant pools. Travelling north, conditions intensified: deep chasms had opened up in the ground under the relentless sun and a great stillness had descended on the land – 'scarcely a bird was seen or heard; no game, native dog, nor the evidence (even of the most ancient date) of a passing human being'. Even larger rivers like the Macintyre had terrestrial plants growing in their sandy beds, indicating a very long period without substantial flows, but most, like the Namoi, still contained deep pools. Notably, despite his commitment to careful scientific documentation Cunningham made no mention of smallpox among Aboriginal people he met on the expedition, less than 18 months before the Darling River outbreak of 1828.

The journal *Cunninghamia*, produced by the Royal Botanic Garden Sydney, continues his legacy, publishing scientific papers on plant ecology and the flora of eastern Australia.

Picken, A. (ca. 1835) *Allan Cunningham (no poet truly but) His Majesty's Botanist in New South Wales* [lithograph]. Courtesy Wikimedia Commons and National Library of New Zealand, A-040-013.

the wind whips into swirling clouds of grey or brown. Most shrub and tree species of the interior have a deep green, olive or greyish colour, a feature linked to the presence of a waxy cuticle on leaves to prevent water loss, dull or silvery colours that reflect solar radiation, and thick, long lived, and frequently concolorous leaves that often hang vertically – all adaptations to low nutrient supply, drought and herbivory.[4] The muted colours add to the dullness of the scene, especially when coated with dust. Birds, mammals and insects disappear or move to remaining sources of water, and a strange silence descends over the land. Perhaps the first European to describe this phenomenon was the explorer–botanist Allan Cunningham when crossing the Liverpool Plains just 18 months earlier, around 350 km to the south-east (Box 9.1).

By time Sturt arrived in late 1828, however, the drought had greatly intensified, leaving the outwash plains that lie between the Bogan, Castlereagh and Darling rivers under the spell of what he called 'the stillness of death'. To him, the landscape was 'dismally brown', 'as dreary as can be imagined' and 'truly melancholy' – a pervasive state that, when viewed from one of the few areas of higher terrain, 'strengthened as we gazed upon it'.[5] One cannot read Sturt's journal without feeling that he was genuinely and deeply affected by the bleak conditions, emotions that resonate with people who have lived in the outback. But Sturt had only arrived in Australia 18 months earlier, and with no prior experience in conditions of the interior, had to contend with the strain of travelling with horses, cattle and men through blistered, drought-stricken terrain, and interacting with melancholy groups of Aboriginal people devastated by disease and hunger. These sentiments peaked on 30 March 1829 when after riding north of the Barwon River near its junction with the Macquarie River (Fig. 9.2),[6] he wrote a famous passage with references to dying trees, gasping emus and starving dingoes, reflecting that 'none who had not like me traversed the interior at such a season, would believe the state of the country'.[7]

> *I am giving no false picture of the reality. So long had the drought continued, that the vegetable kingdom was almost annihilated, and minor vegetation had disappeared. In the creeks, weeds had grown and withered, and grown again; and young saplings were now rising in their beds, nourished by the moisture that still remained; but the largest forest trees were drooping, and many were dead. The emus, with outstretched necks, gasping for breath, searched the channels of the rivers for water, in vain; and the native dog, so thin that it could hardly walk, seemed to implore some merciful hand to despatch it. How the natives subsisted it was difficult to say, but there was no doubt of the scarcity of food among them.*
> – Charles Sturt, Two expeditions into the interior of South Australia, vol. 1, p. 145

GASPING FOR MOISTURE: DROUGHT ON THE DARLING RIVERINE PLAINS

While Sturt's account was obviously written for an audience unfamiliar with such conditions – including Governor Ralph Darling, who authorised the expedition – and therefore embellished with florid imagery, the phenomena he describes leave no doubt that the drought he experienced ranks as among the worst experienced in eastern Australia before the 1895–1903 Federation

Drought. We can tell this because Sturt found numerous watercourses to be either very low or completely dry, particularly on his return journey. This includes not only smaller creeks such as Merri Merri Creek, Marthaguy Creek and other smaller streams in and near the Macquarie Marshes, but also the middle and lower reaches of the Bogan and Castlereagh rivers, the Barwon and, of course, the Darling itself.[8] In the Dubbo and Wellington regions the Bell and Talbragar rivers had also ceased flowing, and although the Macquarie River was flowing into the marshes in early December 1828, long stretches had dried by April 1829.[9]

The Castlereagh River, which rises in the eastern Warrumbungle Ranges, was particularly dry during Sturt's expedition. In March 1829 he encountered the Castlereagh just upstream of the present-day town of Coonamble, where the channel, which was 130 yards (119 m) wide, was completely dry. After searching for some time, they found a small pond 15 yards (14 m) in circumference, around which numerous kangaroos had apparently congregated.[10] Downstream towards its junction with the Macquarie River, pools and lagoons along the course of the Castlereagh were putrid and lacked either fish or wildlife, and a stretch 45 miles (72 km) long was entirely without water.[11] Local Aboriginal populations, most likely the Ngiyampaa (= Ngemba) Wayilwan (= Weilwan) people, had resorted to digging holes in the bed of the river to secure a water supply, and had apparently deserted terrain adjacent to the smaller tributaries of the Castlereagh, which had all run dry.[12]

Further north the drought must have been even more severe, for when Sturt arrived in February 1829 the Darling River above Bourke had been reduced to a series of salty pools amid long stretches of parched riverbed. According to CSIRO colleagues who have studied the hydrology of the Murray–Darling Basin for many years, these pools probably arose from concentration of the salt load in the river water due to evaporation, and incursion of saline groundwater via springs into the bed of the river itself. The latter can be very significant: during the Millennium Drought, for example, the surface waters of the Darling River 55 km downstream of Bourke consisted of 60–99% groundwater with a high salt content.[13] When river flow is high, this groundwater is partially kept at bay by the hydraulic pressure generated by a freshwater lens or aureole that develops in the aquifer that surrounds the river, but during drought this hydraulic gradient reverses, and saline groundwater is able to penetrate the riverbed as springs. Sturt described one not far from Bourke as being a salt-encrusted stream of considerable size, gushing with water sufficient to generate a small flow in the river itself.[14]

The state of the river in 1829 also indicates that severe drought must also have extended across the New England and Liverpool Range regions of New South Wales, which are drained by the Namoi and Gwydir rivers, and the immense catchments of the Barwon–Macintyre and Culgoa–Balonne–Condamine rivers, which, as Sturt correctly inferred, extend well into subtropical Queensland. While the 2017–19 drought demonstrated that the Darling River can be reduced to pools during an acute, basin-wide 3-year drought,[15] the extreme state of the Darling and Castlereagh rivers observed by Sturt, and the presence of saplings in dried riverbeds, suggest that the drought here must have lasted for many years. South-eastern Queensland and New South Wales experienced severe drought in 1824 and 1825, and perhaps the drought-breaking rains that fell across eastern parts of the New South Wales from July 1825 to April 1826 failed to

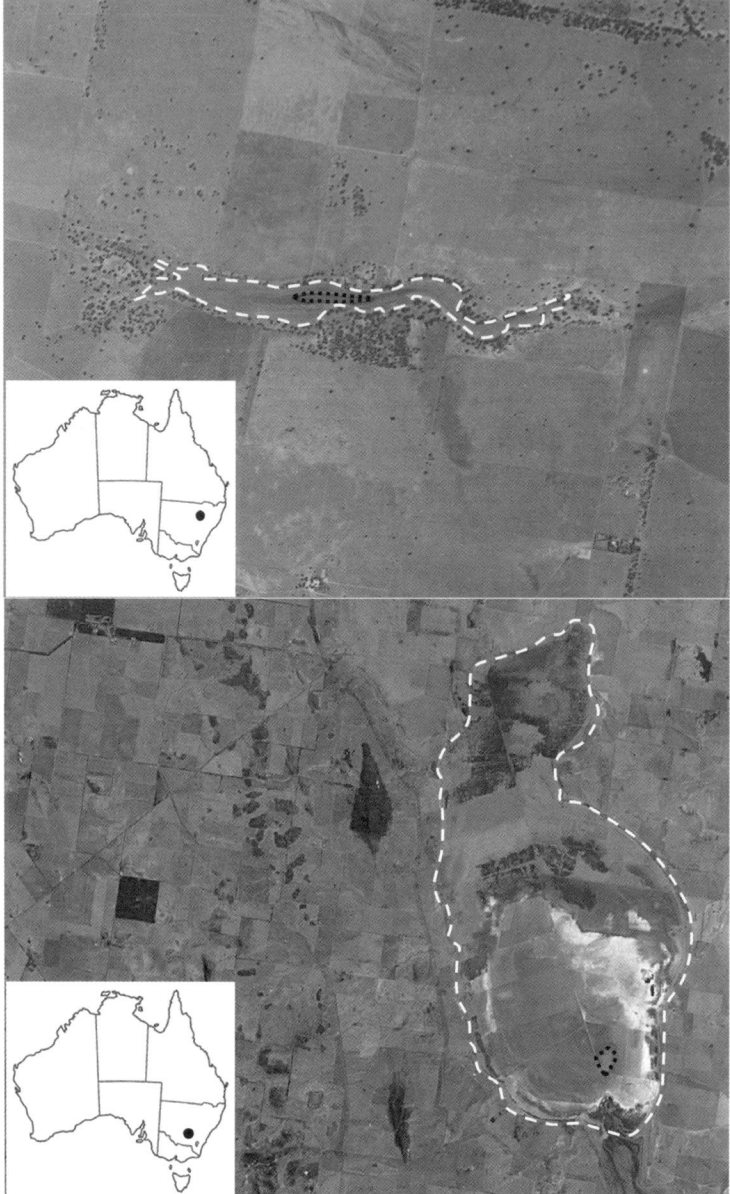

Fig. 9.3: Two ephemeral lakes in inland New South Wales. Top: Buddah Lake, 10 October 1967 (Film number CAC7010, run 1, frame 80); available at Commonwealth of Australia (Geoscience Australia's) Archival Aerial Photography. When visited by Charles Sturt in December 1828 Buddah Lake contained several metres of water; here, after almost 4 dry years, it is nearly dry. Bottom: Lake Cowal (Google Earth, image compiled 31 December 2002). Between December 2000, when the lake covered approximately three-quarters of the lakebed, and the end of 2002, after 2 dry years, Lake Cowal had shrunk to a very small swampy area at the southern end of the lake. It may have looked similar in mid-1828. The white dashed line indicates the approximate extent of each lake when full, while the inner black dotted line indicates the approximate extent of water when the image was taken.

> *There was scarcely a living creature, even of the feathered race, to be seen to break the stillness of the forest.*
>
> – Charles Sturt, Two expeditions into the interior of South Australia, vol. 1, *p. 73*

penetrate this far north. This is conjectural, but Allan Cunningham encountered increasingly dry conditions while travelling north from the Upper Hunter River towards the Darling Downs in 1827, with many watercourses either entirely dry or reduced to pools.[16]

To the south, conditions were definitively less severe because the Macquarie River was still flowing as far as the Macquarie Marshes in late 1828, and Marra Creek contained good water when Sturt camped near it on the night of 31 December 1828.[17] Sturt also described a small, serpentine lake known to local Aboriginal people (probably Wiradjuri) as the 'Buddah' (now Buddah Lake, 11 km south-east of Trangie) as 5–6 feet (1.5–1.8 m) below the normal depth of 4 fathoms (7.3 m) when full. At 300–400 yards (274–365 m) across[18] and more than a mile (1.6 km) in length, it was still a substantial body of water that teemed with both fish and birds. In contrast, it has been known to go dry in past dry periods, including in the early 1900s and again at the end of the 1965–67 drought (Fig. 9.3).[19] According to local Aboriginal people, Lake Cowal, a large ephemeral lake fed by the Lachlan River and Bland Creek, also still retained a swampy body of water in its centre in 1828.[20] Since Lake Cowal is only around 3 m deep and lies in a semi-arid area with a mean annual rainfall of 486 mm compared to pan evaporation of around 1500 mm, it typically dries up in only 2 or 3 years when inflow ceases. This dynamic has repeated regularly in recent years, with the transition between full or high lake levels to complete drying occurring in 1995 to 1998, 2000 to 2002, 2013 to 2015, and 2017 to 2019.

Whatever the exact duration and geographical extent of the drought, the impacts on plant and animals were certainly severe. Sturt often wrote in a poetic style typical of European exploration at the time, but he paints an ecologically convincing story of a blighted landscape characterised by a scorched understorey and stressed and dying trees. For example, he noted that 'minor vegetation', which probably included ephemeral grasses and forbs and possibly smaller perennial species, had entirely perished, and that across the bleak visage of the plains adjacent to the Macquarie Marshes there was 'not a flower in bloom, nor a green object to be seen' (Plate 8). To the north of the Barwon River in the Narran Lakes region, conditions were even more severe, and Sturt travelled all day without seeing a drop of water or blade of grass.[21] This is surprising, because Sturt's expedition pre-dated the introduction of cattle and sheep to the region. Perhaps, even in pre-European times, grazing pressure by native marsupials was sufficient to denude the understorey of some inland ecosystems during the most severe droughts. The lack of ground layer vegetation evidently left the soil exposed, particularly in non-alluvial areas away from the Darling River where there was a 'singular want of vegetable decay',[22] and clouds of a 'minute and penetrating dust' rose as Sturt's party passed by. As pastoralism and livestock numbers grew in decades to come, rising dust was to become a feature that increasingly defined the outback.

Animals attempting to persist in the drought-stricken landscape fared little better. Sturt observed mortality among aquatic species that are particularly prone to the drying out of waterbodies; in the Bogan River 'neither fish nor muscles [mussels] remained', and shallow pools of water on the plains between Mt Oxley and Gundabooka contained dead frogs in a state of putrefaction.[23] On the return journey the Macquarie River had stopped flowing upstream of the Marshes, and water in the channels had dropped so low that the backs of fish had become exposed. Crows, the scourge of helpless animals, congregated around the fish, pecking at them.[24] Massive fish kills have increasingly become a problem in drought-affected parts of the Darling River in recent years, but it is interesting that cases also occurred prior to European settlement. Sturt implies that predation by Aboriginal people may have reduced fish and mussel numbers during extreme drought events, although some must have remained in larger waterbodies.[25]

THE DECEMBER 1828 HEATWAVE

During droughts that last for many years, the land gradually becomes stripped of its moisture, and the atmospheric cooling achieved through *evapotranspiration* (the movement of water from soil, water bodies and vegetation into the air) slowly diminishes. This landscape-level drying is exacerbated by the death or grazing of ground-layer vegetation, which leaves the earth increasingly exposed to solar radiation. Dry soils tend to heat more rapidly and achieve much higher daytime temperatures than damp ones – in experiments that I have conducted in grassland ecosystems the difference can be 5–10°C[26] – and these hot soils drive additional heating in the lower atmosphere, which in turn causes higher evapotranspiration. The impact of positive feedbacks like this on natural systems can be profound: heatwaves continue to build instead of collapsing, and then become more protracted and intense.[27] This explains why many of the worst heatwaves to strike eastern Australia have occurred during multi-year droughts: examples include January 1896, January 1939, February 2004, January–February 2009, and November–December 2019.

Sturt had the great misfortune of setting out to explore the inland riverine plains of New South Wales at just about the worst possible time: in mid-summer during a protracted drought. Even in normal years, temperatures here regularly exceed 40°C for days on end, and in the worst heatwaves they can exceed 45°C. At first, his journey was comfortable enough, the heights of the Blue Mountains and the elevated country around Bathurst providing respite from the burning heat of the plains below. But upon descending into the Wellington Valley, he experienced his first taste of what was to come: here it was so 'exceedingly hot ... that it was impossible to take exercise at noon'.[28] This was probably quite an understatement, for Sturt was no shrinking violet when it came to enduring gruelling physical hardship. Yet further west conditions grew worse, with temperatures becoming intolerable even before the sun reached its highest point in the sky.

While he was probably prone to some exaggeration, the heat was clearly fierce. Sturt made few direct temperature observations during his journey, but at Buddah Lake (Fig. 9.3) he noted that 'At 2 p.m. the thermometer stood at 129° of Fahrenheit [53.9°C], in the shade;

and at 149° [65°C] in the sun'.[29] Above the Macquarie Marshes '[t]he thermometer was seldom under 114° [45.6°C] at noon, and rose still higher at 2 p.m.'. These temperatures were not recorded under standard conditions (53.9°C is much hotter than the maximum ever measured inside a Stevenson screen in Australia of 50.7°C) and cannot be compared directly with modern values. Nevertheless, the fact that Sturt reported that birds coming to drink at the lake were 'gasping', 'almost too weak to avoid us' and 'indifferent to the reports of our guns' indicates a very high level of heat stress. I have observed similar behaviour in the foothills of the Nandewar Ranges, near Narrabri, on 3 January 2014 when the temperature reached 45.6°C, and I suspect that the heat endured by Sturt was similar. He also reports that nights were very hot. During heatwaves in this region temperatures can stay above 37.8°C (100°F) well into the late evening, with minimum overnight readings occasionally exceeding 30°C.[30] The stifling, sleepless conditions add much to the misery endured during the day.

Sturt leaves us with a memorable account of temperatures being so extreme that sugar melted in their canisters and all the dogs that accompanied the men on their expedition were 'destroyed'.[31] Sucrose thermally decomposes at 186°C, but much of the sugar available at the time was in crude brown or muscovado forms, and it is possible that these could undergo changes that resemble melting or dissolving if exposed to somewhat lower temperatures, particularly if they contained impurities such as water, salt or organic acids. The list of supplies obtained from His Majesty's Stores for Sturt's expedition includes tin pots and cases,[32] and these canisters would have been far too hot to touch if directly exposed to the sun on days exceeding 40°C – perhaps hot enough even to 'melt' sugar. It is also not unusual for dogs to suffer heatstroke and die in extremely hot weather, especially when they lack shade or are forced to exercise. While much depends on the breed and condition of the dog and other factors such as humidity, anecdotal reports of dogs dying from heat exhaustion in inland areas seem to occur when temperatures reach 44°C or more, and so I think that conditions at least this hot were experienced by Sturt in 1828.

DUST AND DEVILRY

Contact with Sturt was undoubtedly disruptive to the Aboriginal populations that he encountered along his trek, and this destabilisation grew rapidly as absconding convicts and pastoralists began to move down the Macquarie River towards the Darling River soon after. But while history shows that Sturt's journey marks the first wave of European exploration and occupation in north-west New South Wales, a much more dangerous invader had arrived on the banks of the Darling just a few months earlier. Unlike Sturt's party, which arrived in a cloud of dust, braying cattle and wagons that creaked their tortuous way across the plains, this invader was silent and invisible,[33] and outside of anything experienced by Aboriginal people before. Unaware of the new devilry in their midst, life continued unchanged for a week or two after its arrival, until the first unfortunates were suddenly struck down with violent nausea and fever. Within months, vast stretches of the river that had only months earlier supported thousands of people in permanent villages lay empty and silent in the summer heat.

> *What is the meaning of Boy [Buyi] in your language?*
> *The devil.*
> *What do black fellows mean by devil devil?*
> *Devil devils is its all over small pox like.*
>
> – Mahroot, Dharug man from Cook's River, 1845[34]

Sturt's observation of a 'violent cutaneous disease' among Aboriginal people (probably Barundji) on the Darling River near Bourke on 5 February 1829[35] is now generally regarded as the first European record of the 1828–32 smallpox epidemic that ravaged southern and eastern Australia.[36] By August 1830 it had spread roughly 250–300 km to the south-east along the Castlereagh River,[37] and by October 1830 to the pastoral districts around Bathurst and Wellington. By 1831 it was present in the Liverpool Ranges, the lower Murray, and the coastal districts of northern New South Wales and southern Queensland, and a year later the disease was still spreading south from Port Macquarie.[38] Most early accounts from inland New South Wales, including that of George Clarke, a convict escapee who lived among the Kamilaroi (= Gamilaraay) on the Liverpool Plains from 1827 to 1831, said that the outbreak spread from the north-west and reached the Namoi River in October 1830.[39]

This pattern of spread has long been taken as evidence that the outbreak originated from contact with Macassan trepangers (people from Sulawesi and other parts of the Indonesian archipelago who collected sea cucumber) in northern Australia, particularly since the first cases among Europeans did not occur until contact with infectious Aboriginal people in 1831 near Wellington.[40] However, the *Bussorah Merchant*, which arrived in Port Jackson on 26 July 1828, has also been cited as an alternate source of infection, since convicts from this vessel were assigned to the Upper Hunter region after arrival, and active smallpox had been on board during the voyage.[41]

Whatever the origin of the outbreak, the impact of the disease on local Aboriginal populations can only be described as catastrophic. When the military surgeon Dr John Mair investigated the outbreak among Wiradjuri people in the Bathurst and Wellington regions he found that the case fatality rate (CFR; the ratio of confirmed deaths to confirmed cases) varied from one in five or six (16–20%) to one in three (33%), and recent authors concur that a figure of 25–33% is most likely.[42] At least some Aboriginal people here were apparently resistant to smallpox, either naturally by having survived an earlier epidemic (probably after 1789), or following vaccination by Europeans, but in some locations, such as the Wellington Valley, they were struck down 'numerously and without exception', with but few survivors.[43]

Since smallpox ranks of one of the most brutal diseases ever to strike humanity, most historians have focused on the direct impact of smallpox on Aboriginal populations. But these studies have overlooked a remarkable coincidence – that the first two smallpox outbreaks in eastern Australia (1789 and 1828) both overlapped with periods of extreme drought. I have previously discussed that the spread of smallpox to the Hunter Valley and to the west of the

Blue Mountains very likely coincided with the most extreme phase of the Settlement Drought. Similarly, Sturt's accounts provide unequivocal evidence that the early stages of the 1828 outbreak took place during what had become among the worst droughts of the 19th century. We must therefore take seriously the possibility that drought may have exacerbated the depopulation of the New South Wales interior that took place during both epidemics.

The form of smallpox that struck the Wiradjuri people in central New South Wales during 1830 began with a 2- to 8-day phase of languidity and oppression, growing fever, and development of severe headache and pains in the chest or stomach. This was followed by an eruption of small red spots which expanded from the face, tongue and lips to the head, chest and extremities. Over about a week these grew into pustules roughly the size of a pea and filled with a milky or yellowish fluid. Death usually occurred around 3 days after the eruption due to fever and swelling of the tongue and pain in the throat that prevented swallowing.[44] To the north, reports suggest that confluent smallpox was prevalent, distinguished by the coalescence of pustules on the face followed by development of convulsive fits and a bloody discharge from the mouth. This critical period was followed swiftly either by death or recovery.[45]

Epidemiological and cultural considerations, however, suggest that while some sufferers undoubtedly died simply due to the severity of these symptoms, the impact of the disease was greatly exacerbated by exposure to the elements, and thus partially preventable. This typically occurred when the unfamiliar disease caused the remaining healthy people to move camps frequently or simply to flee, leaving those left to fend for themselves.[46] For example, the Yuwaalaraay (= Euahlayi) people of the Narran River recalled that during the smallpox epidemic, brought by the 'white devils', the healthy would flee the dying in terror, and leave the dead unburied.[47] This response was by no means ubiquitous: among the Kamilaroi, for example, treatment was given to the sick by a traditional healer, and although the disease was known to be infectious, the people did not shun each other on that account.[48] But for those stranded or left behind and incapacitated by disease, the timing of the epidemic during the tremendously hot and dry summer of 1828–29 could not have been worse.

Yet even for those that recovered from the disease, a process that took 2–3 weeks, the immediate threat to survival was not over. Convalescents were often left debilitated by emaciation, ulcers or abscesses, and even blindness in one or both eyes. Such sufferers, even if mobile, would have struggled to participate in hunting or harvesting, which were vital to food acquisition. But perhaps the cruellest feature of smallpox was that the bottom of the feet, which were often nearly covered by pustules, took longer to heal than other parts of the body. In many cases the soles ultimately peeled away and secondary ulcers and infections formed in the scabs, leaving the feet so tender that walking was difficult or impossible for a long time. This effectively left sufferers unable to collect food or move to new locations when necessitated by drought-induced drying of water sources or diminishment of food supplies.[49]

We can also surmise that the challenge of caring for those affected by the disease, already difficult for a people unfamiliar with smallpox and lacking access to European healthcare, must have been greatly exacerbated in drought-stricken areas. According to George Clarke, 20–30 Kamilaroi were struck down at a time on the Liverpool Plains, and nearly 40 people

All about very sick, and tumble down dead like rotten sheep.

– *Aboriginal oral account from the western Victorian Wimmera*[51]

Fig. 9.4: When Thomas Mitchell travelled down the Darling River in 1835, he frequently encountered evidence of the 1828–32 smallpox outbreak, including pock marks on many Aboriginal people and graves of those that had died. So devastating was the epidemic that he thought that the disease had 'almost depopulated the Darling'. This image is taken from Plate 16: Tombs of a tribe, after some great mortality, probably from a disease resembling smallpox. Major T.L. Mitchell del. G. Barnard Lith. J. Graf Printer to Her Majesty. Published by T. and W. Boone, London. State Library of NSW, Dixson Library, FL8635857.

covered in pustules appeared at Port Macquarie in late 1831. The incapacitation of such large numbers of people would likely have placed extreme demands on those unaffected by the disease even at the best of times, let alone during 1828–29, after years of exceptionally severe drought. Indeed, based on the subsistence patterns of the Bagundji (= Paakantyi) of the mid- to lower Darling, it is thought that strenuous effort was likely needed to support the population during bad seasons, although seed stores of native millet (*Panicum* spp., esp. *P. decompositum*), pigweed (*Portulaca oleracea*) and other species were collected for such times.[50]

Finally, eyewitnesses to the epidemic, including Mair, noted that the smallpox outbreak was fatal mainly among adults and the elderly, with children seldom killed by the disease.[52] This differs from the usual pattern of elevated mortality among pregnant women and children seen in smallpox outbreaks elsewhere, and it is possible that these observations were atypical of the epidemic in general. Nevertheless, if taken at face value, the loss of large numbers of adults would have left the remaining children in a highly vulnerable state, especially given the severity of the heat and drought observed during 1828–29. Indeed, Sturt noted that Aboriginal groups near the Castlereagh and Bogan rivers at, or just before the arrival of smallpox, were 'dying fast, not from any disease, but from the scarcity of food', and that near the Macquarie Marshes children were brought to his camp to implore something to eat, every fish and mussel in the river having been collected and consumed.[53]

In following decades Sturt and other explorers and early European settlers commented on the dramatic recent depopulation that had occurred along the Murray and lower Darling rivers, and Aboriginal survivors themselves said that before European arrival entire groups had been virtually or totally wiped out by disease (Fig. 9.4).[54] Two hundred years later, it is difficult to determine with certainty the role that drought played in this process, but we can be confident that it was very likely a key factor in northern and central New South Wales in 1828–29, especially along the Darling, Macquarie and Castlereagh rivers. It may also have played a similar role in the Barcoo, Maranoa, Paroo and other regions of southern Queensland and northern New South Wales where smallpox is known to have occurred, particularly if smallpox did spread from the north coast to the Darling region before 1829 as suggested.[55] Whatever the exact chain of events, the balance of evidence suggests that Aboriginal society here must join the growing list of those devastated through history by the coincidence of drought and disease.[56]

THE INTERIOR REVEALED

Thou hast given me a south land; now give me also springs of water.
  – Mr Barron Field, The Australian, *8 December 1825*[57]

The extreme conditions experienced during 1826–29 revealed to Europeans that droughts in Australia would forever play a central role in shaping the development of the continent. Not since the Settlement Drought had conditions been so dry, but now settlers and explorers witnessed the failure of rains across vast swathes of the continent,[58] and rivers and creeks from Brisbane to Tasmania either dry or reduced to stagnant pools.[59] As early as December 1826, following months of heat, drought and bushfires, the oldest settlers in Sydney considered the conditions to be unprecedented, but when lagoons at Wallis Plains on the lower Hunter Valley dried up in 1829, even local Aboriginal people had no tradition or recollection of such

> *For the last three years, westerly winds have been more prevalent than during any former period of our Colonial history, and have, consequently, brought along with them years of calamitous drought, unprecedented in our annals.*
>   *–* The Sydney Gazette and New South Wales Advertiser, *30 June 1829*

severe conditions.[60] In Sydney water was selling for 3 pence per bucket during 1827, and as shallow wells dried up new ones were being sunk to access deeper groundwater.[61] Crops failed repeatedly from Port Macquarie to Tasmania, and the scarcity and high price of grain quickly led to a general want of food among the working classes of the colony. By the summer of 1827–28 European tenant-settlers and 'famishing wives and children' in the Hawkesbury and other regions were 'scarcely ever in a worse plight'.[62]

Worse still, droughts in inland Australia proved far more severe and protracted than those seen on the coast. Early explorers, such as Allan Cunningham, who set out across the Liverpool Plains in May 1827 hoping to find verdant pastures found instead a parched and barren landscape, dead vegetation, and oppressive silence (Box 9.1).[63] So long had the drought lasted in northern New South Wales and southern Queensland that shrubs were found growing in the channels of watercourses, a phenomenon that we now understand frequently occurs in inland river systems during multi-year droughts.[64] Europeans could no longer ignore the blinding reality that even in higher quality pastoral regions of inland Australia it was not unusual for years to pass with little or no moisture available for agricultural pursuits.

There was, in fact, little excuse of mismanagement of grazing country during the 1826–29 drought, since only 11 years had passed since the last drought had ravaged the Sydney Basin. But much like before, cattle died in droves through starvation or by bogging in drying waterholes, and again grain and fodder had not been stored to mitigate against future reductions in pasture growth. Thus, while the drought was considered by some to be 'an irremediable evil, and incapable of mitigation',[65] others simply blamed selfishness, improvidence and waste for the disaster.[66] Nevertheless, new challenges had also emerged. For the first time trampling and resulting soil compaction from cloven-hooved animals emerged as a major threat to pastures, in some cases rendering the earth incapable of producing anything.[67] Inland pastures, eaten down and left to wither with roots exposed to the sun, failed to recover, forcing herds ever further into the dry interior.[68] Outside of the Sydney Basin grazing by kangaroos and predation by dingoes added to the misery of sheep farmers, ushering in a new appreciation of the way complex ecological interactions between livestock and native animals exacerbate the impact of drought on farming enterprises.[69]

Foreseeable or not, the 1826–29 drought proved a major setback for agricultural aspirations across the colony. Even major operations such as the Australian Agricultural Co.'s million-acre grazing property established at Port Stephens were severely affected (Box 9.2), but the weight again fell most heavily on the poorer classes. Years of failed harvests and crippling debt drove farmers, graziers and traders alike to insolvency, and thousands of convicts, having no-one who could afford to feed and house them, were returned to the government payroll.[70] But despite the vagaries of the Australian climate, stories of 'virgin pastures of the Interior, untrodden and uneaten, ... affected but little by the drought'[71] kept alive an underlying optimism that these pursuits might yet thrive. When the drought finally broke, first in southern coastal districts in mid-1829, and then more broadly in January to April 1830, the rains fell on a people clamouring for new opportunities and path to recovery, and another people, struck down by pandemic, who were unable to resist.

Box 9.2: Drought and the birth of the Australian Agricultural Co.

The Australian Agricultural Co. (AACo) was founded in 1824 by a group of British and colonial investors with a combined capital of £1 000 000. An Act of British Parliament and Royal Charter granted the company the right to 1 million acres (404 685 ha) in New South Wales for the 'cultivation and improvement of waste lands', the primary interest being the production of fine Merino wool for export to Great Britain. While the highly fertile Liverpool Plains were initially recommended, agent Robert Dawson, under pressure to select a site, settled on land extending inland and north from Port Stephens.

Operations began in 1826, but problems began almost immediately. The adoption of a summer lambing season resulted in the loss of lambs due to exposure to heavy rain. Much of the terrain was swampy, causing sheep to stand in mud overnight before being driven to graze on sterile pastures. Large numbers of sheep died from what was thought to be rot, a disease caused by liver flukes that are prevalent in marshes in the region. This disease may have been worsened by the decision to graze the sheep on salt marshes, as was customary on England. Scab, which is caused by mites, also spread across the AACo sheep stations; by mid-1828 not one remained free of the disease.

To make matters worse, a shift towards drier conditions occurred in 1826, and the intensifying drought began to add to the economic woes of the company. By September 1828 the number of sheep had fallen from 15 000 to 10 000, and losses were also high among cattle and horses. The weather was so dry that the unfenced wheat crops barely grew, and

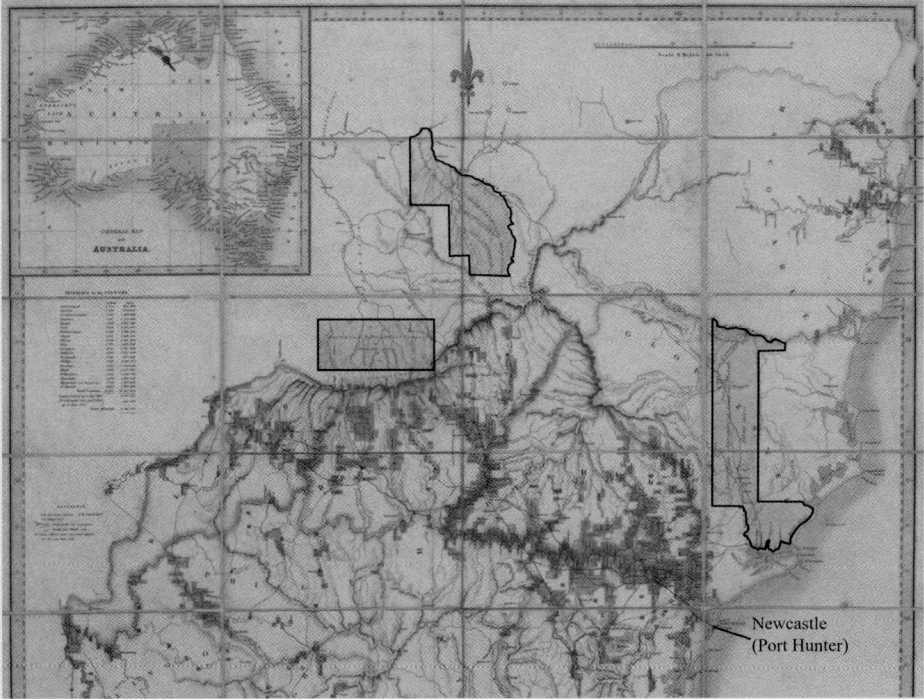

Early holdings of the Australian Agricultural Co. (bold lines), including the Port Stephens estate which, like most of New South Wales, was severely drought affected from 1826 to 1829. National Library of Australia, nla.obj-231316713-1.

> **Box 9.2: (Continued)**
>
> kangaroos and cattle decimated what was left. Conditions on the million-acre property were soon obviously so poor for sheep breeding that plans were set in motion to abandon the grant, and in 1832 two huge estates, Warrah and Goonoo Gonoo, were selected further inland. Today, the AACo is Australia's oldest continuously operating company, owning around 7 million hectares of land, primarily in Queensland and the Northern Territory.

Endnotes

[1] Sturt C (1833) *Two expeditions into the interior of Southern Australia, during the years 1828, 1829, 1830, and 1831: With observations on the soil, climate and general resources of the colony of New South Wales.* vol. 1. Smith, Elder and Co., London, p. 86.

[2] Sturt mentioned that the drought commenced in 1826, but severe drought also extended from the Moreton Bay region of Queensland to Tasmania in 1824 to 1825, and the broader drought period likely extended from ca. 1824–30.

[3] Letter of instructions to Charles Sturt. Sturt (1833), vol. 1, p. 183.

[4] King DA (1997) *Australian Journal of Botany* **45**, 619–639.

[5] Sturt (1833), vol. 1, pp. 29, 34, 113 and 144.

[6] The Barwon River becomes the Darling River downstream of its junction with the Culgoa River.

[7] Sturt (1833), vol. 1, pp. 144–145.

[8] Sturt (1833), vol. 1, pp. 50–52, 73, 119–125.

[9] Sturt (1833), vol. 1, pp. 7–8, 11, 149.

[10] Sturt (1833), vol. 1.

[11] Sturt (1833), vol. 1, pp. 119–125.

[12] Sturt (1833), vol. 1, pp. 137, 140–141. The parts of the Macquarie and Castlereagh river catchments visited by Sturt in 1828–1829 were apparently inhabited by the Ngiyampaa (or Ngemba) Wayilwan people.

[13] Meredith KT *et al.* (2009) *Journal of Hydrology* **378**, 313–324.

[14] Sturt (1833), vol. 1, pp. 95–96. Springs of this kind are known to be particularly prevalent in the Bourke area.

[15] When I visited the Darling River at Wilcannia, NSW in March 2019 it was virtually dry, despite experiencing very substantial flows between July and December 2016.

[16] Lee I (1925) *Early explorers in Australia : From the log-books and journals, including the diary of Allan Cunningham, botanist, from March 1, 1817, to November 19, 1818.* Methuen, London.

[17] Sturt (1833), vol. 1, p. 43. Sturt crossed Marra Creek south of the Big Lagoon (–30.580, 147.196); see Cumpston JHL (1951) *Charles Sturt: His life and journeys of exploration.* Georgian House, Melbourne.

[18] Either Buddah Lake has since contracted in width or Sturt's dimensions were incorrect, for the lake now has a maximum breadth of around 220 m (233 yards).

[19] *The Daily Telegraph* (Sydney), 7 July 1920, p. 10; Geoscience Australia's Archival Aerial Photography film number CAC7010, run 1, frame 80.

[20] *Sydney Gazette and New South Wales Advertiser* (hereafter *Sydney Gazette*), 27 January 1829, p. 2.

[21] Sturt (1833), vol. 1, pp. 113–114.

[22] Sturt (1833), vol. 1, p. 108.

[23] Sturt (1833), vol. 1, pp. 74, 78. The locations of Mount Oxley and Gundabooka are 30.20°S 146.23°E and 30.58°S, 145.75°E respectively.

[24] Sturt (1833), vol. 1, p. 110.

[25] Sturt (1833), vol. 1, p. 113.

[26] Godfree RC *et al.* (2017) *Royal Society Open Science* **4**, 170934.

[27] Mueller B, Seneviratne SI (2012) *Proceedings of the National Academy of Sciences* **109**, 12398–12403.

[28] Sturt (1833), vol. 1, p. 6.
[29] Sturt (1833), vol. 1, p. 15.
[30] Bourke has experienced minimum overnight temperatures as high as 33.7°C on 27 January 2019 (Bourke Airport AWS) and 33.3°C on 11 January 1939 (Bourke Post Office).
[31] Sturt (1833), vol. 1, p. 127.
[32] Sturt (1833), vol. 1, p. 189.
[33] From Campbell J (2002) *Invisible invaders: Smallpox and other diseases in Aboriginal Australia, 1780-1880*. Melbourne University Press, Victoria.
[34] 'Report from the Select Committee on the Condition of the Aborigines with Appendix, Minutes of evidence, and Replies to a Circular Letter, 31 October 1845'. W. W. Davies, Government Printing Office.
[35] Sturt (1833), vol. 1, pp. 89–94. The infected Aboriginal people were likely Barundji.
[36] Dowling PJ (1997) *'A great deal of sickness': Introduced diseases among the Aboriginal People of colonial Southeast Australia 1788-1900*. Thesis submitted for the degree of Doctor of Philosophy of the Australia National University, Canberra; Butlin NG (1985) *Historical Studies* **21**, 315–335.
[37] *The Sydney Herald*, 30 April 1832, p. 4; Campbell J (1983) *Historical Studies* **20**, 536–556.
[38] *Sydney Gazette*, 24 December 1831, p. 2.
[39] Clarke probably lived among Aboriginal people between October 1827 and November 1831; see Boyce D (2013) *Clarke of the Kindur: Convict, bushranger, explorer*. Melbourne University Press, Carlton, Victoria.
[40] Campbell J (1983) *Historical Studies* **26**, 536–556.
[41] Butin (1985); Campbell (2002); although cf. *The Monitor* (Sydney), 4 August 1828, p. 2.
[42] Dowling (1997), p. 95; p. 829 in Carey HM, Roberts D (2002) *Ethnohistory* **49**, 821–869.
[43] *Sydney Gazette*, 29 November 1831, p. 3.
[44] Campbell J (1985) *Historical Studies* **21**, 336–358.
[45] Cumpston JHL (1914) *The history of small-pox in Australia, 1788-1908*. Albert J. Mullett, Government Printer, Melbourne, pp. 152–153.
[46] Cumpston JHL (1914) pp. 147–162; Dowling (1994) pp. 46–83 and Campbell J (1983) *Historical Studies* **20**, 536–556.
[47] Parker KL (1905) *The Euahlayi tribe: A study of Aboriginal life in Australia*. Archibald Constable, London, p. 39.
[48] p. 551 in Campell J (1983) *Historical Studies* **26**, 536–556.
[49] Movement of Aboriginal people to permanent water is described in Sturt (1833), vol. 1, pp. 140–141.
[50] Allen H (1974) *World Archaeology* **5**, 309–322.
[51] Hamilton, JC (1912) *Pioneering days in western Victoria: A narrative of early station life*. Exchange Press, Melbourne, p. 100.
[52] Cumpston (1914), p. 153.
[53] Sturt (1833) *Two expeditions*, vol. 1, pp. 113–114, 137, 149.
[54] Cited in Carey HM, Roberts D (2002) *Ethnohistory* **49**, p. 827; also Campbell J (1985) *Historical Studies* **21**, 336–358.
[55] Campell J (1983) *Historical Studies* **26**, p. 549.
[56] E.g. Acuna-Soto R *et al.* (2005) *Medical Hypotheses* **65**, 405–409.
[57] The quote here by the drily named Barron Field comes from Joshua 15:19 (King James Version): 'Give me a blessing; for thou hast given me a south land; give me also springs of water.'
[58] *Sydney Gazette*, 30 June 1829, p. 3.
[59] There are many references to low water levels beginning in 1824 and extending through 1829. A few include Cunningham A (1824) and Cunningham A (1829), cited in Steele JG (1972) *The Explorers of the Moreton Bay District 1770-1830*, University of Queensland Press, St Lucia, Qld; Lee (1925); *Historical Records of Australia (HRA)*, series 3, vol. 5, pp. 850–854 (26 January 1827); *HRA*, series 3, vol. 5, p. 857 (27 March 1827; *The Australian* (Sydney), 10 March 1827, p. 2; *Sydney Gazette*, 27 January 1829, p. 2, *Sydney Gazette*, 6 February 1830, p. 3; Sturt (1833), vol. 1; *Colonial Advocate and Tasmanian Monthly Review and Register* (Hobart Town), 1 March 1828, p. 46; Hovell WH, Hume H (1837) *Journey of discovery to Port Phillip, New*

South Wales in 1824 and 1825, by W. H. Hovell and H. Hume, Esquires. Second edition. (Ed. William Bland). James Tegg, Sydney.

[60] *Sydney Gazette*, 30 June 1829, p. 3.

[61] *Colonial Times and Tasmanian Advertiser* (Hobart), 2 March 1827, p. 2; *The Australian* (Sydney), 3 March 1829, p. 3.

[62] *The Australian* (Sydney), 21 October 1824, p. 3; *The Australian* (Sydney), 23 February 1826, p. 2; *The Monitor* (Sydney) 17 February 1827, p. 5; *The Monitor* (Sydney), 11 February 1828, p. 4; *Colonial Advocate and Tasmanian Monthly Review and Register* (Hobart Town), 1 August 1828, p. 293.

[63] Cited in Lee (1925).

[64] Cunningham A (1829) cited in Steele (1972), p. 319.

[65] *The Sydney Monitor*, 13 June 1829, p. 3.

[66] *Sydney Gazette*, 24 December 1828, p. 3.

[67] *The Australian* (Sydney), 21 September 1827, p. 2.

[68] *Sydney Gazette*, 30 June 1829, p. 3; *The Australian* (Sydney), 27 March 1827, p. 2.

[69] *The Sydney Monitor*, 4 October 1828, p. 5; *Sydney Gazette*, 9 June 1829, p. 2.

[70] *The Sydney Monitor*, 6 September 1828, p. 5 and 13 June 1829, p. 3.

[71] *The Sydney Monitor*, 13 December 1828, p. 3.

10
Depressed times

The droughts of this country are certainly insurmountable, and therefore a serious drawback to agricultural speculations

– The Colonist *(Sydney), 23 March 1837*

IN OCTOBER 1839, CHARLES AND LOUISA ANN MEREDITH, HAVING RECENTLY arrived in Sydney after a 4-month voyage from England aboard the *Letitia*, were bumping by cart over the rutted and dusty road to the growing central-western town of Bathurst, NSW. Charles, a well-connected man who, since 1821, had spent time in Tasmania and New South Wales, was at home in the Australian bush. For Louisa, however, the contrast between the harsh climate and culture of her new home and her English upbringing could not have been starker. An exceptionally well-educated and erudite young woman who had minor celebrity status in England as an author at just 23 years of age, Louisa (Fig. 10.1) was unimpressed by the filthy public-houses, slovenly people and monotonous cuisine encountered in the colony. It was even less ready for her: in an era when women seldom played a major role in public political discourse, Louisa's sharp eye for detail and acerbic pen frequently fell on the inhabitants of her new home, much to the ire of the educated urbanites and wealthy graziers who dominated the cultural scene.

For the moment, however, things seemed positive. In 1834 Charles had been given £2000 by his father to start his own business and was now a squatter of considerable means, with extensive sheep

Fig. 10.1: Louisa Anne Meredith (1812–95), ca. 1850, photographer unknown. Louisa arrived in New South Wales in 1839, already a successful writer and poet in England, later becoming one of the first women to forge a career as an author and artist in Australia. Unfortunately, with her finances devastated by the depression of the early 1890s, she wrote in a letter to Sir Henry Parkes: 'I was born under an evil star, or put myself under one, in quitting England in the first instance – and my life long struggle against fate – is vain.' Her works, including *Notes and sketches of New South Wales* (1844), *My home in Tasmania* (1852) and *Some of my bush friends in Tasmania* (1860) provide some keen insights into early colonial life and the plants and animals of Australia. Image: Wikimedia Commons and Libraries Tasmania, 613811.

interests in the favoured Murrumbidgee region of southern New South Wales. The novel plants, animals and landscapes were fertile grounds for Louisa's creative talents and passion for writing, a welcome change from the cramped and monotonous voyage. The climate even seemed agreeable, and upon arrival at Port Jackson in late September 1839, Louisa commented on the bright sunshine, warmth and the remarkable clarity of the atmosphere, which in contrast to the 'diffused effect of the English landscape' left even remote objects with a clean and distinct outline.[1] It was not until later, when crossing the waterless tracts of the Blue Mountains, that she fully realised that such conditions are not always viewed with same enthusiasm by Australians, who know well the perils of clear skies. Disaster was looming, not just for the Meredith family, but for the whole colony, wealthy and poor alike.

EXPANSION AND ENTERPRISE

Here in these wilds reigns the liberty of rags and the freedom of dirt.
– George Hamilton, pioneer and overlander[2]

The years that followed the breakup of the drought of 1826–29 was a period of unparalleled expansion of agricultural investment and settlement in south-eastern Australia. Optimism returned with the rain, providing new opportunities for colonial and English capitalists to benefit from the wealth generated by the booming Australian wool industry. Colonial banks, old and new, poured foreign and overseas capital into new pastoral businesses, while British banks also entered the fray, all enjoying relaxed controls on lending. A 'speculative credit nexus' soon developed, in which merchants would speculate on future wool markets through money lending, since command of increasing quantities of wool underpinned profitability, and pastoralists speculated on both land and future wool markets, because wool prices were favourable and credit easy to secure.[3] Woolgrowers were therefore incentivised to maximise income by rapidly expanding their flocks and selling surplus sheep to new speculators entering the market. In this way pastoralism became the engine of the colonial economy, and speculation came to focus on the pastoral unit of production and the expanding frontier.

From the outset this synergy between British colonial mercantile and penal interests was inconducive to agricultural improvement and economic durability, and, equally importantly, to the long-term management of pastoral properties. From a legal perspective, the tenure of frontier squatters was insecure, and so there was little incentive for fixed capital investment, technical improvement or animal breeding. Additionally, much land was controlled by absentee or commercial capitalists from the settled districts, like Charles Meredith, or wealthy British emigres, thus depriving new agricultural districts of the homesteading families that would have had a greater interest in such pursuits.[4] Thus, what capital accumulation did occur was often primitive in nature, labour intensive and, for many, only possible given access to a steady flow of convict workers, which was supported by powerful government and private interests (Fig. 10.2), and foreign investment.[5] Stations established in the more isolated areas, 6–8 weeks travel by bullock dray from export terminals, were also heavily dependent on high wool prices, access to credit, and availability of transportation. But while this house of cards remained standing, expansionist colonial pastoralism as a world-economic industry system remained viable.

Whatever the squatters may have lacked in commercial nous, however, they more than made up for in vigour. Prior to 1830, colonial spread had been slow: the settled frontier in New South Wales at that time was still restricted to a roughly 120 000 km² area extending from Port Macquarie to Jervis Bay and east to the major rivers of the southern and central slopes (Fig. 10.3). Major sheep stations dotted the Murrumbidgee as far as west as Jugiong, but further north the Liverpool Ranges had impeded the spread of stockmen from the Hunter Valley, which in 1826 contained some 50 000 sheep, and few Europeans had been beyond the southern Liverpool Plains.[6] On the back of favourable rainfall and the tracks of explorers, the squatters quickly burst these shackles, and rode the wave of speculation along the coast and down inland rivers into the deep interior of the continent. Two main routes were used: the larger down the Murrumbidgee and over the south-western slopes of New South Wales, and the smaller north from the Hunter Valley towards Queensland (Fig. 10.3). In 1836 Hawdon's party had successfully established the Overland Route to South Australia, which effectively linked these colonies, and by 1839, 694 stations and sheep holdings dotted the landscape in a 1700 km–long chain from New England to South Australia.[7] In just 10 years the most favourable land in south-east Australia had all been occupied, the main exceptions being south-east Queensland, Gippsland and parts of the western and southern Riverina.

Fig. 10.2: Newspaper article from 1836 reminding country residents of the date of application to have convicts assigned to them as agricultural workers or mechanics. Interestingly, the Bank of Australia which mainly catered to the upper class, failed in 1843, while the Bank of New South Wales, which was dubbed the 'convicts bank', survived.

The rapid expansion of the frontier posed new environmental problems for early squatters. First, stations and runs were now being established well within the semi-arid zone for the first time, in locations as far west as Hillston, Hay and the northern Bogan and Macquarie rivers. Here, rainfall is scant (360–420 mm annually), summers hot, and moisture availability drives the productivity of the landscape. Agricultural scientists often measure this in terms of the ability of a system to produce vegetative growth, or its *above-ground net primary productivity*

(ANPP). In semi-arid Australian rangelands ANPP typically averages 500–1000 g of dry biomass per m² per year (g m^{-2} yr^{-1}), compared to 30–300 g m^{-2} yr^{-1} in arid regions globally (Australia deserts, being quite wet, have ANPP values towards the top end of this range) and 2000–2500 g m^{-2} yr^{-1} in wet tropical forests.[8]

The problem of low rainfall is itself not insurmountable, for dryland farming is undertaken profitably in many parts of the world that are as dry or drier, including in fallow-based systems of south-western Western Australia.[9] The key, however, is that the timing of rainfall in these regions is predictable, which allows farmers to plan for and manage seasonal water deficiencies. In western New South Wales there is no such luxury: rainfall is tremendously erratic, and much of what does fall is either lost to run-off during intense storms or to evaporation. Even during 'average' seasons the soil is frequently dry enough to retard plant growth, and during droughts it can become so parched that native vegetation dies and creeks dry up entirely. Thus, lacking any infrastructure to gather and store water or fodder, squatters were now truly at the mercy of the fickle inland weather.

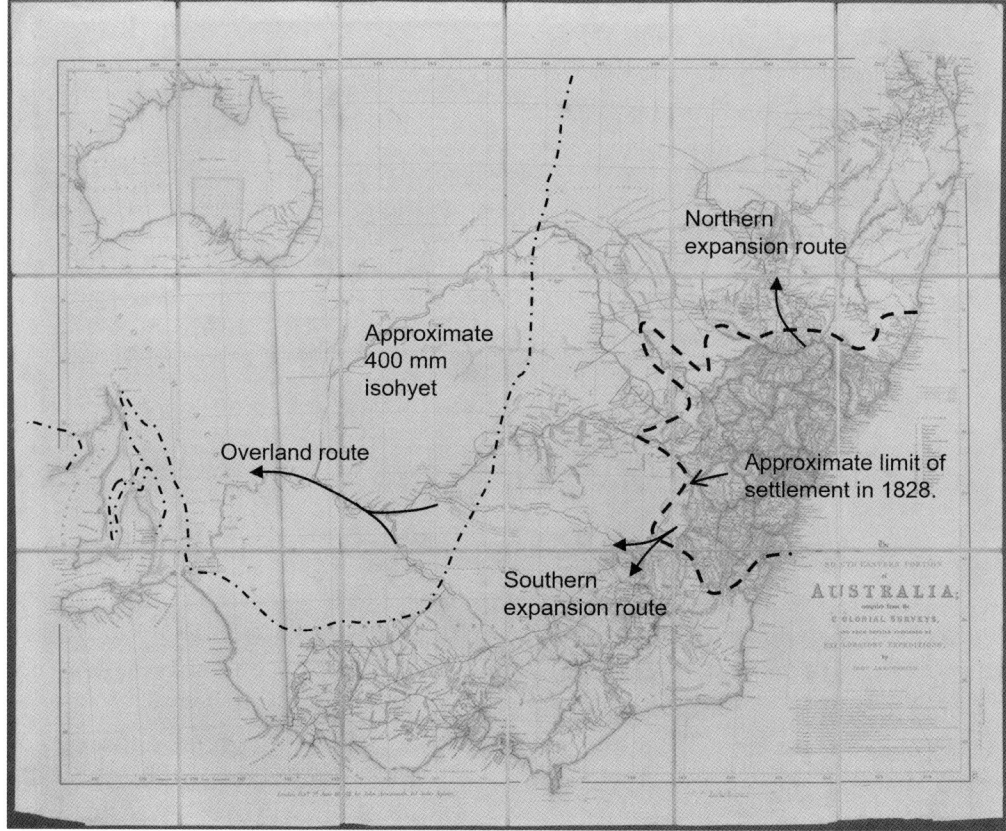

Fig. 10.3: Map of settlement in south-eastern Australia in 1842, showing the rapid expansion that occurred after 1828. From Arrowsmith J (1842) *The south eastern portion of Australia / compiled from the colonial surveys and from details furnished by exploratory expeditions.* John Arrowsmith, London. National Library of Australia, nla.obj-231341828.

The second problem was that as the supply of available land closer to major settlements dried up, squatters were forced to establish new holdings in far-flung areas remote from towns (Fig. 10.4). While this provided opportunities for those with few means, access was difficult. Roads, if they existed, were in terrible condition and, especially in cracking clay country, impassable after rain. This limited access to supplies, and people in the outer stations often lived in a state of severe deprivation. In 1839, Louisa Meredith commented on how due to a 'devastating and ruinous drought' at Bathurst 'meat was lean to starvation, and flour liberally adulterated with various cheaper ingredients; vegetables there were none' and that 'butter and milk had long been but a name'.[10] Given these circumstances, she was unable to form a 'very high opinion' of the towns' beauties or advantages. But she would have found conditions on the outer stations far worse: a traveller passing through the Lake Cargelligo region in the same year found stockmen and other workers in a 'miserable condition; no sugar, no tea, very little bread, and less meat'. The livestock proprietors, probably absentee owners of higher social standing, had apparently not yet sent their *half-yearly* supplies to the station, which lay more than 300 km west of Bathurst down the Lachlan River.[11]

Fig. 10.4: Doyle, J. T. (ca. 1862-1863). *'Prospecting' at an out station. Dingoes, or Wild Dogs of the Bush prowling round the sheep fold* [watercolour]. While both romantically and humorously portrayed here, life on remote runs was difficult, dangerous and often lonely. Dingoes were indeed a major threat to sheep, which led to their persecution after settlement. State Library of NSW, Mitchell Library, PXA 1983.

Although some libertarian-minded people rejoiced in the freedom from social formalities and rules that frontier life provided, stockmen and farm hands willing to endure such conditions were scarce, especially given the chronic shortage of workers that prevailed at the time. To secure their services squatters had to pay salaries that were considered exorbitant compared to the cost of labour in England: shepherds could earn £25–60 annually during the wool boom, and additionally received a ration in the form of wheat or flour, beef or mutton, and milk, while married couples earned £10–20 more than single men. Shearers earned £1 per 100 sheep and could shear 60–80 sheep per day, and skilled workers like mechanics could make 10 shillings per day (half a pound).[12] One newspaper article from Geelong in 1840 lamented that labourers were asking 'the enormous sum of 12 shillings per day' just for digging ground for an orchard.[13]

An even more challenging problem was to get their products, especially wool, to market, and the only effective form of haulage available was drays pulled by bullock teams. Bullock drays typically travelled at only 1–3 miles per hour (1.6–4.8 km/hour), depending on the quality of roads, and in most areas could only be expected to cover 9–15 miles per day (14–24 km/day, up to 40 km/day under ideal conditions). Driving bullock teams was also arduous and often dangerous work,[14] and so 'bullockies' were well paid, even compared to other farm workers. Frontier stations, which were 3–6 weeks or more from Sydney by dray, therefore faced the threefold problem of poor roads, large distances and high driver wages, a 'tyranny of distance' that dramatically reduced the profitability of their wool-growing operations. While wool prices remained high, squatters could cover these costs. But an even more slender thread held this entire system together – the availability of grass and water.

Today, in an era of mass livestock movement on diesel-powered road trains, it seems strange that transportation was once fundamentally limited by rainfall. But bullock and horse teams required large amounts of food, and while in some cases fodder could be hauled for these requirements, in most cases it was simply obtained by grazing along roadsides or at campsites. The number of drays moving about the landscape numbered in the thousands – one traveller passing from Bathurst to Sydney in December 1834 passed more than 150 loaded with wool.[15] Squatters also sold excess livestock to speculators, and from the early 1830s vast mobs of sheep and cattle were continually in transit around the pastoral districts from southern Queensland to South Australia. The scale of movement was enormous – 50 000 sheep and 2000 cattle and horses arrived in the Port Phillip region in just 2 months during mid-1837, while mobs of 5000–10 000 sheep and hundreds of cattle were driven to Adelaide almost weekly across the famous overland routes.[16] Collectively, the dependence of transport on animals pushed food supplies along travel routes to the breaking point.

Fortunately, the breakup of the 1826–29 drought in mid- to late 1829 had ushered in conditions that were perfect for pastoral expansion, and by 1830 all western rivers contained more than enough water for livestock movement. In December 1829 to January 1830, Charles Sturt found that both the Murrumbidgee and Darling rivers were flowing swiftly to their junctions with the Murray River, indicating that major tributaries of their upper catchments must have also been running for several months.[17] The Darling itself measured 12 feet (3.66 m) deep and flowed so strongly that its silty waters remained separated from the clear waters of the Murray for a considerable distance downstream of their confluence (Fig. 10.5). Sturt estimated the velocity of the Murray itself to be 2.5 knots (1.29 m/s) and, although I think this may be an overestimate, flow was obviously fast.[18] Crucially, however, most rivers appeared to have remained below bank level, thus allowing stock movement downstream while avoiding challenges associated with flooding.[19]

Interestingly, Sturt noted that, despite the recent rains, the lower plains of the Riverina near the junction of the Lachlan and Murrumbidgee rivers were still relatively dry in December 1829. Days passed without his party seeing so much as a blade of grass, and furnace-like north-north-easterly winds blew up clouds of dust.[20] However, the plains and rolling hills on the upper slopes to the east had fared better, and Sturt encountered luxuriant vegetation along the plains that flank the Murrumbidgee from Jugiong to Wantabadgery, upstream of modern Wagga Wagga.[21] At the same time, grass grew in abundance across the tablelands, Hunter Valley, Sydney Basin and Southern Highlands.[22] Wet conditions continued the following year, with significant

Fig. 10.5: *Junction of the Darling and Murray River*, pencil drawing, after the engraving in Charles Sturt's *Two expeditions into the interior of southern Australia*, vol. 2. In January 1830 the Darling was flowing swiftly where it entered the Murray, indicating that significant rain must have fallen in northern New South Wales and southern Queensland during the previous 6 months, breaking the 1826–29 drought. However, when visited by Thomas Mitchell in 1836, when rivers with more southerly catchments were flowing strongly, the Darling was very low, suggesting that flooding rains that fell that year may not have reached the northern interior. National Library of Australia, nla.obj-135204400.

flooding affecting the Hawkesbury, Nepean, Hunter and Hasting River catchments on several occasions between April 1830 and May 1831, the latter of which apparently reached into the interior.[23]

What then ensued was an interesting period of rapid oscillation between wet and drought conditions as the El Niño–Southern Oscillation shifted between neutral and El Niño phases. The winter of 1831 was cold and mostly dry, by October the ground near Bathurst was 'as firm and hard as in the heat of the summer', and 2 months later the Sydney Basin was parched.[24] In March 1832 rain fell across the colony, and rivers and creeks in the Macquarie and Nepean catchments rose high enough to disrupt travel but dry conditions again took hold in mid- to late 1832 with the onset of a weak El Niño event, and by summer crops and pastures had failed from the Hunter to the Illawarra and west to Bathurst.[25] Strong ENSO-positive conditions developed in 1833 and soon severe drought was being reported across most of New South Wales, with widespread failure of crops and pastures precipitating a serious decline in the condition of livestock.[26] However, this short but sharp drought also soon broke and from December 1833 to September 1834 periods of heavy rain and localised flooding extended from Tasmania to New South Wales.[27]

THE ACCURSED COUNTRY

this is an accursed country, drought and flood and fire being leagued in fatal hostility against the hapless settler.
 – The Sydney Gazette and New South Wales Advertiser, *27 March 1832*

So far, squatters and their creditors had been able to withstand the volatile conditions, since demand for wool remained high and financial liquidity was in abundance. But circumstances again took a turn for the worse. The summer of 1834–35 began dry, and this time conditions intensified so rapidly that within less than a year thousands of livestock lay dead from starvation and bogging in drying waterholes across the colony.[28] The unusual speed with which this drought developed suggests that the periods of adequate rainfall that had fallen during 1832 and 1834 had masked groundwater deficiencies that been gradually building up during the intervening dry years. This idea is supported by the rapid decline in the depth of Lake George that set in after 1823 (Fig. 3.3); by March 1835 Lakes George and Bathurst had both shrunk to only around 'half their dimensions' on their way to drying out completely in 1839.[29] When water deficiencies build up in a punctuated manner over many seasons like this, it can be difficult to define the onset and duration of drought, but here I follow the view of William Stanley Jevons, who argued in his *Some data concerning the climate of Australia and New South Wales* in 1859 that 1835 marked the beginning of a 5-year period that was to become one of the worst droughts in Australia's history.

The climate in 1836 also began poorly, with hot, dry summer winds laying waste to crops and pastures from the Liverpool Plains to Victoria.[30] On the Monaro Plains the dry conditions, exacerbated by fires that burnt pastures, huts and stockades, caused stock keepers to abandon their stations and take up new holdings much further afield – a process that sped up frontier expansion in much the same way as it had in 1813–15 and 1826–29.[31] The developing drought was briefly interrupted in the autumn of 1836, when heavy rain fell across much of the colony, especially in the south and on the coast.[32] According to James Dunlop, superintendent of the government observatory at Parramatta, rainfall there measured over 8 inches (200 mm) in each month between April and July 1836 before again tailing off.[33] In March to July 1836 Thomas Mitchell found the Murrumbidgee, Loddon, Murray and numerous other rivers and creeks in Victoria all flowing strongly or in flood, but the Lachlan River remained virtually dry and the Darling greatly reduced at its junction with the Murray, indicating that drought may have persisted in the central and northern inland.[34]

Perhaps the coldest and wettest winter ever experienced in south-eastern Australia then set in. Three to six inches (7–15 cm) of snow fell in Hobart, a very rare event, and a light fall even graced downtown Sydney for the only time in recorded history. Higher altitudes received much more: the central and southern highlands of New South Wales received 1–2 feet (0.3–0.6 m) of snow, and a 4-day storm in the Monaro Plains left snow lying 3–4 feet (0.9–1.2 m) deep and thousands of sheep dead, which must have been a devastating blow for those who had moved their stock into more remote, higher pastures during the drought and fires of the previous year.[35] Further strains were placed on agricultural profitability by heavy rain and snowmelt that

> *This drought was a circumstance, however, which could not fail to strike the visitors of Port Phillip very forcibly after the confident assertions of immunity from this Australian plague*
>
> – The Australian *(Sydney), 21 April 1837*

left interior rivers swollen and roads little better than a quagmire, blocking movement of cattle and goods.[36] Thus, despite prolific growth of pastures and crops in the following spring and summer, the net result of these conditions was that the physical and commercial ties that had stimulated unprecedented expansion of the past 6 years had become increasingly precarious.

Further trouble began with a return to dry conditions early in the summer of 1836–37 and the development of a strong El Niño event that persisted through 1838.[37] Although crop and pasture failure were first observed across the coast and inland of central New South Wales in December 1836,[38] the shift in conditions must have been much more general, since by early autumn 1837 settlers at the newly established colony at Port Phillip, contrary to their expectations, were complaining of a drought 'as determined, and as fatal to vegetation, as any similar drought of one season's duration in our own part of the Colony'.[39] Tasmanian agriculturalists were also not spared; drought had set in by May 1837 and by the end of winter had become severe enough to damage crops.[40] Further west, the interior of Western Australia also apparently suffered below-normal rainfall, the spring cropping season being the driest since the colony was established.[41]

By early 1838 conditions in New South Wales were becoming critical. Creeks had run dry from the Sydney Basin to the Monaro Plains, and feed for stock was in short supply everywhere. The southern interior of New South Wales was particularly hard-hit, and now presented a scene of utter desolation.[42] Near Marulan, 25 km east of Goulburn, conditions were considered the worst in the 18–20 years since the area was first settled, which included the great 1826–1829 drought.[43] An account from Lake Bathurst, probably written in early 1838, reported that the leaves of forest trees were withering, and that honeysuckle, mimosa and 'light wood' (probably *Banksia* spp., *Acacia dealbata* and *A. implexa*) were 'almost dead' and which 'by their shrivelled unhealthy appearance, make the dreariness complete'.[44] Similarly, trees reportedly began to die in the vicinity of Lake George in 1837, perhaps compounded by disease.[45] In my experience native trees and shrubs growing on ridges and steeper slopes in this region can show signs of severe water stress after 18 months to 2 years of severe drought, which implies that in this region soil water must have been generally declining since 1835, and that 1837 was exceptionally dry indeed.

The drought peaked in both severity and extent in late 1838 and 1839, a behemoth that spread across more than 1 000 000 km² of New South Wales, Tasmania and Victoria (Fig. i.1). Whether it matched the scale of the great droughts of the 20th century is difficult to determine, since much of the continent still remained beyond the reach of Europeans, but drought and crop failure were reported at York and Northam, WA in October 1838, and in 1839 the state of the River Torrens at the Adelaide Plains was so disgusting that 'a loaf of bread has been offered for a draught of water'.[46] On the other hand, many early settlers seemed to be simply unaware that sharp seasonal drying is a normal characteristic of the Mediterranean climate experienced in these regions. For example, observations made over ensuing years revealed that the River Torrens normally dries down to pools in the summer months, and by 1846, contrary to early experience, at least some argued that South Australia had yet to experience a major drought.[47] Evidence of abnormally dry conditions in Western Australia and Adelaide in 1838–39 remains equivocal, apart perhaps from a brief period in the autumn of 1839,[48] although overlanders driving cattle from the Murrumbidgee to South Australia in May 1839 described the whole country as suffering greatly from drought, suggesting that at that time the entirety of the continent east and south of Adelaide was affected.[49]

In much of central New South Wales to northern Victoria the drought was probably the worst experienced since settlement. Major watercourses including the Nepean, Lachlan, Murrumbidgee, Macquarie, Hume and Ovens ceased flowing, sometimes for the first time in European experience, while ephemeral waterbodies such as lakes Bathurst, George and Cargelligo dried up completely.[50] From Wellington to the Monaro, cattle and sheep died in the thousands from lack of feed and water, the remaining pools in drying rivers and creeks choked with putrid carcases,[51] some of which were dead or diseased animals intentionally thrown into waterways for disposal.[52] Nearby, conditions on the Lachlan were so severe that 'suffering much from the drought, the trees of the forest withered, and great numbers are dying along acres together for want of moisture'.[53] This unequivocal eyewitness record of drought-induced tree dieback (the first apart from those mentioned above in 1837–38 and possibly Sturt's observations on the Castlereagh in 1829) clearly places the 1835–39 period, in terms of the severity of hydrological drought, among the worst experienced in New South Wales since European settlement.

Further north, conditions were little better. John Henderson, Captain of the 78th Highlanders, provided a vivid description of the Hunter Valley and Liverpool Plains during his travels there in January to February 1839. From his account, published in *Excursions and adventures in New South Wales* (1854) we know that drought had reduced most rivers in the region to mere chains of pools, and that little or no feed was available for livestock. At Darlington, near Singleton, the Hunter River was completely dry as far as the eye could see, apart from one stagnant, green pool which emitted 'the most abominable effluvia' courtesy of the putrefying carcases of bullocks that lay in the water. Local settlers had to travel many miles to get fresh water for domestic use.[54] Over the Great Dividing Range, the fertile Breeza Plains, which once supported a vast sea of waving native grasses, had been reduced to dirt, and the Mooki, Peel and Namoi rivers, which to this day remain the lifeblood of the region, contained little but drying pools and abandoned

> *No grass, or herbage of any kind, was to be seen; no dew-drops sparkled among our horses' feet, 'like orient pearls at random strung;' the dust rose at every footfall, and the few lean gaunt cattle that crossed our path, as they returned to the forest from slaking their thirst in the Mooki, seemed to the dreamy midnight fancy of a tired traveller, to stalk the plain - the ghosts of the departed, or the demons (alas! poor wretches, they were only the victims,) of the great drought.*
>
> – *John Henderson, February 1839, near Breeza, NSW*[55]

Fig. 10.6: Bullock dray with wool from Yandilla, 1898. Bullock teams were used to haul wool from outstations to markets and ports in the main cities on a massive scale beginning in the 1830s. This photograph was taken during the 1895–1902 Federation Drought, but conditions were similarly barren in 1837–39, when vast numbers of bullocks that died while hauling lay strewn along roadsides across the colony. Difficulties in moving wool from remote stations due to a lack of feed and water along transport routes contributed to the collapse of the speculative pastoral bubble of the 1830s. Queensland State Archives, ITM1109458.

outstations. Henderson's haunting portrayal of starving cattle on the plains, victims of the 'great drought', still resonates with anyone who has seen this country in a similar state.

Henderson's comments on the plight of beasts of burden are no less confronting. All along the route from Singleton to the Liverpool Plains bullocks lay dead, the task of hauling drays loaded with 10 or more bales of wool during the heat of summer over the Great Dividing Range being too much to ask. Louisa Meredith observed a similar ghastly scene on her trip across the Blue Mountains later in the same year, where carcases in various states of decay dotted the road where animals had fallen, some still covered in hide but others reduced to skeletons (Fig. 10.6).[56] Heavily travelled roadsides across the colony contained similar scenes of carnage; 50 dead oxen were seen between Cowpastures and Sydney in 1838.[57] Perhaps these animals were the lucky ones: the wretches that remained were forced to labour over the steep slopes of the ranges and burning plains without food or water, lashed from stem to stern by unfeeling bullock drivers. As Louisa put it, 'the patience and docility of the ox are justly proverbial, but unfortunately colonial drivers are less gifted with these virtues'.[58]

ECONOMIC COLLAPSE

'Water, oh God!' is the only cry.
– The Colonist *(Sydney), 24 August 1837*

Exactly which economic and social drivers were responsible for the collapse of the speculative bubble in eastern Australia in 1840–43 remains the subject of some debate. Internally, profitable

Box 10.1: Commodity prices, drought and the 1840s Depression

The drought of 1835–39 set the stage for the economic depression that developed in the early 1840s. Throughout the 1830s speculators had driven up the value of land and livestock to very high levels and pastoralism had spread rapidly into the interior of the continent. The onset of severe drought, particularly after 1837, resulted in a rapid rise in commodity prices, the importation of which drained liquidity from the colony, but also a decline in the price of sheep and cattle. Wool prices also fell, due in part to a glut in the English market. Falling profitability in the wool industry after years of severe drought then led English investors to withdraw finance from the pastoral sector. The final boom in land price speculation was over by the end of 1840, and following a collapse in the price received for farm produce, a severe financial depression ensued that lasted several years. Data compiled from Panza L, Williamson JG (2017) 'Australian squatters, convicts, and capitalists: Dividing up a fast-growing frontier pie, 1821–1871', Discussion Paper No. 2017–02, ANU, Canberra; Butlin NG, Ginswick J, Statham P (1987) The economy before 1850. In: Vamplew W (Ed.), *Australians, historical statistics*, pp. 102–125. Fairfax, Syme & Weldon Associates, Broadway, NSW; and Butlin SJ (2002) Foundations of the Australian Monetary System 1788–1851. University of Sydney Library, Sydney.

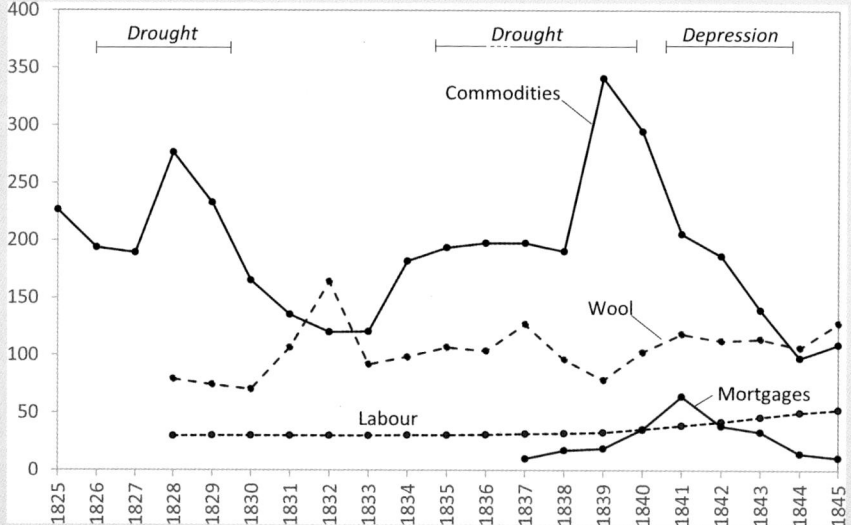

Value of commodities, wool, labour and mortgages. Commodities = NSW Commodity Price Index with 1848–50 = 100; Wool = Relative Wool Price Index with 1861 = 100; Labour = free unskilled average annual earnings in pounds per year; Mortgages = value of country mortgages in New South Wales in tens of thousands of pounds.

pastoral expansion had probably reached a temporary limit due to excessive transportation costs, declining availability of quality land, and the rising costs of equipment, supplies and labour.[59] The latter was under particular pressure both due to the rate of expansion and the impending cessation of mass convict transportation to New South Wales, which ultimately took place in 1840.[60] At the same time, the price of wool, which had reached as high as 2s. per pound in 1832 and 1836–37, fell sharply in 1838–39 (Box 10.1) due to oversupply in England[61] and declining fleece quality as inferior sheep were sold on to speculators rather than being culled from the breeding pool. While squatters and their creditors may have been able to withstand these conditions, *ceteris paribus*, the onset of drought in 1835, and the extreme conditions experienced in 1838–39, proved decisive.

The drought had devastating ramifications for the colonial economy. First, with stock losses from starvation and diseases like catarrh[62] mounting and travel from the outer runs impeded by shortages of feed and water, the profitability of speculative agricultural concerns nosedived. News of this trickled back to England, and, along with a growing glut of imported goods brought to the colony by British merchants, triggered the withdrawal of capital from the domestic market and a major call-in of funds by English banks.[63] This loss of liquidity, reportedly amounting to more than £300 000 in the 4 years to 31 December 1840, was exacerbated by an increase in the importation of wheat and other commodities to cover crop failures in New South Wales and Tasmania. To make matter worse, wheat prices skyrocketed from between 5s. and 8s. per bushel to more than 15s. in little more than a year; by mid-1839 prices of 20s. to 28s. per bushel were seen in Sydney.[64] This, along with shortages and 'appalling' prices of meat, tea, potatoes and other commodities led to severe distress among prisoners, who resorted to begging for food in interior settlements; the poor; and, unusually, the working classes, particularly those with large families.[65]

It was under these dire circumstances that Charles Meredith, after his gruelling trip across the Blue Mountains, left his wife in Bathurst to visit his holdings on the Murrumbidgee in late 1839. Fresh off the boat from England, he must have been shocked to see the landscape scorched bare from years of rainless skies, and the thousands of sheep that lay dead on his journey south. It seemed that he had little option but to sell his stake in the struggling industry, and fortunately two buyers were found – one being the court registrar John Edye Manning, a man with large land holdings in New South Wales and Port Phillip and an active life as a member of numerous public committees and associations in the colony.[66] In Bathurst, Louisa found every article of food to be extremely dear, and vegetables, butter and milk unprocurable. Fodder for livestock was both extremely scarce and exorbitantly priced, and starvation had left the local horses with so little strength that families could no longer use them for haulage and transport. The struggle was exacerbated by the high price of firewood required during the cooler months, the closest forests being miles from the town, and by furnace-like winds that frequently swept into their house with 'the force of a hurricane'. Both were undoubtedly relieved when after Charles returned from the Murrumbidgee they left the town in a cloud of insects and dust to spend Christmas in Sydney.[67]

The last fling of the bubble occurred in 1839–40 with a spectacular inflation of land prices, particularly in Adelaide and Melbourne, financed in part by newly formed colonial banks and companies based on the transferral of British funds for loan on mortgages in Australia. Established banks in England and the colony also underwent increased capitalisation, with generous lines of credit given to a small cohort of Sydney merchants who used the funds to purchase overpriced rental accommodation, although by this time the rate of economic

Fig. 10.7: Artist unknown, detail from Political Sketches by B.B. No.1., *Raising the wind or Sydney in 1844*, published by E. B. Barlow. This political cartoon was one of a series that were all highly critical of land regulations proposed by Governor George Gipps. After the collapse in the value of sheep many turned to boiling down the carcases of sick and old animals to produce tallow, which is a rendered form of mutton or beef fat. This cartoon portrays Gipps's efforts to raise revenue by charging new licence fees for the use of crown land as akin to rendering down squatters for the fruits of their labours, which was a popular sentiment given the financial destruction caused by extreme drought in 1837–39 and the subsequent economic depression. Dixson Library, State Library of NSW, FL991394.

expansion had already slowed.[68] Ironically, it was not drought but the joyous sound of rain that fell on dusty iron roofs across the interior in October 1839 that ultimately led to the final economic collapse. As the drought broke, favourable conditions returned, leading to a rapid drop in the price of wheat, flour and other farm commodities in November 1839. The collapse in prices continued the following year, a double blow for farmers and flour millers who had already been economically ruined by drought.[69]

The first insolvencies were of flour millers, but a more serious problem arose when indebted merchants and traders were forced to auction off goods at a loss and press for payments from account holders. Almost overnight the pressure for cash spread to banks and other lenders, and in 1842 a flood of more than 600 insolvencies claimed merchants, wealthy business owners and agents.[70] British investment in the colonies dried up, and crown land sales collapsed, which in turn led to the failure of multiple banks – including the Bank of Australia – and severe contraction among those that survived. Unemployment rose to 10% or more, which led to the formation of the Mutual Protection Society in 1843, the first union in Australia to protect the interest of the working and middle classes. Back on the land farmers, faced with supporting flocks of sheep that had become nearly valueless, turned to boiling down vast numbers of older, diseased and surplus animals for tallow fat (Fig. 10.7), in what was called at the time 'sheep-boiling mania'.[71] Slowly, economic conditions improved, and the suffering caused by one of the three worst depressions in Australian history eased.[72]

Many, however, never fully recovered from the destruction of their livelihoods that took place in 1841–43. We again pick up the story of the Merediths, who, after arranging sale of Charles's assets on the Murrumbidgee, moved into a run-down house called 'Homebush' on the Parramatta River. Soon after the birth of their first son on 1 July 1840 came news that Charles had lost everything: apparently John Edye Manning had purchased his sheep on promissory notes, transferred them to his wife and declared himself insolvent, while a second purchaser died.[73] Whatever the specifics, insolvency proceedings published on 21 April 1842[74] suggest that Charles had a claim on the Manning estate of £1740, a small fortune given that the average farm worker's wage at the time was around £40 per year and professionals were paid £100–£300. Charles received only 4d. on the pound (one-sixtieth of the value), all of which was paid to his agents, which in the economic chaos also failed. Despite Louisa's subsequent success as a one of Australia's first female authors, and her husband's role as a police magistrate and politician, the family joined the growing list of people devastated by the vagaries of the Australian pastoral industry.

Endnotes

[1] Meredith LA (1973) *Notes and sketches of New South Wales during a residence in the colony from 1839 to 1844*. Ure Smith, Sydney, in association with the National Trust of Australia (NSW), p. 35.

[2] From 'Experiences of a colonist forty years ago', written by George Hamilton (1812–1883); excerpt published in *The Illustrated Adelaide News*, 1 March 1878, p. 6.

[3] McMichael P (1980) Crisis in pastoral capital accumulation: A re-interpretation of the 1840s depression in colonial Australia, p. 19. In *Essays in the political economy of Australian capitalism. Volume 4*. (Eds EL Wheelwright and K Buckley) pp. 17–40. Australian and New Zealand Book Company, Sydney.

[4] McMichael (1980), p. 18.

[5] Roberts S (1968) *History of Australian land settlement*. Macmillan of Australia, Melbourne, pp. 120–130.

[6] Movement of stock into the Liverpool Plains began on a large scale in the early 1830s; see Roberts (1968), pp. 166, 170.

[7] Roberts (1968), p. 166.

[8] Hadley NF, Szarek SR (1981) *Bioscience* **31**, 747–753; Liu X et al. (2017) *The Rangeland Journal* **39**, 387–400.

[9] Schillinger WF et al. (2006) Dryland cropping in the Western United States. In *Agronomy Monograph No. 23, Dryland Agriculture*. (Eds GA Peterson, PW Unger, WA Payne) pp. 365–393. Madison, WI: American Society of Agronomy, Crop Science Society of America, Soil Science Society of America; Chen C et al. (2023) *Agricultural Systems* **204**, 103561.

[10] Meredith (1973), p. 85.

[11] *The Sydney Herald*, 17 April 1839, p. 2. Stations had been established at Lake Cargelligo by 1838 and at Hillston by 1839, possibly earlier.

[12] Dingle, T (1984) *The Victorians. Settling*. Fairfax, Syme & Weldon Associates, McMahon's Point, NSW, p. 31; *Colonial Times* (Hobart), 4 August 1840, p. 5; *The Sydney Herald*, 20 August 1835, p. 2.

[13] *Australasian Chronicle* (Sydney), 17 December 1840, p. 3.
[14] E.g. *The Colonist* (Sydney), 8 January 1835, p. 3.
[15] *The True Colonist Van Diemen's Land Political Despatch, and Agricultural and Commercial Advertiser* (Hobart Town), 23 December 1834, p. 4.
[16] *Launceston Advertiser*, 22 June 1837, p. 3; *The South Australian Colonist and Settlers' Weekly Record of British, Foreign and Colonial Intelligence* (London), 16 June 1840, p. 230. Charles Bonney and Joseph Hawdon first overlanded from a station on the Hume River to Adelaide in 1838.
[17] Sturt C (1833) *Two expeditions into the interior of Southern Australia, during the years 1828, 1829, 1830 and 1831: With observations on the soil, climate and general resources of the colony of New South Wales. Volume 2*. Smith, Elder and Co., London, pp. 80–84, 108–112.
[18] Cf. Mallen-Cooper M, Zampatti BP (2018) *Ecohydrology* **11**, e1965; Mallen-Cooper M et al. (2011) Managing the Chowilla Creek Environmental Regulator for fish species at risk. Report prepared for the South Australian Murray-Darling Basin Natural Resources Management Board, doi:10.13140/RG.2.1.5086.9847.
[19] Sturt (1833), vol. 2, p. 7.
[20] Sturt (1833), vol. 2, pp. 59, 66.
[21] Sturt (1833), vol. 2, p. 35.
[22] *The Sydney Gazette and New South Wales Advertiser* (henceforth *Sydney Gazette*), 29 October 1829, p. 2; 3 November 1829, p. 2; 19 November 1829, p. 2.
[23] *The Australian* (Sydney), 9 April 1830, p. 3; *The Sydney Monitor*, 21 April 1830, p. 4; 12 January 1831, p. 3; 25 May 1831, p. 2; 7 May 1831, p. 3; 24 March 1832, p. 2; *Sydney Gazette*, 20 November 1830, p. 3; 4 November 1830, p. 2; 9 August 1831, p. 3.
[24] *Launceston Advertiser*, 19 October 1831, p. 4; *Sydney Gazette*, 5 November 1831, p. 2; *The Australian* (Sydney), 23 December 1831, p. 2.
[25] *Sydney Gazette*, 5 April 1832, p. 3; 5 April 1832, p. 3.
[26] *The Australian* (Sydney), 4 October 1833, p. 2; *Sydney Gazette*, 19 October 1833, p. 2; 5 September 1833, p. 2. A history of El Niño and La Niña events extending back to 1525 is provided in Gergis JL, Fowler AM (2009) *Climatic Change* **92**, 343–387.
[27] *Sydney Gazette*, 14 January 1834, p. 2; 26 June 1834, p. 3; 7 October 1834, p. 2; *The Sydney Monitor*, 24 January 1834, p. 2; 4 March 1834, p. 3; *The Tasmanian* (Hobart), 25 April 1834, p. 7; *The True Colonist Van Diemen's Land Political Despatch, and Agricultural and Commercial Advertiser* (Hobart Town), 9 September 1834, p. 4.
[28] *The Cornwall Chronicle* (Launceston), 19 September 1835, p. 2; *The Hobart Town Courier*, 9 October 1835, p. 4; *The Australian* (Sydney), 25 September 1835, p. 2; *The True Colonist and Van Diemen's Land Political Despatch, and Agricultural and Commercial Advertiser* (Hobart Town), 4 September 1835, p. 3; *The Sydney Herald*, 21 September 1835, p. 3.
[29] *The Sydney Monitor*, 11 March 1835, p. 2.
[30] *Sydney Gazette*, 29 March 1836, p. 2; *The Sydney Monitor*, 2 January 1836, p. 2; *The Australian* (Sydney), 12 February 1836, p. 2.
[31] *The Sydney Monitor*, 11 March 1835, p. 2.
[32] *The Colonist* (Sydney), 7 April 1836, p. 4; *Sydney Gazette*, 7 May 1836, p. 2.
[33] *Sydney Gazette*, 21 November 1837, p. 4.
[34] Mitchell TL (1996) *Three expeditions into the interior of eastern Australia: With descriptions of the recently explored region of Australia Felix, and of the present colony of New South Wales*. Eagle Press, Maryborough, Victoria. Reprint. Originally published: 2nd ed., rev., T. & W. Boone, London, 1839.
[35] *The Sydney Herald*, 7 July 1836, p. 2; *The Sydney Monitor*, 6 July 1836, p. 3; 24 August 1836, p. 3; *The True Colonist Van Diemen's Land Political Despatch, and Agricultural and Commercial Advertiser* (Hobart Town), 10 June 1836, p. 7; *The Colonist* (Sydney), 30 June 1836, p. 4; *The Australian* (Sydney), 12 August 1836, p. 2.
[36] *The Cornwall Chronicle*, 9 July 1836, p. 3; *Commercial Journal and Advertiser* (Sydney), 13 August 1836, p. 3; *The Colonist* (Sydney), 29 September 1836, p. 6; *The Australian* (Sydney), 30 September 1836, p. 2.
[37] Gergis JL, Fowler AM (2009) *Climatic Change* **92**, 343–387.

[38] *The Sydney Herald*, 26 December 1836, p. 2; *Sydney Gazette*, 12 January 1837, p. 2.

[39] *The Australian* (Sydney), 21 April 1837, p. 2.

[40] *The Tasmanian* (Hobart Town), 5 May 1837, p. 7; *The True Colonist Van Diemen's Land Political Despatch, and Agricultural and Commercial Advertiser* (Hobart Town), 25 August 1837, p. 680.

[41] *The Perth Gazette and Western Australian Journal*, 3 Dec 1836, p. 808. The authors may have been referring to the establishment of towns further inland in the early 1830s.

[42] *The Sydney Monitor*, 22 December 1837, p. 2.

[43] *The Sydney Herald*, 9 April 1838, p. 2. The Goulburn Plains were first reached by Europeans in 1818 and then soon settled.

[44] *The Sydney Monitor*, 11 April 1838, p. 4.

[45] Russell HC (1877) *Climate of New South Wales: Descriptive, historical, and tabular*. C. Potter, Sydney, NSW, p. 136.

[46] *The Colonist* (Sydney), 6 March 1839, p. 3; *Sydney Gazette*, 19 March 1839, p. 2; *The Perth Gazette and Western Australian Journal*, 27 October 1838, p. 170.

[47] *The Royal South Australian almanack and general directory for 1846*. J. Stephens, Adelaide, p. 81.

[48] *Southern Australian* (Adelaide), 24 April 1839, p. 3; *The Sydney Monitor and Commercial Advertiser*, 15 February 1839, p. 2; 8 March 1839, p. 2.

[49] *Southern Australian* (Adelaide), 29 May 1839, p. 1; *The Colonist* (Sydney), 8 May 1839, p. 4.

[50] *The Australian* (Sydney), 14 February 1839, p. 2; *The Sydney Monitor and Commercial Advertiser*, 28 January 1839, p. 2; 2 March 1839, p. 2; *Sydney Gazette*, 9 March 1839, p. 2; *The Sydney Herald*, 18 March 1839, p. 3; 17 April 1839, p. 2.

[51] *The Australian* (Sydney), 23 March 1839, p. 3; 6 November 1838, p. 2; 25 May 1838, p. 2; *The Colonist* (Sydney), 27 March 1839, p. 2, 11 April 1838, p. 2; 22 August 1838, p. 2; *The Sydney Herald*, 17 April 1839, p. 2.

[52] *Southern Australian* (Adelaide), 4 September 1839, p. 4.

[53] *The Sydney Herald*, 17 April 1839, p. 2.

[54] *The Sydney Monitor and Commercial Advertiser*, 22 March 1839, p. 2.

[55] Henderson J (1854) *Excursions and adventures in New South Wales*. Second edition, vol. 1, Saunders and Otley, London, pp. 188–189.

[56] Meredith (1973), p. 71.

[57] *Bent's News and Tasmanian Register* (Hobart Town), 9 November 1838, p. 3; *The Australian* (Sydney), 8 November 1838, p. 2.

[58] Meredith (1973), p. 71.

[59] A detailed account of the factors that led to the Depression is provided in Chapter 10 of Butlin SJ (2002) *Foundations of the Australian Monetary System 1788–1851*. University of Sydney Library, Sydney, Australia.

[60] Mass transportation ended in 1840 although due to pressure from the squatting lobby a small number of convicts were brought to NSW until 1850.

[61] *Colonial Times* (Hobart), 16 January 1838, p. 6.

[62] *Sydney Free Press*, 20 November 1841, p. 2; *Sydney Gazette*, 19 August 1841, p. 4.

[63] *Sydney Free Press*, 20 November 1841, p. 2.

[64] *Launceston Advertiser*, 29 March 1838, p. 3, *The Hobart Town Courier*, 9 November 1838 p. 2; *Port Phillip Patriot and Melbourne Advertiser*, 19 August 1839, p. 4.

[65] *Colonial Times* (Hobart), 27 August 1839, p. 276; *Bent's News and New South Wales Advertiser* (Sydney), 13 July 1839, p. 2.

[66] Rae-Ellis V (1990) *Louisa Anne Meredith: A tigress in exile*. St. David's Park Publishing, Hobart; sources therein (e.g. p. 89).

[67] Meredith (1973), p. 85.

[68] *The Austral-Asiatic Review, Tasmanian and Australian Advertiser* (Hobart Town), 9 September 1842, p. 2; Butlin (2002).

[69] Butlin (2002) cites flour prices of £75 per ton in September 1839, £33 in November, and £20 a year later, and Sydney wheat prices of 22s per bushel in August-September 1839, 13s. 6d. in November, and 6s. 11d.

in September 1840. These prices are broadly consistent with the drop in the NSW Commodity Price Index provided on p. 108 of Butlin NG, Ginswick J & Statham P (1987) Chapter 7: The economy before 1850. In *Australians, historical statistics* (Ed. W Vamplew), pp. 102–125. Fairfax, Syme & Weldon Associates, Broadway, NSW. Declines in other commodities over this period were less severe, but still typically in the order of 25–50%.

[70] Details of insolvencies are provided in Butlin (2002).

[71] *Geelong Advertiser*, 3 July 1843, p. 1; *The Correspondence of John Cotton, Victorian Pioneer, 1842-1849, part 1*, p. 47. Volume XXVIII of Australian Historical Monographs (Ed. G. Mackaness), Review Publications Pty. Ltd., Dubbo, Australia.

[72] The two others were the 1890s depression and the Great Depression of the 1930s.

[73] Rae-Ellis (1990) provides details.

[74] *Australasian Chronicle* (Sydney), 21 April 1842, p. 2.

11
Transitions

The drought has burnt up Bathurst Plains,
The river's bed is dry,
Mankind and brute weep blood for rain,
It comes not to their cry.
– By 'The Hermit of Geelong', The Sydney Gazette and
New South Wales Advertiser, *19 March 1842*

THE STUDY OF HISTORICAL EVENTS IS OFTEN DESCRIBED AS BEING AKIN TO peering through a glass darkly, hoping that some glimmer of reality, however imperfect, will emerge from the murky depths of time.[1] This is a treacherous path because we must often resort to forming opinions based on only a fragmentary record of either the events themselves or the motivations of the people involved. But a second problem that receives less attention than it deserves is that each of us is fundamentally prone to making systematic errors in the way in which we view time itself. We typically underestimate the length of long intervals of time but overestimate short ones, a phenomenon known as Vierordt's law. We thus tend to forward telescope distant but important events, compressing them into a much shorter perceived timeline, and focus too much on the significance of events that occur within our lifetime.

For scientists seeking to understand past environmental change this cognitive distortion of time is compounded by the fact that, compared with the processes that shape landscapes, our lives are infuriatingly short. Naturally, we tend to focus on iterative and pervasive features of ecosystems that can be easily and relatively quickly measured, such as the demographic processes of seed production and dispersal, or species interactions like predation and competition. By studying these we hope to gain insight into patterns of change that unfold over century to millennial timescales, such as the movement of plant species across the landscape at the end of an ice age. This concept can be traced back to the theories of *gradualism* and *uniformitarianism* developed by the geologists James Hutton (1726–97) and Charles Lyell (1797–1875), and later to Darwin's theory of evolution by natural selection.

The problem, however, is that humans have understood for millennia that rare but extreme perturbations can result in rapid and irreversible environmental and social change. This concept is embodied in many religious beliefs, but the first to formally codify it into scientific theory was the French scientist Georges Cuvier (1769–1832) in what is now known as *catastrophism*. Catastrophism fell out of favour during much of the 19th and 20th centuries, largely because uniformitarianism provided incontrovertible insights into the study of evolution and geology over deep time, but in recent decades it has been revitalised by mounting evidence of stepwise or punctuated change in the organisation and trajectory of environmental systems from around the globe. It is now understood that many ecosystems can rapidly transition between alternative states if pushed beyond certain biophysical thresholds ('tipping points') by disturbance (Box 11.1). Crucially, these transitions can be difficult or impossible to predict and even more difficult to reverse.

I raise these points because we must now turn our attention to a burning question in ecology: what role do extreme events and natural disasters play in driving sudden and long-term changes to Australia's soil, flora, fauna and even human societies? And could these be just as important as processes that gradually play out on much longer timeframes? Many Australians became acquainted with this idea for the first time in 2019–20, when catastrophic bushfires and drought ravaged eastern Australia. Millions of hectares of woodland and forest burned, billions of animals lay dead, and, when flooding rains returned, creeks and rivers turned black with ash and mud. The scale was so large, and resources so strained, that there was little we could do to prevent, or even mitigate, the environmental damage.

However, Australia is no stranger to disasters, and we know that some of these, especially the great droughts of the late 19th to mid-20th centuries, caused rapid and permanent transformation of Australian ecosystems on an even larger scale. We will return to this theme in future chapters, but here I want to focus on droughts that took place earlier, and on three processes: erosion and loss of hydration of soil profiles, the collapse of native animal assemblages and the destruction of native vegetation. Each has contributed to the impoverishment of Australian biodiversity and all are major ongoing threats today. Could these earlier droughts have been instrumental in setting these changes in motion?

The first reason to suspect so is that they occurred simultaneously with an abrupt change in the way that Australian landscapes were managed: Aboriginal fire and land use patterns were in rapid decline and European settlement was growing explosively into ever-drier country along a 1500-km-long frontier from the Darling Downs to South Australia. But it was the introduction of hundreds of thousands of European livestock that proved decisive (Fig. 11.1). Today, red and eastern grey kangaroos and the southern cassowary are the

Fig. 11.1: King H. (ca. 1887). *Photograph of a painting of a herd of cattle crossing a river, Australia.* Vast herds of cattle were driven across the settled inland towards the expanding frontier during the 1830s–60s. Most travelled along watercourses, where feed and water were more reliable. National Library of Australia, nla.obj-149691998.

Box 11.1: When catastrophes unfold

The science of catastrophes has advanced significantly in recent decades and it is now understood that many ecosystems can undergo sudden shifts between different stable states. This figure shows one conceptualisation of this phenomenon, based on the *fold catastrophe* framework. The bottom of the figure shows how shifting environmental conditions can result in an abrupt shift in a given ecosystem from a healthy to a degraded state. Consider a healthy ecosystem experiencing a severe ongoing drought, which is represented as a shift from favourable to unfavourable external conditions (trajectory A to B). For a while, the system is stable, but once a critical tipping point is reached (point B), a very small stimulus can cause the system to rapidly shift into a degraded state (transition B to D). The difficulty, from a conservation perspective, is that even a return to favourable conditions (say, increased rainfall, trajectory D to C) may not cause the system to return to a healthy state unless another tipping point (C) is reached (transition C to A).

The 'ball and cup' models at the top are another way to illustrate these ecosystem shifts. If conditions are extremely favourable the ball (i.e. ecosystem) always settles in a healthy state (model I), while if they are very unfavourable (model V) the ecosystem always becomes degraded. In average conditions the ecosystem can exist either in highly stable healthy or degraded states (model III). Close to the tipping points small perturbations can cause an ecosystem to undergo rapid degradation (model IV) or recovery (model II). Restoration projects, which overcome the intrinsic stability of degraded systems, can facilitate ecosystem recovery. However, a critical problem in semi-arid and arid rangelands is that the process of degradation can result in the irreversible loss of plant or animal species and much of the soil profile, which took millennia to develop, and only partial recovery is possible. Most Australian ecosystems exist in this state today. Interested readers can find a more detailed discussion in *Frontiers in Ecology and Evolution* **8**, 561101 and references therein.

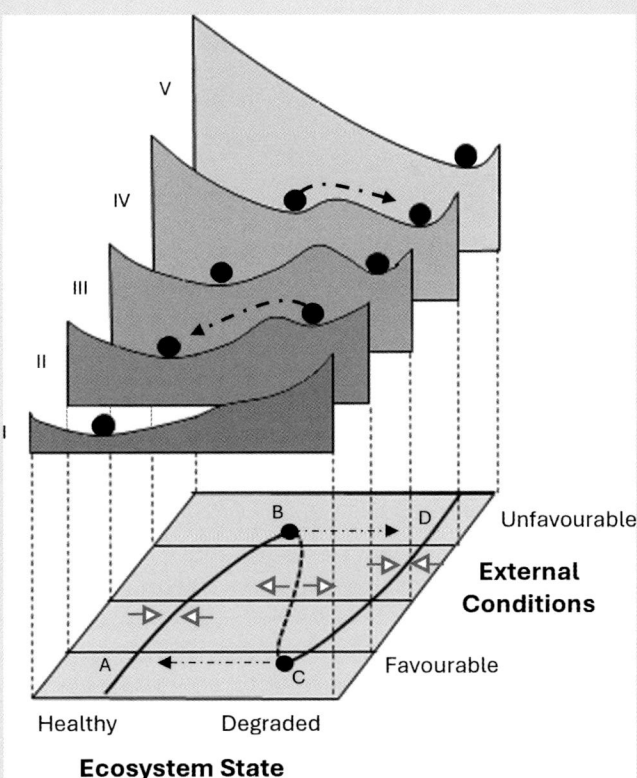

largest terrestrial Australian animals, weighing up to 70–90 kg, dwarfed by beef cattle and stock horses, which at maturity typically weigh 400–1200 kg. No animals of this size have existed in Australia for tens of thousands of years, when colossi like *Diprotodon optatum*, which stood 2 m at the shoulder and weighed up to 3000 kg, and many other megafauna roamed the landscape.[2]

Ungulates, and particularly cattle, also had hard hooves that could pulverise the fragile soil far more than the elongate, padded feet of kangaroos and wallabies (the term 'macropod', which includes these marsupials, literally means 'large foot') and the tridactyl-toed feet of the emu.[3] Indeed, large cattle, based on their weight and foot area, exert a static pressure of 100–200 kilopascals on the ground, and even more when walking; this is higher than even unloaded rubber-tyred tractors and some tracked vehicles.[4] By compacting the soil and destabilising river banks by trampling, cattle are more than just large animals – they are agents of geomorphological change, particularly when grazing at high densities.[5]

A COUNTRY LAID BARE

Sheep have golden feet, and wherever the print of them appears, the soil is turned into gold.

– Inquirer *(Perth), 26 May 1841, p. 5*

The deterioration in soil structure began soon after settlement. Squatters, whose land tenure was insecure, had much to gain from running as many sheep as possible and less interest in responsibly managing the grass supply for the long term.[6] Within just a few years habitats that once boasted grass so thick that it screened kangaroos from the 'keen eye of the Aboriginal hunter' now could scarcely 'afford shelter or shade for the grasshopper'.[7] Soon, grasslands and pastures had become enveloped by 'swarming numbers of sheep, trampling over the ground eternally, and devouring the foliage, especially in dry seasons, like locusts or caterpillars'.[8] When drought arrived across southern New South Wales in 1835–39 herbage died to the roots, and the landscape lay bare for several consecutive years.[9] In northern New South Wales drought-breaking rains failed in 1841, leaving some 60 000 km² of the Liverpool Plains as bare as a 'turnpike road'.[10] Under pressure like this, native grasslands dominated by tall, warm-season species, many tended by Aboriginal people for their seed production, began to die out, replaced by boom-or-bust weeds, year-round native grasses or non-native species like couch grass that were encouraged for their ability to stabilise soils.[11]

That drastic and irreversible change was now sweeping the landscape was clearly understood as early as 1838 by James Dunlop, superintendent of the Government Observatory at

It is a great mistake to suppose that our pasture lands can last for ever under the present system of overstocking, of keeping the grass constantly bare to the roots, and never suffering it to recover itself or run to seed.

– Sydney Chronicle, *12 December 1846, p. 3*

Parramatta, who noted that prolonged trampling was leaving soil profiles around the colony heavily compacted, damaged by pug marks, and lacking porosity to store water.[12] Ponds, rivers, streams and springs were now rapidly becoming ephemeral, and permanent water scarcer,[13] and worse still run-off dramatically increased during high-intensity rainfall events, leaving prime, agricultural land with deeply incised gullies or washed away entirely (Fig. 11.2). Agricultural scientists have shown that heavy grazing can reduce soil water infiltration by 50% or more and, by removing litter and vegetation, increase soil loss more than 200-fold. When ground cover falls below 70–75%, which is typical during drought, soil losses become much worse.[14] The cultivation of hilly landscapes for feed crops further increases the potential for catchment-level soil degradation, especially when levee banks or terraces are not constructed to slow water, or in times of drought when seed fails to establish.[15]

Fig. 11.2: Woodbury W. (1856). *Cleared land and erosion on the banks of the Ovens River, near Beechworth, Victoria* [picture]. The severe impact of grazing and clearing is evident in this early photograph, taken around the same time as native mammals were disappearing from the landscape. The country had been heavily overgrazed, leaving no cover, and extensive cattle tracks can be seen in the foreground. Erosion has already commenced on the bank of the river. National Library of Australia, nla.obj-151253993.

Reports of serious soil erosion rapidly emerged across the colony. By 1830, 6 inches (15 cm) of topsoil had washed off the Baulkham Hills, north of Parramatta, and in 1833 heavy rains were causing erosion in gardens nearby.[16] By the mid-1830s to 1840s, soil was washing off cultivated hilly country of the southern Sydney Basin near Campbelltown, on slopes in the Hunter Valley, and in coastal areas of the Illawarra,[17] while heavy rains that fell during October 1844 in overgrazed pastures and hilly terrain near Hobart caused major damage to the fertile alluvial soils favoured by early settlers. Gully and other soil erosion was forming on the Torrens River and other lowland areas of South Australia as early as 1847.[18] The rapid increase in sedimentation of creeks and rivers across eastern Australia after European settlement has been well documented by soil scientists and geomorphologists. Well-studied examples include Sandy Hook on Coxs River, where sedimentation increased between 1809 and 1858 due to disturbance and gully erosion in the upper catchment,[19] and on the Southern Tablelands, where broad-scale erosion and channel incision began soon after the termination of the 1835–39 drought.[20] Crucially, these and other similar studies have noted that these changes occurred during a 'drought dominated regime' that prevailed across eastern Australia from around 1820 to the mid-1850s.[21]

The state of the landscape also gave rise to a new phenomenon that has plagued Australia ever since: tremendous anthropogenic dust storms. Apart from numerous early accounts of dusty roads, the first major storm of which I am aware occurred on 17 February 1839, in the

Yass region, when strong winds ahead of a cold front caused 'the sand and dust to aris[e] in such thick dark clouds, that the traveller could scarcely see which way he was directing his steps'.[22] Another vast storm, probably containing driving sand and dust, was seen across Victoria in March 1839.[23] Data collected from a peat mire in the Snowy Mountains indeed show a marked increase in dust deposition during the 1830s[24], although since western New South Wales and Victoria were unstocked this peak in activity was far smaller than that associated with the 1895–1946 eastern Australian 'dust bowl' era.[25] Nevertheless, mass aeolian mobilisation of the soil was at least occurring on a local scale, in places where it had not done so for thousands of years.

Changes in the abundance and diversity of aquatic animals were taking hold around the same time. Soil erosion and sedimentation reduced water quality, and nitrogen, phosphorus and other nutrients began to wash into waterways (1000 kg of sheep, cattle and horse liveweight produces around 40–86 kg of manure and 10–39 kg of urine daily).[26] In addition, human and industrial waste was a growing problem near towns. The supply of these nutrients normally limit plant and algal growth, and so contaminated run-off can cause eutrophication of waterways that results in blue–green algae blooms, hypoxia (low oxygen) and other problems. To make matters worse, the Hunter and Lachlan river catchments and other central, southern and coastal New South Wales waterways were choked with dead cattle and sheep during 1838–39,[27] as were the Muckoi (= Mooki) and Namoi rivers in northern New South Wales during 1841–42.[28] Waterholes act as crucial 'island' habitats in dryland river systems during drought, and so the long-term viability of populations of fish and other aquatic taxa is very sensitive to reductions in water quality.[29] The fact that dying livestock were forced to concentrate at these very same drought refugia must have greatly accelerated the decline.

We have little direct evidence that these processes impacted on aquatic life, but it seems very likely, since putrefaction of water was becoming a major concern to people who relied on inland waterways for domestic use, due to the increased likelihood of disease.[30] A major fish kill event seems to have occurred in the Murrumbidgee River in early 1839, when the river fell so low that many waterholes dried up completely, and where 'fish [almost certainly Murray cod, *Maccullochella peelii*] weighing from thirty to forty pounds [13.5–18 kg] may be seen lying in a putrid state in the bed of the river'.[31] Reports that dead and diseased sheep were being thrown into the Murrumbidgee around this time suggest that water contamination might have played a role.[32] Another fish kill was observed in Campbells River, south of Bathurst, in the drought of 1833, again ostensibly due to extremely low water levels and excessive heat.[33]

A PRACTICAL ERROR OF GREAT MAGNITUDE: DROUGHT AND THE SALTBUSH PLAINS

> *And all the world became a withered void*
> *Of hardened barrenness – ye little know*
> *The terrors of a drought.*
>
> – The Geelong Advertiser, *17 March 1848*

After inland routes had become established along the main westward-flowing river systems in the mid-to late 1830s the great plains of the Riverina and lower Murray–Darling Plains also incrementally fell under the hoof. Here, graziers encountered shrublands of saltbush, black bluebush and cottonbush that dominated some 77 000 km² of clay and aeolian sandplain country.[34] In the gaps between the shrubs grew wallaby grasses, spear grasses and highly palatable native grasses and forbs such as Darling clover, the most nutritious pastures of which were quickly exploited (Plate 9). Initially, sheep and cattle grazing was restricted to the main river frontages where livestock had access to water, but these soon began to show signs of deterioration. Initially, the more remote waterless backcountry, where travelling stock were (often unsuccessfully) driven 70 miles or more (113 km) between watering points,[35] was only opportunistically grazed, but in the wetter years that followed the breakup of the drought in late 1839 it too experienced a rapid increase in cattle and sheep numbers. Still more flooded in from the European-settled districts.

The influx and then concentration of livestock, however, could scarcely have come at a worse time, for the megadrought that had prevailed since the early 1800s had not finished with the region yet. Rainfall deficiencies returned, first along the eastern coast and central districts from Goulburn to Moreton Bay in 1846,[36] and by the summer of 1847–48 even the colony of Port Phillip was heading into a drought of 'unparalleled severity'.[37] On the Riverina this return of drier weather was initially of little concern to graziers because their livestock, having exhausted the supply of grasses and forbs, could turn their attention to the tremendous standing biomass of halophytic (salt-loving) saltbushes. The forage of these species is high in salt and betaine and so sheep that feed on them require increased access to fresh water, but on balance their relatively high protein levels make them an excellent drought reserve.[38] But once ephemeral water sources failed, the vast herds crowded to the main river frontages, and by early 1847–48 the middle and upper stretches of the Lachlan River and the recently settled lower Darling River frontages had been rendered bare,[39] apart from protected pastures on some stations and a few more favourable locations.[40] Losses of sheep and cattle were heavy.[41]

Heavy rain across eastern New South Wales and Victoria in late 1848 broke the drought there, but in 1849 drought conditions continued to intensify in the western and northern districts, and stations from the Barwon River to the lower Murray and Darling rivers were stripped of their ground cover. Conditions were so severe that 'native' troopers sent to investigate conflict between Aboriginal people and European settlers in mid-1849 found it difficult to prevent their horses from starving.[42] Through 1850 drought conditions still gripped virtually the entire inland from the Burnett River in Queensland to the lower Murray; the Darling was again reduced to a chain of shallow pools and the surrounding countryside was in a deplorable and

> *The suffering that has been inflicted on us is awful; flocks have been driven to the river from the plains in order that the shepherd, as well as the sheep, might get water. You can have no idea of the present state of the plains, behind Darlot's; we have the Lahara [sic] in miniature. The blacks say that even their 'oldest inhabitant never saw such a thing.'*
> – Port Phillip Gazette and Settler's Journal, *18 March 1848, p. 2*

wretched state for want of rain.[43] Conditions did not improve until later in the year, after some 3–4 years of drought.

For the vegetation of the saltbush plains, the coincidence of these events proved disastrous. The precise pattern of destruction remains unclear, but palatable shrubs and grasses were quickly destroyed by repeated grazing, which probably caused a pulse in resources available to less competitive species, especially fast-growing, nitrophilous (nitrogen-loving) forbs and grasses. Introduced species from Europe were being used to improve native pastures in the east by the 1820s, and by the 1840s a veritable potpourri of annual and perennial grasses and forbs was available for purchase around the colonies.[44] Many were introduced intentionally, but others simply as impurities in seed and hay, or by hitching a ride on (or in) livestock. But so too did terrible species like the Bathurst burr (*Xanthium spinosum*; Fig. 11.3) and Scotch thistle (*Carduus nutans*), which were well established by the 1840s and had become widespread in southern New South Wales and Victoria by 1850.[45] By this time these and a vast array of other invaders were beginning to transform the plains of the south-east into something approximating the composition that we see today.[46] Interestingly, the introduction of nitrogen-fixing clovers and medics, which improved soil fertility for new plant and animal invaders, is an example of a feedback process known as invasional meltdown;[47] one of the beneficiaries was the European rabbit, a topic to which we will return in later chapters.

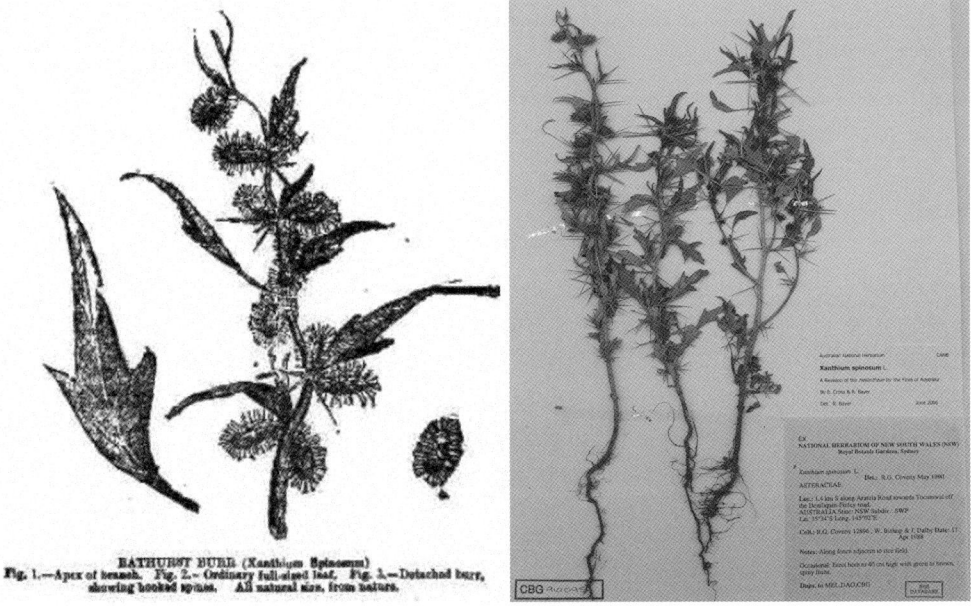

Fig. 11.3: Weeds like the dreaded Bathurst burr (*Xanthium spinosum*) spread rapidly across inland eastern Australia during the mid-1800s. The burrs, illustrated at left, adhered to wool and hair, and so were easily dispersed long distances by livestock. The image at right is of a mounted specimen collected from the southern Riverina in 1988 and now housed, along with thousands of other records, at the Australian National Herbarium. Such specimens are an invaluable resource for those interested in reconstructing the introduction history, spread and even genetics of invasive species. Left: image from *Weekly Times* (Melbourne), 20 July 1895. Right: photograph by Robert Godfree.

Yet another drought battered eastern Australia during 1858–59, by which time more than a million sheep and 200 000 cattle roamed the Riverina and hundreds of thousands more had been moved to newly established runs along the Darling River frontages, now serviced by a steamer trade.[48] In South Australia new graziers had been pushing increasingly further afield from the Murray River and Adelaide, and north from Port Augusta into true desert country beyond the end of the Flinders Ranges. Cattle and sheep were again driven to shrinking waterbodies, damaging springs and watering points, and ultimately dying of starvation and thirst on an immense scale.[49] The saltbush receded as the squatters advanced, but until 1860 the 'waste lands' away from permanent water had escaped sustained and heavy grazing pressure, especially during dry times, and as late as 1863 these areas were thought to still measure around 100 000 km² in extent across the Riverina alone.[50] Significant tracts of relatively intact saltbush-dominated vegetation also persisted in South Australia, where livestock were mainly limited to scattered waterholes and ephemeral lakes and creeks.

This reprieve, however, proved short-lived. The sinking of successful wells in the Victorian goldfields and near Adelaide showed water-hungry graziers that subterranean water was within reach.[51] By the early 1860s, wells up to 120 feet (37 m) deep were under construction from the Murray to far north-western New South Wales, an operation that was so successful that by 1864 entire runs had become fully serviced by permanent water from wells and tanks, each capable of supporting tens

Fig. 11.4: The establishment of permanent water sources in the 'backcountry' increased stock numbers and the consequences of overstocking during drought. Top: horse-driven whim at McCoy's Well, SA ca. 1870. These watering points suffered intense grazing and stock pressure. State Library of South Australia, B 45776. Middle: the Kallara no. 2 artesian bore, Kallara Station, near Tilpa, NSW. Open water sources like this allowed more intensive sheep grazing, but later provided a crucial water supply to migrating rabbits. Photo: Robert Godfree. Bottom: dam construction on Mt Moriah Station, 40 km north-east of Toowoomba, Qld, ca. 1870. Excavating dams on this scale was labour-intensive and time-consuming; this one took approximately 6 months to complete. Photo courtesy Toowoomba City Library.

of thousands of sheep (Fig. 11.4).[52] Progress was similarly rapid in South Australia: in 1865 it was reported that 63 stations in the Flinders and Gawler ranges had spent £128 738 on procuring water, which opened the remaining country.[53] The construction of dams, tanks and wells continued, so that by the mid-1870s most of the back blocks were fully watered, even in drought years, and carrying staggering numbers of sheep – more than 5 million were being shorn in the Riverina alone by 1876.[54]

The end was now assured: when drought returned in 1864–65 the vast stands of saltbush were eaten down to stumps by starving sheep from the Narran Lakes in northern New South Wales to the Murray River.[55] Judging by comments made by the English author Anthony Trollope in 1873, the saltbush that remained in the back country had now entering a death spiral, and the periodic return of drier weather during the 1870s and 1880s, and particularly the drought of 1875–77, delivered the coup de grâce. With the last of the saltbush annihilated and trampled to dust,[56] nothing was left to stop the mobilisation of the sand and soil that had lain hidden from the elements for millennia. Although at the time this was understood to be a 'practical error of great magnitude',[57] the full ramifications would not be experienced for another two decades.

LOST THINGS

A little pale ghost, it seemed hardly to touch the ground; it floated along – eerie, effortless.

– HH Finlayson describing the desert rat kangaroo in 1932, just before its likely extinction[58]

Plants were not the only life forms devastated by the establishment of European grazing systems across the south-east of the continent. Australia also has the unenviable record of having lost 11% (28 out of 273) of its endemic land mammal species since 1788, the highest number of any country globally. Losses in western New South Wales have been particularly severe: of the 61 mammal species thought to have been present at the time of European settlement only 37 still exist there, while seven are now nationally extinct and a further 17 have been regionally extirpated.[59] Losses in adjacent parts of South Australia and the Flinders Ranges were even more severe.

The reasons why these species disappeared have been the subject of debate for decades, but recent research has shown that this process unfolded over a period of about a century, and that many factors, including the arrival of rabbits, predators and disease, were involved.[60] Nevertheless, it is a striking fact that a large number of species declined soon after the arrival of livestock, and before the arrival of the rabbit in the 1870s–80s and the fox in the 1880s–1900s. Indeed, three mammal species were last recorded in the NSW Western Division in the 1840s, while a further nine were last collected during the Blandowski Expedition to the lower Murray and Darling rivers in 1856–57. Another 13 species disappeared in the 1880s–90s, but again some of these were in serious decline by the mid-1850s. For example, according to the great zoologist Gerard Krefft, the enigmatic pig-footed bandicoot had become exceedingly rare on the Murray and lower Darling by 1857, just 21 years after being described by Thomas Mitchell in 1836, and several other species had already disappeared from the Victorian side of the Murray River.[61]

Krefft noted that in 1857 the pig-footed bandicoot was 'disappearing as fast as the native [Aboriginal] population', and that 'the large flocks of sheep and herds of cattle occupying the country will soon disperse those individuals which are still to be found in the so-called settled districts', and similarly that the 'burrowing *Bettongia* has long retreated before the herds of cattle with which the plains bordering on the Murray are now stocked'. The greater stick-nest rat, which builds large, communal nests out of sticks, stones and grass in perennial chenopod shrubland and woodland, had already 'retreated before the herds of sheep and cattle across the Murray', and on the south side only a few empty nests remained (Plate 10). I think that these comments indicate that many native species probably succumbed to the removal of shrubs and grasses that provide cover (e.g. western quoll, brush-tailed bettong), competition for food (e.g. crescent nailtail wallaby), and by the trampling and erosion of the soil, which destroyed the habitat of fossorial species (e.g. burrowing bettong).[62]

Some researchers have questioned whether livestock could have been responsible for a collapse in native mammal numbers within only 5–10 years of their arrival. However, as I have shown, vast herds of cattle and sheep were being driven across the region by the late 1830s by overlanders, and the severe droughts of the 1840s were probably pivotal, when stock stripped the river frontages and watered plains of the western Riverina and Murray bare of vegetation. The 1858–59 and 1864–66 droughts devastated similar habitat but on a much larger scale, including most of the Darling River country and isolated springs, clay depressions and waterholes from northern New South Wales to northern South Australia. These areas were highly productive parts of the landscape and essential drought refuges for native animals.

Thomas Mitchell provided a vivid description of the impact that cattle were having on these habitats and the Aboriginal people that depended on them while travelling near the Macquarie River in 1846, writing:

> *the water in these holes had been recently drunk, and the mud trampled into hard clay by the hoofs of cattle. Thus it is, that the aborigines first become sensible of the approach of the white man. These retired spots, where nature was wont to supply enough for their own little wants, are well known to the denizens of the bush ... Cattle find these places and come from stations often many miles distant, attracted by the rich verdure usually growing about them, and by thus treading the water into mud, or by drinking it up, they literally destroy the whole country for the aborigines, and thereby also banish from it the kangaroos, emus, and other animals on which they live. I felt much more disgusted than the poor natives, while they were thus exploring in vain every hollow in search of water for our use, that our 'cloven foot' should appear everywhere. [Fig. 11.5]*[63]

Cats, which often closely followed, or in some cases preceded European settlement, also probably played an important role,[64] because the loss of vegetation would have subjected small to mid-sized native mammals to dramatically increased predation pressure, a phenomenon that

Fig. 11.5: Calvert, S. (ca. 1877–78) *The drought – sheep at a dry creek*; from engraving by J. W. Curtis. By the late 1840s the damage caused by cattle and sheep to river frontages and permanent water sources in semi-arid regions was affecting the viability of wildlife and the Aboriginal people that depended on them. Image courtesy National Library of Australia, nla.obj-135892738.

has been linked to more recent mammal declines in northern Australia.[65] Given the chronology of settlement, I think that cats were probably important predators in open woodland and grassland habitats in the eastern colony during the 1830s and 1840s – the last white-footed rabbit-rat recorded from Victoria in 1839 was killed by a settler's cat! – but that in western New South Wales and South Australia they probably contributed to the final destruction of many species that had already declined sharply in range during the 1840s–60s. This process probably occurred in the 1870s–80s, when cats were becoming more numerous.[66]

The final blow for many species that survived longer, like the burrowing bettong, was delivered by the rabbit plagues that began in the 1880s and the arrival of what I think had proved to be the most devastating predator (apart from humans) ever to reach Australia – the European red fox – in the 1890s (Fig. 11.6).[67] Foxes are among a select group of few species to undertake 'surplus killing' of prey, a behaviour that may explain the collapse of native mammal populations across the continent.[68] In southern New South Wales their arrival coincided with the 1895–1903 Federation Drought, and with this added predation pressure populations of several mid-sized native animals driven to low levels during this time were simply unable to recover. By 1910 the landscapes witnessed by Sturt, Mitchell and thousands of generations of Indigenous people before them, were swept away forever, their unique animals but a memory.

Fig. 11.6: Many native animals were unable to withstand the onslaught of feral foxes and cats, especially when drought and overgrazing left them exposed to predation. Top: Egersdörfer, H. (1891). *The fox in Victoria* (detail) [wood engraving]. State Library of Victoria, IAN01/07/91/9. Middle: a wild cat defending her young against an enemy. *Australian Town and Country Journal* (Sydney), 1 March 1890. Bottom: foxes killed on Lake Victoria Station, ca. 1910. Fox populations rose rapidly after their arrival in most inland areas. Museums Victoria, MM 7121.

THE DYING HEART

The fruitful land is turned into barrenness; the rivers of water are dried up, and the fire has devoured the pastures of the wilderness.

– Empire *(Sydney), 6 January 1866*[69]

Let us return now to South Australia, where after 1840 an emerging synergism between early explorers and pastoralists and relatively easy access across the coastal plains north of Adelaide to the Flinders Ranges contributed to an unbelievably rapid expansion of the pastoral frontier. By 1861, when Burke and Wills lay dying of malnutrition on Cooper Creek, runs had been established near Murnpeowie, less than 250 km away, where rainfall averages well under 200 mm per year. In New South Wales the establishment of massive runs on the Darling River during the 1850s[70] stimulated further expansion into the interior, so that stations now dotted the fringes of the Simpson, Tirari, Strzelecki and Sturt Stony deserts. Europeans had finally reached the most inhospitable parts of the continent, far to the north of Goyder's famous rainfall line separating cropping from pastoral areas (Box 11.2). It is here, during 1864–66, that the first true collapse and reorganisation of an entire *socio-ecological* system – the set of biophysical and social factors that interact to structure societies and the ecosystems in which they exist – caused by drought and European agriculture was witnessed.

It is often argued that early European settlers in the semi-arid and arid interior were oblivious to the severity of the climate of the interior or the need to carefully manage stock numbers in anticipation of future dry conditions. And the undeniable fact is that many of the earliest squatters in these regions had never directly experienced the full ferocity of a central Australian drought, lulled by the swaying grass, verdant saltbush and lakes covered in vast flocks of birds that can be seen in good seasons. After reading 10 chapters of this book, though, perhaps you are beginning to question whether this narrative may not be wearing a bit thin. As we have seen, the dangers of being overstocked during drought had been experienced first-hand by settlers in every decade from the 1813–15 drought through the 1820s and 1830s, as was the permanent landscape degradation that had taken place during the 1840s and 1850s in the Riverina and Darling River districts. These events were remembered by many, including drovers that passed through these regions with large mobs of sheep and cattle bound for western New South Wales, Victoria and South Australia. Settlers in the northern runs had even experienced a significant drought in 1859–60, when waterholes dried up and cattle and sheep died in significant numbers.[71]

Even in the true desert country, which had only just been settled, the nature of drought was also widely understood thanks to accounts written by early European explorers. The best known of these was the expedition led by Charles Sturt, who crossed some of Australia's driest terrain between Menindee and Birdsville in 1844–46. Sturt's famous account of being trapped for 6 months near a waterhole at Depot Glen, 30 km north of Milparinka, which was published in all major newspapers around the country in 1847–49, left no doubt about the severity of the conditions they experienced or the prospects for settling in the region. Apart from the waterless state of the country, Sturt reported fierce temperatures during January and February 1845, including a shade thermometer reading of 132°F (55.6°C).

> **Box 11.2: Goyder's Line**
>
> One of the legacies of the terrible drought of 1864–66 was the development of the first serious cartographic attempt to explicitly link rainfall with potential agricultural productivity in Australia. The task was left to the Surveyor-General of South Australia, George Woodrofe Goyder. Following the devastating losses of livestock in the north and subsequent establishment of a parliamentary inquiry into the state of the northern runs, Goyder was tasked with travelling north in November 1865 to establish the portion of the country where drought still prevailed. While the original intent was to determine the regions that might benefit from a relaxation of the terms of the Crown leases held by pastoralists, the 'rainfall line' penned by Goyder (Plate 18), which came to be known as 'Goyder's Line', was to have more serious implications for the development of agricultural aspirations in the southern part of the state.
>
> Members of the commission of inquiry quickly concluded that Goyder's Line could also be used to delineate the northerly boundary of the cropping (agricultural) zone; this was consistent with the ethos that central planning was the best way to direct future agricultural development in South Australia. However, pressure soon mounted for new agricultural land to be released along the agricultural frontier, and opponents of the restriction posed by Goyder's Line argued that it was placed too far south. Despite Goyder's objections, land far to the north of his line, where he considered rainfall to be too unreliable for growing crops, and the soil too easily blown away, was thrown open. Initially, this decision seemed correct, for crop production reached record levels through the 1870s. However, dry weather returned in 1881–82 and 1882–83, crops failed, and the expansion was over. Many properties were abandoned while others converted to grazing or mixed farming operations, and Goyder's remarkable intuition about the productivity of the South Australian landscape, which he gained through direct observation and experience, was proved correct. During the mid- to late 20th century advances in crop genetics and farming practices allowed cropping to once expand significantly north, but Goyder may again have the last laugh, for global circulation models predict that rising temperatures and declining rainfall under climate change may soon reimpose the hard limit to cropping that he defined more than 150 years ago.

While this reading was not taken under standard conditions, he probably experienced temperatures at the upper limit of those which occur in the arid zone, perhaps 47–49°C (Fig. 11.7). In his journal of the expedition,[72] he famously recounts that during the day bullocks pawed the ground to get to the cool soil beneath, the upper leathers of people's shoes were 'burnt as if by fire', dogs died or lost the skin off the soles of their feet, and riders could not keep their feet in the stirrups. The heat was so intense that 'every screw in our boxes had been drawn', handles made of horn split into fine laminae, lead dropped out from pencils, human hair and wool on sheep ceased to grow, nails became as brittle as glass, flour lost more than 8% of its original weight, and a bottle of citric acid became fluid.

More ominously, Sturt encountered large stands of coolabah (*Eucalyptus coolabah*) that had been killed by drought in the vicinity of Lake Lipson (Lake Lady Blanche) and Lake Etamunbanie between Cooper Creek and Birdsville. One was 5 miles (8 km) across, presumably containing hundreds or thousands of trees. The flooded plains on which they grew had been

Fig. 11.7: The expedition of Charles Sturt into some of the driest terrain in Australia in 1844–46 proved conclusively that no wet, fertile region, let alone an 'inland sea', existed in central Australia. He also recorded widespread tree death, probably caused by severe drought in the late 1830s and early 1840s, and extremely hot weather. Left: Photograph of Depot Glen, near Milparinka, NSW in 1935. Sturt was trapped here for many months. State Library of South Australia, B 7138. Right: Sleap, F. A. (1891). *The Stony Desert – Sturt's third expedition* [engraving]. State Library of Victoria, IAN 01-01-91 SUPP P.5.

'torn to pieces by cracks of four, six and eight feet deep'. If these trees had died recently, as Sturt implies, it probably happened during the drought of 1836–42.[73] No one who has read Sturt's account can fail to understand the implications of what he saw: coolabah is extremely drought tolerant, having both deep roots that reach the lower water table and an extensive network of shallow roots that can harvest water from light rainfall events. It takes several years of extremely low rainfall to kill mature trees.

Other explorers also witnessed the impact of this drought, or possibly an earlier one, on inland vegetation. In 1846, Thomas Mitchell noted that along the Barcoo River in western Queensland a recent drought had 'killed much of the brigalow [*Acacia harpophylla*] scrub so effectually, that the dead trunks alone remained on vast tracts, thus becoming open downs'.[74] In 1858, Augustus Charles Gregory, while searching for Ludwig Leichhardt's lost 1848 expedition, also observed that trees on the back country near the Barcoo and Thompson rivers had been annihilated by drought. Later he argued that the 'character of the vegetation indicated excessive droughts' and that the south-western quarter of Queensland was 'unfit for occupation, for, though in favourable seasons there might in some few localities be abundance of feed for stock, the uncertainty of rain and frequent recurrence of drought renders it untenable, the grasses and herbage ... are swept away by the hot summer winds, leaving the surface of the soil completely bare'.[75] Media coverage of Gregory's expedition in 1858 explicitly mention the annihilation of vegetation in the region by drought, although Mitchell's despatches, which were published in newspapers in 1846–47, did not.[76]

But none of this prevented the dramatic intensification of sheep grazing through the 1850s, and by 1861 the South Australian sheep flock comprised more than 3 million sheep.[77] According to Henry T Morris, the Chief Inspector of Sheep and Estimator of Runs, runs in the far north and both western and eastern plains were already overstocked, and noted that 'if their grazing

> *At eight miles we descended to a flooded plain, scattered over with stunted box-trees [E. coolabah], the greater number being dead, and I may remark that we generally found such to be the case on lands of similar description; a fact, it appears to me, that can only be accounted for from the long-continued drought to which these unhappy regions are subject.*
> – Charles Sturt, October 1844, *Narrative of an expedition into central Australia*

capabilities are overtaxed, they will retrograde very rapidly ...'. In 1860, livestock to the west of the Flinders Ranges were dying in hundreds around the few permanent watering points, and the plains were 'almost as bare as a roadway'.[78] But memories become shorter during good times; numbers rose again during the exceptionally favourable year of 1863 and by 1864 the entire region was overstocked.

The real explanation for this lack of planning involves a combination of economic, social and physical drivers. Wool growing was highly profitable, wealthy squatters could make substantial sums of money by doing little more than putting sheep on the land, and the general view was that dry years were the exception rather than the rule, and that stock numbers could be held near the level that could be sustained during wetter years.[79] Severe practical limitations also existed for squatters who wished to retain a store of feed for dry times, for transportation costs were high, most runs were unfenced, public stock routes typically passed through the most well-watered and productive country, and few had alternative land further south with which to withdraw surplus stock.

The collapse happened suddenly. Rainfall tapered off in late 1863 across the north of state and deficiencies intensified greatly as the normal winter rains failed in 1864.[80] Reports began to flow in of the death of livestock and even kangaroos in September, and by early 1865 the mortality of livestock was described as 'appalling'. The cost of cartage exploded from 9d. to 3s. 2 1/2d. per ton per mile since teamsters were unwilling to cross the waterless and grassless wastes from Port Augusta, and most hauling bullocks were dead from disease and starvation. By the end of 1865 a report prepared by the Northern Runs Commissioners estimated a loss of 235 000 sheep (out of 827 000) and 28 000 cattle in the year ending September 1865, with many stations losing 80–90% of their animals (Plate 11). Only those fortunate owners who had sold before their food supply ran out were able to salvage anything from the situation.[81]

The ceaseless pounding of hooves on the bare earth and hundreds of thousands of starving mouths quickly devastated the fragile vegetation, as it had done in western New South Wales. By early 1865, extensive tracts of saltbush were scorched and failing, and by the end of the year little or nothing was left alive. The light and sandy soils, stripped of all cover, soon blew up into tremendous dust storms the likes of which had never been seen before in the colonies. Riding on strong north-westerly winds, great clouds of dust, sand and pebbles swept across the landscape for hours, blackening the sky and reducing visibility to a few yards. Yards and fences were buried. Dust and sand poured into the more simply built station houses, and later had to be removed by wheelbarrow. In some places the entire upper soil profile, which provided an essential role in retaining nutrients and water, disappeared.[82]

At the time of European settlement, the Flinders Ranges and adjacent plains supported an even more diverse assemblage of terrestrial mammals than western New South Wales. But here, too, the destruction of vegetation, watering points and soil proved devastating. Already in decline through the 1850s, species like the golden bandicoot, greater bilby, western quoll, crescent nailtail wallaby, both stick nest rats and two different hare-wallabies, were driven close to extinction. In May 1865, an eye-witness account from the plains east of the Flinders Ranges reported that kangaroo rats (probably burrowing bettongs but possibly brush-tailed bettongs and/or desert rat kangaroos) were 'lying about dead all over the country'.[83] These and many other species held on in small, scattered populations for a couple more decades, but by the early 1900s more than two dozen species, or roughly two-thirds of those originally present, had disappeared. The 1864–66 drought is now understood as a seminal point in the collapse of these species, but even at the time there was a sense that the country would never recover.[84]

The drought also devastated the populations of larger and more common animals. For example, John Bristol Hughes, after travelling from Port Augusta to Yudanamutana in the far northern Flinders Ranges in 1865, observed 'wallabies, euroes, and kangaroos' lying dead in all directions, and that '[t]hose that remain alive are so reduced in number that they have become a rarity in districts that generally teemed with them'.[85] The species in question were likely to be the red kangaroo, euro, and one or more wallabies, perhaps including the crescent nailtail wallaby, yellow footed rock wallaby, tammar wallaby and smaller hare-wallabies.[86] Hughes also noted that '[t]he natural severity of the drought is greatly aggravated by the flocks and herds of the squatters, which have utterly consumed or trodden out every vestige of grass or feed within miles of water', and thus competition with livestock for food was a crucial factor.

> *Euroe no food now – big one tumble down – that fellow all bone, no butter [fat] – by-and-by all die.*
>
> – South Australian Register, *12 July 1865*
>
> *There is not a hill, a creek, or an area of any kind which the aborigines may retire to or look upon as their home or place of resort.*
>
> – South Australian Register, *2 August 1865*[87]

The collapse of native mammal numbers and decline in water sources brought the Adnyamathanha people of the northern Flinders Ranges into direct conflict with squatters. Reduced to starvation and driven from waterholes, they had little option but to beg squatters or work as shepherds for food, steal, or prey on sheep and cattle. The first option proved a temporary solution at best: even if willing, squatters were increasingly unable to provide the necessary rations due to soaring freight costs and mounting financial losses.[88] Those Aboriginal people who killed livestock often had their possessions destroyed, or worse, were captured and flogged, and then, depending on circumstances, were secured by a collar and chain riveted around the neck and marched tens to hundreds of kilometres across the drought-stricken landscape to stand trial. Others were shot in skirmishes that arose over stock, food or water, or even in unprovoked circumstances.[89] The injustice of this situation was not lost on everyone – a few people became vocal advocates for extending natural rights to dispossessed Aboriginal people – but the Frontier Wars and the incarceration for stealing cattle were to continue for decades, with tragic consequences for Indigenous society.

Adnyamathanha people have described the disappearance of mammal species as being in essence a loss of part of the meaning of existence, since the extinction of totem species severed the sense of oneness with the land and the living environment that had been passed down for millennia.[90] Other authors have covered the implications of European settlement and the loss of animal species on Aboriginal life in far greater depth than it is possible to do here, but I think that the events documented here reveal the extent to which the 1864–66 drought accelerated and deepened these processes. The collapse was sudden, permanent and catastrophic, and the impoverishment of both the natural world and human society that resulted surely ranks among the most devastating that has ever occurred on the planet.

Endnotes

[1] From 1 Corinthians 13:12: βλέπομεν γὰρ ἄρτι δι' ἐσόπτρου ἐν αἰνίγματι, meaning 'For now we see through a glass darkly'.

[2] Table 2.1 in Johnson C (2006) *Australia's mammal extinctions. A 50 000 year history*. Cambridge University Press, Cambridge.

[3] The clade Ungulata mainly contains the hooved mammals, and includes horses, cattle, pigs, sheep, camels and deer.

[4] Blunden BG *et al.* (1994) *Australian Journal of Soil Research* **32**, 1095–1108; Greenwood KL, McKenzie BM (2001) *Australian Journal of Experimental Agriculture* **41**, 1231–1250.

[5] Trimble SW, Medel AC (1995) *Geomorphology* **13**, 233–253.

[6] *The Hobart Town Advertiser*, 23 October 1840, p. 2.

[7] *The Sydney Gazette and New South Wales Advertiser* (hereafter *Sydney Gazette*), 22 November 1838, p. 3.

[8] *Geelong Advertiser*, 25 June 1850, p. 1.

[9] *The Australian* (Sydney), 23 April 1839, p. 2; *The Sydney Herald*, 17 April 1839, p. 2; 9 April 1838, p. 2; *The Colonist* (Sydney), 8 August 1838, p. 3, *The Sydney Monitor*, 22 December 1837, p. 2.

[10] *The Sydney Monitor and Commercial Advertiser*, 15 December 1841, p. 2.

[11] Garden DL *et al.* (2000) *Australian Journal of Experimental Agriculture* **40**, 225–245; *Sydney Gazette,* 30 June 1829, p. 3; *The Sydney Monitor and Commercial Advertiser*, 26 December 1838, p. 2; *Port Phillip Gazette*, 20 April 1842, p. 4.

[12] Soil compaction caused by livestock is reviewed in Drewry JJ, Cameron KC, Buchan GD (2008) *Australian Journal of Soil Research* **46**, 237–256.

[13] *Sydney Gazette*, 22 November 1838, p. 3; *South Australian Register* (Adelaide), 11 April 1840, p. 5.

[14] Greenwood KL, McKenzie BM (2001) and references therein.

[15] *Sydney Gazette*, 30 June 1829, p. 3.

[16] *The Sydney Monitor*, 7 April 1830, p. 2, and 1 May 1833, p. 3.

[17] *The Sydney Monitor*, 20 June 1835, p. 2; *The Maitland Mercury and Hunter River General Advertiser*, 15 November 1845, p. 4; *Australasian Chronicle* (Sydney), 17 September 1839, p. 1; *The Sydney Herald*, 12 December 1833, p. 2.

[18] *The Hobart Town Advertiser*, 15 October 1844, p. 3; *South Australian Register* (Adelaide), 28 July 1847, p. 4; *South Australian* (Adelaide), 16 July 1847, p. 3.

[19] Rustomji P, Pietsch T (2007) *Geomorphology* **90**, 73–90. Cattle arrived here around the time of the 1813–1815 drought.

[20] Wasson RJ et al (1998) *Geomorphology* **24**, 291–308; Eyles RJ (1977) *Australian Geographer* **13**, 377–386.

[21] Warner RF (1997) *Catena* **30**, 263–282.

[22] *The Colonist* (Sydney), 2 March 1839, p. 2.

[23] *The Austral-Asiatic Review, Tasmanian and Australian Advertiser* (Hobart Town), 2 April 1839, p. 6.

[24] Marx SK et al. (2014) *Journal of Geophysical Research: Earth Surface* **119**, 45–61.

[25] Cattle SR (2016) *Aeolian Research* **21**, 1–20.

[26] Hubbard RK, Newton GL, Hill GM (2004) *Journal of Animal Science* **82** (E. Suppl.), E255–E263.

[27] *The Sydney Monitor and Commercial Advertiser*, 31 October 1838, p. 2; 22 March 1839, p. 2; 27 December 1839, p. 2; *The Sydney Herald*, 17 April 1839, p. 2.

[28] *Southern Australian* (Adelaide), 17 December 1841, p. 3; *The Sydney Monitor and Commercial Advertise*r, 15 December 1841, p. 2.

[29] Balcombe SR et al. (2006) *Marine and Freshwater Research* **57**, 619–633; Sheldon F et al. (2010) *Marine and Freshwater Research* **61**, 885–895.

[30] *Colonial Times* (Hobart), 12 January 1836, p. 6; *South Australian Gazette and Colonial Register* (Adelaide), 9 February 1839, p. 3; *Port Phillip Gazette*, 13 May 1840, p. 3.

[31] *Sydney Gazette*, 9 March 1839, p. 2.

[32] *Southern Australian* (Adelaide), 4 September 1839, p. 4.

[33] *Sydney Gazette*, 14 March 1833, p. 2.

[34] Leigh JH (1994) Chenopodium shrublands. In *Australian Vegetation*. Second Edition. (Ed. RH Groves) pp. 345–367. Cambridge University Press, Cambridge.

[35] *The Argus* (Melbourne), 28 September 1858, p. 1 and 14 October 1865, p. 5.

[36] *The Sydney Morning Herald*, 13 February 1846, p. 3; *Morning Chronicle* (Sydney), 30 May 1846, p. 3; *The Maitland Mercury and Hunter River General Advertiser*, 20 June 1846, p. 2.

[37] *The Moreton Bay Courier* (Brisbane), 18 March 1848, p. 2.

[38] Wilson AD (1966) *Australian Journal of Agricultural Research* **17**, 147–153.

[39] *The Sydney Morning Herald*, 12 January 1847, p. 2; Heathcote RL (1965) *Back of Bourke. A study of land appraisal and settlement in semi-arid Australia*. Melbourne University Press, Carlton; cited in Lunney D (2001) *The Rangeland Journal* **23**, 44–70.

[40] *The Sydney Morning Herald*, 17 February 1846, p. 3; 27 May 1848, p. 2; *Port Phillip and Settler's Journal*, 18 March 1848, p. 2; *South Australian Register* (Adelaide), 30 September 1848, p. 4.

[41] *Port Phillip Gazette and Settler's Journal*, 16 February 1848, p. 2; 5 February 1848, p. 2; *Geelong Advertiser*, 10 March 1848, p. 1; *The Sydney Morning Herald*, 27 May 1848, p. 2; *The Maitland Mercury and Hunter River General Advertiser*, 11 March 1848, p. 4.

[42] *The Moreton Bay Courier* (Brisbane), 29 June 1850, p. 4.

[43] *The Sydney Morning Herald*, 19 September 1850, p. 3; 27 November 1850, p. 3; *The Moreton Bay Courier* (Brisbane), 21 September 1850, p. 2.

[44] *Hobart Town Gazette and Van Diemen's Land Advertiser*, 5 October 1822, p. 2; *Sydney Gazette*, 15 April 1824, p. 4; 19 August 1815, p. 2.

[45] *The Sydney Morning Herald*, 23 May 1845, p. 3; 30 April 1850, p. 2; *Bathurst Free Press*, 27 April 1850, p. 2; *Geelong Advertiser*, 9 August 1849, p. 1.

[46] *Hamilton Spectator*, 1 October 1873, p. 3; *Australian Town and Country Journal* (Sydney), 24 June 1871, p. 796; 20 January 1883, p. 20; *The Sydney Mail and New South Wales Advertiser*, 16 September 1871, p. 903; 25 August 1877, p. 231; *The Australasian* (Melbourne), 28 December 1872, p. 825; *The Ballarat Star*, 22 July 1868, p. 2.

[47] The concept of 'invasional meltdown' was introduced in Simberloff D, Von Holle B (1999) *Biological Invasions* **1**, 21–32; recently reviewed in Braga et al. (2018) *Biological Invasions* **4**, 923–936. The decline of native grasslands in southern Australia and roles is discussed in Lunt I, Barlow T, Ross J (1998) *Plains wandering. Exploring the grassy plains of south-eastern Australia*. Victorian National Parks Association

and the Trust for Nature, East Melbourne, Victoria. The roles of grazing and nutrient enrichment in remnant woodlands and grassland are discussed in Close DC, Davidson NJ, Watson T (2008) *Biological Conservation* **141**, 2395–2402 and Lunt ID *et al*. *Australian Journal of Botany* **55**, 401–415.

[48] Lunney D (2001) *Rangeland Journal* **23**, 44–70; *South Australian Gazette and Mining Journal*, 26 October 1850, p. 3; *Adelaide Times*, 9 December 1853, p. 3; *The Armidale Express and New England General Advertiser*, 9 January 1858, p. 3.

[49] *Bendigo Advertiser*, 7 April 1858, p. 2; *The Sydney Morning Herald*, 17 September 1858, p. 2; 6 August 1858, p. 8; *Empire* (Sydney), 6 March 1858, p. 5; 20 March 1858, p. 3; *South Australian Register* (Adelaide), 19 July 1859, p. 3; 30 December 1859, p. 3.

[50] *Geelong Advertiser*, 31 August 1863, p. 3.

[51] *Sydney Morning Herald*, 13 March 1865, p. 8; *Argus* (Melbourne), 14 October 1865, p. 8; 27 January 1864, p. 6; *South Australian Gazette and Colonial Register* (Adelaide), 11 May 1839, p. 2.

[52] *Argus* (Melbourne), 10 October 1864, p. 3; 28 September 1865, p. 3.

[53] *The South Australian Advertiser* (Adelaide), 15 November 1865, p. 2.

[54] *Australian Town and Country Journal* (Sydney), 3 June 1876, p. 895.

[55] *Empire* (Sydney), 9 July 1866, p. 5; *The Brisbane Courier*, 3 January 1866, p. 3; *The South Australian Advertiser* (Adelaide), 15 November 1865, p. 2.

[56] *The Australasian* (Melbourne), 21 June 1873, p. 6; *Leader* (Melbourne), 26 May 1877, p. 6; Moore CWE (1953) *Australian Journal of Botany* **1**, 548–567. Oldman saltbush had become rare at least 5–10 years before the arrival of the first rabbits.

[57] *The Australasian* (Melbourne), 16 February 1884, p. 23; *Australian Town and Country Journal* (Sydney), 30 December 1882, p. 1268; 12 September 1885, p. 548; *Evening News* (Sydney), 5 August 1882, p. 7.

[58] *The Australasian* (Melbourne), 26 March 1932, p. 41.

[59] Lunney D (2001) *Rangeland Journal* **23**, 44–70.

[60] Woinarski JC, Burbidge AA, Harrison PL (2015) *PNAS* **112**, 4531–4540.

[61] Grefft G (1866) *Transactions of the Philosophical Society of NSW* **1**, 1–38.

[62] Dickman CR *et al.* (1993) *Biological Conservation* **65**, 219–248.

[63] Mitchell TL (1848) *Journal of an expedition into the interior of tropical Australia, in search of a route from Sydney to the Gulf of Carpentaria*. Longman, Brown, Green and Longmans, London.

[64] Abbot I (2002) *Wildlife Research* **29**, 51–74.

[65] Dickman *et al.* (1993) *Biological Conservation* **65**, 219–248; Lawes *et al.* (2015) *PloS One* **10**, e0130626; references therein.

[66] Menkorst PW (2009) *Proceedings of the Royal Society of Victoria* **121**, 61–89. Cats might have been uncommon in the lower Murray-Darling region until at least the 1880s.

[67] Lunney D (2001) *Rangeland Journal* **23**, 44–70.

[68] Short J, Kinnear JE, Robley A (2002) *Biological Conservation* **103**, 283–301.

[69] Cf. Joel 1: 20, the King James Version of which reads: 'The beasts of the field pant for You, for the streams of water have dried up, and fire has devoured the pastures of the wilderness.'

[70] Lunney D (2001) *The Rangeland Journal* **23**, 44–70.

[71] *South Australian Register* (Adelaide), 19 July 1859, p. 3; *Mount Alexander Mail*, 2 March 1860, p. 6.

[72] Sturt C (2001) *Narrative of an expedition into central Australia: Performed under the authority of Her Majesty's Government, during the years 1844, 5, and 6, together with a notice of the Province of South Australia, in 1847*. Corkwood Press, North Adelaide; first published in London by T and W Boone, 1849.

[73] The drought in central Australia probably ended sometime around 1841–42.

[74] Mitchell (1848).

[75] Gregory AC, Gregory FT (1884) *Journals of Australian explorations*. James C. Beal, Government Printer, Brisbane.

[76] *The Argus* (Melbourne), 8 November 1858, p. 1; *The Australian* (Sydney), 10 December 1846, p. 2; Exploring expedition under Sir Thomas Mitchell, New South Wales Government Gazette, 7 December 1846, Issue no. 102 (Supplement), p. 1532.

77 Appendix A in Bowes KR (1962) *Land settlement in South Australia 1857–1890*. PhD Thesis, Australian National University.
78 *The South Australian Advertiser* (Adelaide), 15 July 1861, p. 2.
79 Bowes (1962).
80 *South Australian Register* (Adelaide), 29 April 1865, p. 3; *The Adelaide Express*, 13 June 1864, p. 2.
81 *Adelaide Observer*, 10 September 1864, p. 4; *South Australian Register* (Adelaide), 20 January 1865, p. 2; 15 November 1865, p. 2.
82 *South Australian Register* (Adelaide), 20 January 1865, p. 3; 25 May 1865, p. 2; 15 November 1865, p. 2.
83 *South Australian Register* (Adelaide), 25 May 1865, p. 2; most likely the burrowing bettong (*Bettongia lesueur*).
84 Tunbridge D (1991) *The story of the Flinders Ranges mammals*. Kangaroo Press, Kenthurst, NSW and references therein.
85 *Adelaide Observer*, 22 July 1865, p. 4.
86 Tunbridge (1991).
87 The first quote recounts the destruction of the euro (*Osphranter robustus*), a critical food source of the Adnyamathanha people, during the 1864–66 drought; the second was made by JB Hughes, an early advocate of justice for Aboriginal people, and indeed those 'whether born with a white, a brown or a black skin', in South Australia.
88 *Adelaide Observer*, 22 July 1865, p. 4.
89 One example is reported in *South Australian Register* (Adelaide), 26 January 1864, pp. 6–7.
90 The deep connection between Aboriginal people and mammal species in the Flinders Ranges is described in detail in Tunbridge (1991).

Part 4

Hell's half century

Ah, wherefore tarries the rain?
   – Anonymous, The Capricornian *(Rockhampton), 19 October 1895*[1]

12
The Titan's grip

the entire continent of Australia, excepting a fringe of coastline, is now held in the grasp of a malignant Titan.
— The Yass Courier, *23 September 1902, p. 2*

WHEN I WAS COMPLETING MY ENVIRONMENTAL SCIENCE DEGREE AT Macquarie University in 1994, northern New South Wales was in the grip of a fairly serious drought. Not unusually, it had begun as a slow burn: 1991, 1992 and 1993 had all been all drier than normal, reflecting persistent El Niño conditions,[2] but after near-record low rainfall in the autumn and winter of 1994 things were shaping up very badly indeed. Our dams had all gone dry for the first time in over a decade and, lacking any feed, local landholders were being forced to sell, or even slaughter, large numbers of livestock. Most of Queensland, northern New South Wales and smaller areas of Tasmania and South Australia had been drought-declared under the newly developed 'drought exceptional circumstances' framework,[3] and when arriving in Sydney to complete my studies I was greeted with a massive dust storm that stripped 20 million tonnes of topsoil from the western plains. The sunsets in New Zealand were said to be spectacular.[4] As a budding 20-year-old scientist, I found these events fascinating, particularly since I had only seen one other drought of equal severity, the brutal year of 1982–83 that had savaged eastern Australia with dust, fire and economic hardship (Plate 2).[5]

One day, when I was discussing the dry conditions with an older farmer at Maules Creek, near Narrabri, he remarked that his father, who farmed the same country, had told him that 'back in the thirties and forties it never rained', the implication being that *this* drought wasn't bad at all. With all the arrogance of youth, I assumed that I was being treated to yet another story from the 'olden days' about just how much tougher things were in the past. I didn't give it much credence until my neighbour, who had grown up in the same area in the 1920s–30s, said the same thing, and, to boot, that the then summers were also much hotter. Interest piqued, I went to the CSIRO field station at Myall Vale, west of Narrabri, got my hands on some dot matrix printouts of rainfall data, and my journey into climate history had begun.

When investigating claims like these I like to begin by looking at simple plots of rainfall over time to see if any obvious patterns emerge. And, thanks to the efforts of meteorologists and a small army of dedicated farmers and enthusiasts, Australia has a good long-term rainfall dataset compiled from station records and from which daily rainfall has been modelled back to 1889.[6] Let us then consider the graph of mean annual rainfall across the continent from 1889 to 2022 in Fig. 12.1. Although rainfall varies considerably over time and there have been extremely dry years in most decades, including the driest ever year in 2019, there does seem to be a preponderance of dry years before the early 1970s and very wet years since. The 10-year moving average rainfall, which smooths out some of the variability, is also consistently low before about 1973. At first glance is seems that perhaps the 'old timers' might be right after all.

Detecting long-term trends in highly variable or 'noisy' rainfall datasets can be quite challenging, but one of my favourite strategies is to use cumulative rainfall departure (CRD) curves, which are calculated by subtracting the mean annual rainfall from each year and, starting at the beginning of the time series, sequentially adding the resulting annual deviations (or residuals). Put simply, a downward trend in a CRD curve occurs during runs of years that are generally drier than the long-term (1889–2022) average, and a rising trend during ones that are wetter. As it turns out, the CRD for Australia (Fig. 12.1) confirms the presence of a nearly continuous half-century-long dry epoch that lasted from around 1895 until 1946, and a second period that that lasted from the 1950s until 1972. During these two dry epochs the country as a whole missed out on close to 4 years' worth of average rainfall (ca. 1800 mm total). Between 1973 and 2000 most of this deficiency has been erased, and since 2001 this general pattern has continued, although with a mix of severe droughts and wet periods.

If we now look at the spatial pattern of rainfall deficiencies that prevailed over the 1895–1946 and 1947–72 dry epochs, we can see that both occurred on an incredible scale (Plate 12). The first, and arguably worst, affected virtually the entire country: the interior suffered the most, where rainfall was 10–30% below the long-term average, but chronic dry conditions also extended well into south-eastern Australia and the northern tropics. During this period, which for reasons that will become clear in later chapters has been called the 'eastern Australian dust bowl', rainfall was doled out less generously by the jealous skies, wet years were infrequent and fleeting, and tremendous droughts swept the continent with depressing regularity. At least five have gone down in history as among Australia's worst: the Federation, WWI, 1920s, 1930s and World War II droughts (Table i.1, Plates 1 and 2, Fig. 12.1). The second epoch, 1947–72, was of similar severity, but with a footprint focused in central and northern Australia, more obscure. Nevertheless, it featured the great quasi-decadal central drought of 1958–65 and others that affected the northern tropics (1951–54) and eastern Australia (1964–67).

Determining the start and end of a particular drought is, to some extent, a matter of opinion, particularly since dry periods tend to blur together during long stretches of chronically low rainfall. Indeed, in the scientific literature you will sometimes see dates different to those given above, especially in more regional studies. Perhaps it is more interesting to consider these droughts as collectively constituting a *megadrought* period, a term reserved for decadal to multi-decadal droughts of great intensity, and normally those that cause serious agricultural disruption and societal struggle in afflicted civilisations. I believe that there have been two periods in post-settlement times that clearly meet these criteria: ca. 1820–50 and 1895–1946 (–1972 in central Australia), and we might add the long-term drying trend that has taken place in south-western Western Australia since the 1950s (Plate 12).

By the end of the following chapters I hope to convince you that the great drought that took place in the first half of the 20th century was not only the worst in post-settlement history, but among the most severe to occur globally in the last two centuries. No historical work could

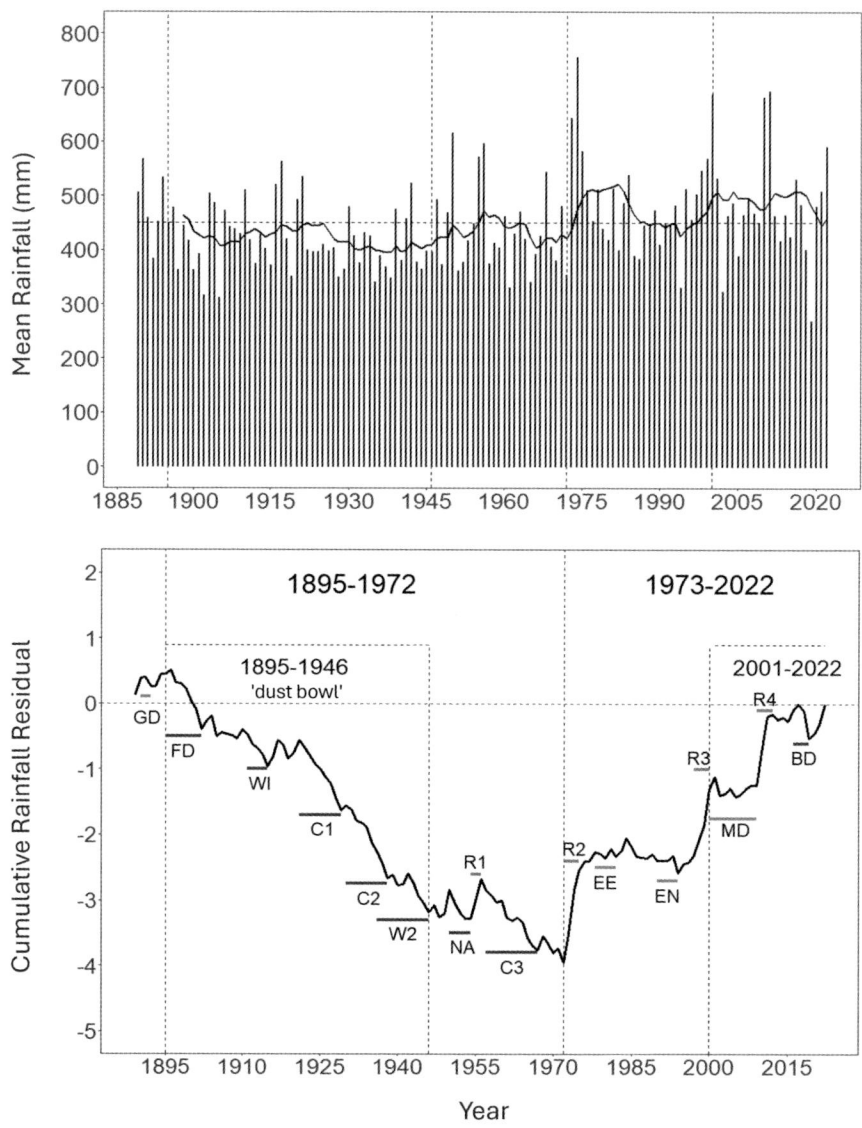

Fig. 12.1: Mean Australian rainfall, 1889–2022, based on interpolation of long-term station rainfall records. Top: annual mean rainfall relative to the 1889–2022 average (horizontal dashed line) and the 10-year moving (running) average (dark solid line). Major rainfall epochs are delineated by vertical dashed lines at years 1895, 1946, 1972 and 2000; see text for description. Bottom: cumulative rainfall departure (CRD) for mean Australian rainfall, 1889–2022, scaled to (divided by) the mean annual continental rainfall of 450 mm. Major rainfall epochs are delineated by vertical bars as above. Continental- and subcontinental-scale droughts are indicated below the CRD line (dark and light horizontal bars respectively) and wet phases as horizontal bars above the CRD line. Major droughts are as follows: GD = 'Great Drought', FD = Federation, WI = World War I, C1 = Central I 1920s, C2 = Central II 1930s, W2 = World War II, NA = Northern Australian, C3 = Central III and Eastern 1950s–60s, EE = late 1970s–early 1980s, EN = early to mid-1990s, MD = Millennium, BD = 'Big Dry'. Note that the 1930s drought, which was particularly severe in central Australia, overlaps with the World War II drought of the eastern states. Wet periods are as follows: R1 = 1955–56, R2 = 1973–76, R3 = 1998–2000, R4 = 2010–11.

ever do full justice to the breadth of environments and social systems impacted nor to the sheer misery endured by people during 'hell's half century'. Instead, I have focused on bringing to light some of the most compelling themes and storylines during four key phases of the megadrought: the explosion of rabbit numbers across inland in the 1880s–90s, the collapse of inland pastoralism during the 1895–1902 Federation Drought, the development of the eastern Australian 'dust bowl' era of 1895–1946 and its continuation in the centre of the continent through the 1960s, and the devastation of Aboriginal and European settler societies of central Australia that reached its peak in the 1920s–30s. For, in an era of general inanition, these were the hungriest of times.

Endnotes

[1] From the poem 'Drought' published in *The Capricornian* (Rockhampton), 19 October 1895, p. 9.
[2] Allan RJ, D'Arrigo RD (1999) *The Holocene* **9**, 101–118.
[3] White DH, Karssies L (1999) *Water International* **24**, 2–9.
[4] O'Neill G (1994) 'Dust storms a national disaster with a $100m price tag'. *The Canberra Times*, 27 May, p. 5.
[5] Purtill A *et al*. (1983) *Quarterly Review of the Rural Economy* **5**, 3; Rawson RP, Billing PR, Duncan SF (1983) *Australian Forestry* **46**, 163–172; *The Canberra Times*, 9 February 1983, p. 3.
[6] In the following analyses I use SILO long-term data hosted by the Queensland Department of Environment and Science (DES) and available at https://www.longpaddock.qld.gov.au/silo/.

13

The living drought

No person can tell what he will do when driven by hunger.
– Alexander Pierce, Supreme Court of Van Diemen's Land (as reported in the
Hobart Town Gazette and Van Diemen's Land Advertiser, *25 June 1824)*

IN NORTHERN CENTRAL NEW SOUTH WALES, THERE IS RIVER COUNTRY, GOOD country, and then there's *rabbit country*. There has always been a perceptible level of cultural cringe associated with living in rabbit country, which, even today, still resonates with those graziers who endured the rabbit plagues of the 1890s–1940s. This runs deeper than the simple fact that the hilly, 'hard' country favoured by rabbits tends to be less fertile and profitable than the black, basalt-derived soil of the Liverpool Plains. Here they represent both a curse and a saviour of desperate times, of a country stripped bare, of starving sheep, and of 'underground mutton'. In those days, the foothills of the Nandewar Ranges to the east of Narrabri, where I lived as a boy, were most definitely *rabbit country*.

Fig. 13.1: Left: steel-jawed traps like this Victor No. 10 model were a mainstay for rabbit control across the inland for nearly a century. *Gippsland Times*, 3 May 1934. Right: rabbiting at Wellington, NSW, 22 January 1944. Mitchell Library, State Library of NSW, FL9547216, and Courtesy ACP Magazines Ltd.

Stories of the great rabbit plagues still endure in the consciousness and, for a few, in memory. When my parents bought their small farm on Bullawa Creek, below Mount Kaputar, in 1976, rabbits still thrived along sandy creek lines in huge warrens with dozens of holes, in the evening moving out in a wave into the adjacent paddocks like some military formation. They had a special fondness for windrows left over from chain clearing of timbered country, and by carefully picking your position you could sometimes hit three rabbits emerging from them with a single rifle shot. During the 1980s they became so numerous that they invaded gardens and burrowed under houses, sheds and tank stands. During family trips I would amuse myself by counting the number as roadkill, once recording 146 on a drive between Narrabri and Maitland (370 km) in the Hunter Valley.

But these populations were much, much smaller than those that existed before the introduction and spread of the myxoma virus in the 1950s. I first became aware of this when Tom Gunter, a grazier from Bullawa Creek and the Upper Horton area near Mount Kaputar, told me that during the Great Depression he would kill hundreds of rabbits for skins each day simply by hitting bushes with a stick containing a nail in the end. I can still remember as a small child my dad digging up rabbit warrens long since abandoned, and most old farm sheds at the time had dozens, or even hundreds, of the old steel-jawed rabbit 'spring' traps hanging on the walls gathering dust (Fig. 13.1). Today, in western New South Wales, where both the Czech and Korean strains of calicivirus (CAPM 351RHDV and RHDV1 K5) have been extremely effective in rabbit control, there are even fewer. I haven't seen a rabbit on my parent's property for years, and the warrens have fallen in, a once vast empire in ruins.

With the benefit of two centuries of hindsight, one is left with a sense that the dramatic population growth and spread of the European rabbit that took place after they were released on Thomas Austin's Victorian property Barwon Park in 1859[1] was inevitable. Surprisingly, though, the odds were stacked against them from the start, because ecologists have found that only around one in 10 species introduced into a new region becomes established, and of these around one in 10 (or 1% of those introduced) become invasive (a pattern known as the 'tens rule').[2] Indeed, by the 1850s small rabbit populations were already scattered around the continent, but these showed no signs of becoming a biological juggernaut, held in check as they were by overwhelming numbers of predatory birds and mammals – 'fresh delinquents supplying the place of the ancient rogues'.[3]

But these later arrivals were forged in a hot evolutionary furnace; in their Mediterranean home they had developed the ability to burrow and form communal warren complexes, a legendary capacity to reproduce, and a cryptic pelt colouration perfectly matched to semi-arid and arid landscapes. Invasion ecologists might say that they were *pre-adapted* to novel environments that exist outside their native range.[4] When settlers began to exterminate native quolls, eagles, hawks, crows and dingoes in the thousands using strychnine-laced baits in the 1860s, rabbits were perfectly placed to take advantage of this situation, and soon numbered in the tens of thousands. In 1866 more than 14 000 rabbits were shot on Barwon Park,[5] but warrens still filled beyond capacity, and displaced younger animals ventured ever deeper into

> *Hark! I hear the tramp of thousands*
>
> – *Bret Harte*, 'The Rèveille'

new terrain.[6] Within 10 years a dark evil stain[7] was spreading across the Australian landscape, the main front moving towards Swan Hill and north-western Victoria, and another, probably derived from a secondary introduction, through hilly country to the north-east of Adelaide and the Flinders Ranges.

The 40 years that had passed since Thomas Mitchell first traced the Darling River to its junction with the Murray in 1836 had not been kind to the native plants and animals of the southern inland. Along the ranges and slopes the clearing of forests and woodlands had paved the way for the establishment of thousands of hectares of introduced cool season crops, grasses and forbs that provide a nutritious dietary food supply during the normal winter–spring feed gap in natural pastures (Fig. 13.2).[8] Squatters and farmers persecuted virtually all predators, and by 1863 few dingoes survived south of the Murrumbidgee River,[9] quolls were being exterminated,[10] and 'eaglehawks' (mainly wedge-tailed eagles; *Aquila audax*) received no quarter (Fig. 13.3) even though sheep and lambs comprise less than 10% of their typical diet.[11] Migrating rabbits found shelter in the burrows of larger native mammals, including common, southern and northern hairy-nosed wombats, burrowing bettongs and bilbies, each capable of excavating deep into the hard Australian earth.[12] Vast warrens formed, and stupendous numbers spilled out into ever more remote parts of the landscape.

After the break of the 1864–66 drought and a brief relapse into drought conditions during 1868–69, a decade-long period of generally wetter conditions commenced over most of south-eastern Australia (Plate 13). Most of the inland pastoral zone was plunged into drought in 1875–77, but despite significant agricultural losses conditions in southern New South Wales were not as severe as those experienced in the 1830s, 1840s and 1860s,[13] and enough feed was present to allow rabbits to colonise the lower Darling, Lachlan and Murrumbidgee River regions by 1881. Many pastoralists, including the phenomenally successful James 'Hungry' Tyson, believed that the tougher conditions of semi-arid and arid New South Wales would limit the worst damage to their southerly neighbours, and laughed at the prospect of rabbits becoming established on Tupra Station, north-west of Hay, arguing that the saltbush would cause them to die of thirst. But by 1886 every vestige of bluebush, saltbush and cotton bush had been destroyed, and the rabbits lived on.[14]

Not everyone was so naïve. David Brown (Fig. 13.4),[15] the manager of the prized million-acre Kallara Pastoral Holding near Tilpa, saw the devastation wrought on properties lower down the Darling, and recognised that the spread of the rabbit was a matter of life and death for their industry. Since purchasing the property in 1875, the owners of Kallara, Charles and Suetonius Officer, had spent a staggering £65 000 on property improvements (hundreds of times the annual wage of a typical farm worker), and were in no mood to allow this investment to disappear down the mouths of the grey horde approaching from the south. After rabbits

Fig. 13.2: The transformation of the mallee country from vast tracts of wilderness to cultivated paddocks is beautifully captured in this series of sketches. Facilitated by innovations such as the stump-jumping plough, these changes aided the spread of rabbits and left the land susceptible to wind erosion and drift during periods of drought when overgrazed. *Illustrated Australian News* (Melbourne), 1 July 1892.

arrived at Curranyalpa Station, less than 20 km from Kallara, Brown became a tireless activist, thinking deeply about both the ecological and anthropogenic factors that were contributing to the rabbit plague. Ultimately, though, the herculean efforts of graziers to poison, trap and fence the newcomers did little or nothing to slow the tide.[16]

The first real check came not from men but from the climate. In 1888 the mean rainfall for the New South Wales fell to only 13.4 inches (340 mm), a record 43% below the average at that time, with severe deficiencies experienced across the Western Division (Fig. i.1). Native

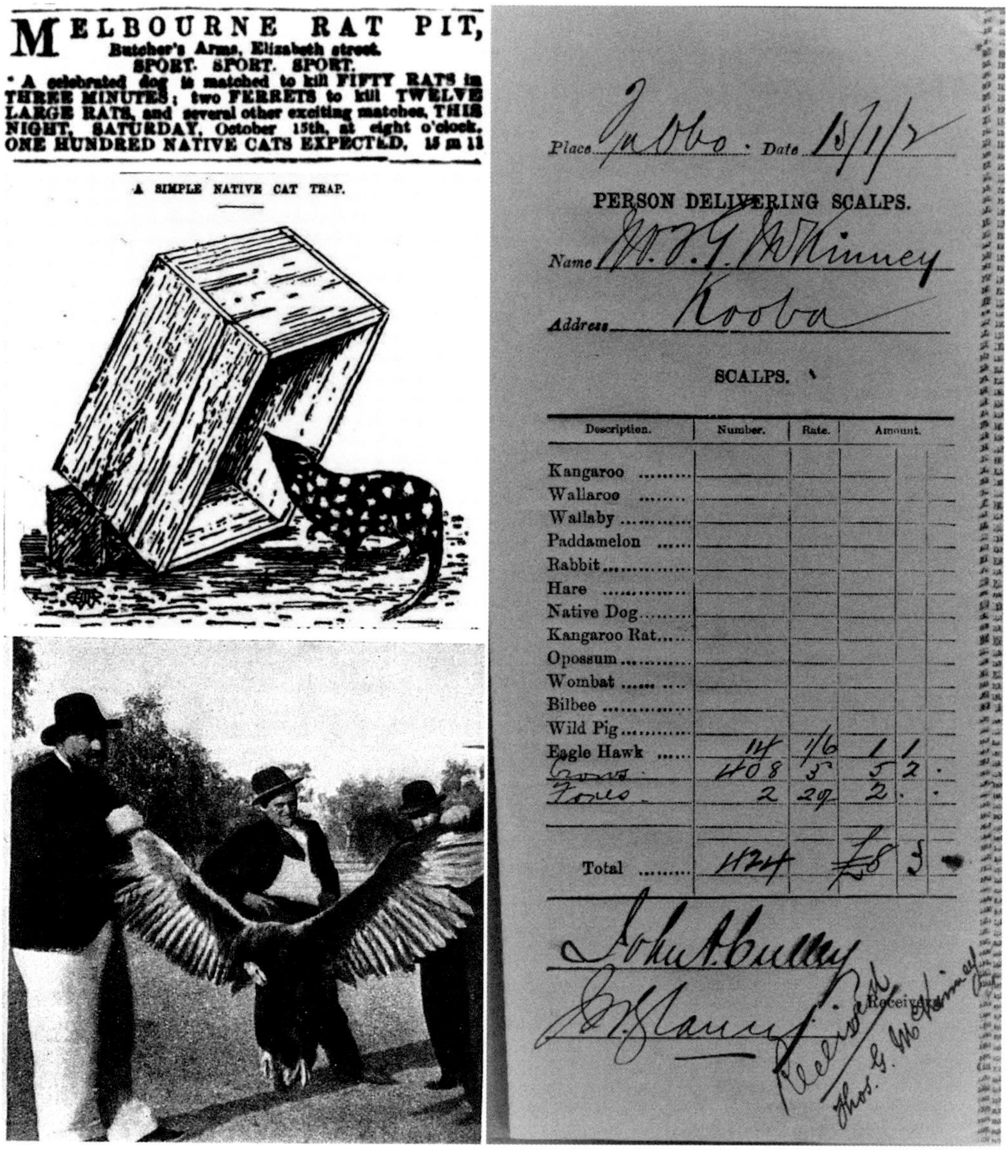

Fig. 13.3: Top left: vast numbers of quolls were killed in organised animal fights. *Bell's Life in Victoria and Sporting Chronicle* (Melbourne), 15 October 1859. Centre left: quolls were often targeted for destruction. *Western Mail* (Perth), 17 March 1899. Bottom left: a huge 'eaglehawk' shot during a tour of drought-stricken western New South Wales by the Minister for Lands, William Patrick Crick, in April 1902. Right: report of scalps of noxious animals delivered from Tubbo Station, near Darlington Point, NSW, January 1902. Note the large numbers of wedge-tailed eagles killed, and the presence of foxes in the area. Courtesy Charles Sturt University Regional Archives, Wagga Wagga.

trees withered and died in the Riverina, New England and Sydney Basin regions of New South Wales, as well as in Queensland and Tasmania, and drought conditions were severely felt across inland South Australia and Victoria.[17] Pastures across the inland, groaning under the weight of rabbits, sheep and cattle, were eaten down to dirt, and millions of rabbits died of starvation and thirst, in some places becoming so rare that it was said to be nearly impossible to procure a pair for the table.[18] In 1890–91 exceptionally wet conditions left the western landscape 'clothed in herbage as green as a leek'[19] (Fig. 13.5) and drowned rabbits across the Murray–Darling Basin in the tens of thousands, their bodies floating down the rivers with the rest of the flotsam or huddling together above the floodwaters on the branches of leaning river red gums.[20]

But this was merely a stay of execution, for numbers were already rebounding from the 1888 drought, particularly in country with lighter soils (containing a high proportion of sand relative to clays).[21] Rabbit warrens grew enormous, often hundreds of metres to kilometres long and containing thousands to millions of animals,[22] earning inland New South Wales the distinction of being referred to as 'one great rabbit warren' that extended continuously from the Victorian to Queensland borders.[23] Hyperbole aside, assuming rabbit densities averaged around six per acre (15 per hectare) over some 38 000 000 ha of infested land in 1892, there were probably in the order of 500–600 million rabbits in New South Wales alone at this time, and conservatively perhaps 1–1.2 billion rabbits across the south-east of the continent.[24] Even relatively low numbers (e.g. one per hectare) of rabbits can cause significant ecological change, but at these densities they become the dominant herbivores in the landscape, capable of completely removing edible vegetation in all but the wettest years.

Fig. 13.4: David Brown, manager of Kallara Station, near Tilpa, with his family. Brown was heavily involved in efforts to prevent rabbits from overrunning Kallara Station in the 1880s and in the development of subsequent control programs. His insights into ecology and behaviour of rabbits and station management in general were ahead of their time. Photo courtesy of Julie and Justin McClure.

The crucial moment came in late 1891 and into 1892 when a terrible drought that had gripped tropical Australia dipped into northern and central New South Wales, causing pastures to brown and then fail. Adult rabbits typically need access to pasture with a water content of 55% or more to remain adequately hydrated,[25] and when forced to persist on dry, senescent grasses and forbs, they experience a *negative water balance* (water losses due to excretion and evaporation exceed that acquired from the pasture they consume), limiting their ability to thermoregulate. In hot weather their bodyweight slowly declines, metabolic processes become impaired, and, after 1 or 2 months, the animal dies. Finding water therefore becomes a matter of life and death.

In drier country lacking permanent watercourses rabbits driven by terrible thirst and dehydration swarmed to dams, bores and wells, often arriving so weak that they fell in and drowned. In January 1892, for example, a 45-feet-deep (13 m) stock well at Poolamacca

Fig. 13.5: Top: thick grass and luxurious gardens on Sussex Station, east of Cobar, ca. 1890. By 1891 the rabbit plague had arrived. Mitchell Library, State Library of NSW, 94R5p5Z1. Bottom: floodwaters at Dunlop Station, 1886, showing inundation of the woolshed and stands of river red gum along the Darling River. In 1890, floodwaters in places extended for 30 miles (48 km) or more from the Darling, isolating towns and many stations. Photo: Charles Bayliss. National Library of Australia, nla.obj-140616884.

Station, north of Broken Hill, had become filled to the top with dead rabbits that had fallen in attempting to get at the water, the collective weight of the carcases having 'consolidated the whole mass into a putrefying jelly and forced the well water level with the surface, allowing the rabbits to drink in safety'.[26] Such macabre events greatly exacerbated the challenge posed by an ongoing water famine at the time in the Broken Hill region, where tens of thousands of litres of water were arriving by train to alleviate the shortage. Rules were put in place to provide this water to the needy on an equitable basis, although the poorer classes apparently still suffered the worst deprivation, some reportedly using bathing water to make tea (Fig. 13.6).[27]

It was not unusual to see rabbits, driven by terrible thirst and dehydration, crowding around any watering point in the thousands (Fig. 13.7), and mobs could even be seen sitting in lakes or even feeding on the water weeds growing along the banks of rivers with their heads just above the waterline.[28] Those surviving on drought-stricken pastures found that any better quality feed was also sustaining immense flocks of sheep that had expanded on the back of favourable conditions during the 1870s, mid-1880s and early 1890s, and the dry feed that was available contained very little digestible energy.

Soon, livestock and rabbits became locked in a grim contest of starvation for the last scraps of vegetation that retained a skerrick of moisture and nutrition. Unconcerned with the human struggle against aridity, rabbits flocked to station houses, raiding the flowers and vegetables and stripping the trees of their fruit. Others converged on towns. In the mid-1890s a vast horde invaded Broken Hill looking for water and food, rendering 'the night hideous by the noise of their gambols and the air unwholesome with their effluvium'.[29] In scenes reminiscent of some medieval plague the dead were removed daily in their thousands by cart. Others would wait beneath kurrajong trees (*Brachychiton populneus*), pouncing like beggars waiting for alms on the

Fig. 13.6: McFarlane, J. (1892) *Broken Hill – The Water Famine* [wood engraving]. State Library of Victoria, PCINF IAN 01-02-92 P.9.

Fig. 13.7: During droughts of the 1890s–1940s, when rabbits were in plague proportions, vast numbers would crowd around any available source of water to stem dehydration. Top: note the second row of fencing around the edge of the water which was often used to exclude rabbits or to prevent livestock from bogging in this farm dam. Photo: Robert Godfree, from Stead DG (1935) *The rabbit in Australia*. Winn & Co., Sydney. Bottom: netting was used near watering points during drought to concentrate and kill rabbits more effectively than trapping and broad-scale poisoning. Courtesy Mitchell Library, State Library of NSW, FL976508, and ACP Magazines Ltd.

Fig. 13.8: The great rabbit plagues and droughts of the 1890s caused incredible destruction to shrubby vegetation across inland Australia. Top left: a specimen of *Capparis mitchellii* (wild orange), a nutritious, drought-hardy shrub, collected near Nymagee, NSW. Courtesy CSIRO Australian National Herbarium, Canberra. Top right: wild orange bushes ringbarked 3–4 m above ground level. *The Sydney Mail and New South Wales Advertiser*, 20 February 1892. Bottom: it was not unusual to find large numbers of rabbits dead in the branches of shrubs, hung by the neck after losing their footing. This rare photograph was taken near Wilcannia, NSW, during the height of the plague. *Australian Town and Country Journal* (Sydney), 13 February 1892.

leaves which would occasionally fall to the ground, eat the leaf litter under river red gums, steal from tea billies and horses' feed boxes, and even eat the lower portions of hayricks.[30]

All that remained after grasses and forbs were long gone were edible shrubs, the bark of which can have a water content of 50% or more – just enough to keep rabbits alive.[31] The most popular species were wild orange (*Capparis mitchellii*; Fig. 13.8), mulga (*Acacia aneura*) and sandalwood (probably *Santalum lanceolatum*), but less palatable species were also eaten.[32] Stems were scientifically ringbarked to a height of 3–10 feet (1–3 m) above the ground, leaving entire patches of scrub bleached white.[33] Stronger rabbits climbed any bushes and shrubs that contained footholds, but many lost their footing on the slippery and narrow branches, resulting in one of the most gruesome images to ever come from the drought stricken west: dead rabbits hanging by their necks or limbs in the forks of dying bushes, as if a victim of execution (Fig. 13.8).[34] When nothing under the sun was left available, rabbits would ringbark the roots of trees and shrubs below the ground level.[35] Even those that escaped the gnawing army were quickly undermined and excavated by wind, their roots spreading out like cobwebs over the shifting soils.

Eventually, starving and thirsty, and frequently harried by landowners with traps and dogs, rabbits began to move, often in a massive, synchronous wave across the landscape. Whether this behaviour reflected an attempt to reduce the risk of predation, which is known as the *selfish herd* theory,[36] to improve navigational performance through the benefits of collective decision-making,[37] simple panic[38] or some other factor such as funnelling by landscape features remains unknown, but whatever the cause, the migrations involved were immense. One of the most extraordinary was seen by a mailman travelling across drought-stricken desert terrain near Lake Hope in northern South Australia in the drought year of 1895, where a mob of rabbits about 4 miles (6 km) wide and as thick as they could run together, all heading south-west towards Kopperamanna.[39] Another tremendous migration, described as being ca. 150 miles broad, was seen travelling north-west at Tinga Tingana Station in 1893.[40]

Unfortunately, migrating rabbits usually found no refuge, and nowhere was crueller than the vast Riverina plains of 'Hay and hell and Booligal' fame,[41] a place where even the horses looked 'forward to the time when death would cut the traces and relieve them from the yoke of the collar'.[42] There, left with no access to green feed, water or shade, rabbits became 'half-baked' – so stressed that they took little notice of passers-by, and could be killed with sticks[43] or even with the boot. In 1892 wagon drivers on the blackened and blasted plains between Wilcannia and the Lachlan River would use the whip to clear them from their path before proceeding.[44] In many places telegraph posts and fence posts were the only sources of shade

> *As they get thicker and thicker, they raise great clouds of dust when travelling, just like a flock of sheep.*
>
> – *A description of rabbits migrating on the Riverina*, Kilmore Free Press, *1 January 1903*

Fig. 13.9: Top left: the Mungindi to Narrabri rabbit fence in 1906. Top right: a rabbit migration trapped by a netting fence at Cockburn Railway, SA, early 1892. State Library of South Australia, PRG 280/1/2/463. Bottom left: countryside eaten out by rabbits near the transcontinental rail line, 1912. State Library of South Australia, PRG 280/1/14/735. Bottom right: rabbits feeding in a barren landscape, ca. 1900. There are more than 100 rabbits on the near side of the fence alone. State Library of South Australia, PRG 280/1/2/453.

available for many miles, and on exceptionally hot days a row of panting, half-dead rabbits would sit in the shade of each post, shifting with the sun as it moved slowly around through the day, like a living sundial.[45] The end came when they finally ran into netting fences that had been installed on stations beginning in the 1880s (Fig. 13.9), wandering up and down until they died from exhaustion, a torment intensified by large flocks of crows that feasted on the dead and dying animals. Kilometre after kilometre the ground turned white with the bones of the dead.[46]

By the mid-1890s, rabbits had swept millions of acres of the outback as clean as a broom. On the most hard-hit stations of the Darling and Riverina more than 95% of edible shrubs stood dead, no more than skeletons in a graveyard-like landscape.[47] The shrub lands which once 'reigned in all their hideous glory for mile after mile, dense and gloomy and weird',[48] stabilising inland soils for tens of thousands of years, now stood defeated, replaced by open sandy plains, 'windswept and worried, churned up as though they had been rich alluvial diggings worked by a myriad of eager miners'.[49] Exposed to the elements, the country now stood awaiting the arrival of the next great drought to finally slip its skin and transform into a new, harder visage. And arrive it did.

Endnotes

[1] The history of rabbit introductions in Australia has been reviewed in Rolls E (1984) *They all ran wild: The animals and plants that plague Australia.* Angus and Robertson, Sydney.

[2] The 'tens rule' has been an important hypothesis in the field of invasion ecology since it was first mentioned in Williamson MH, Brown KC (1986) *Philosophical Transactions of the Royal Society London Series B* **314**, 505–522, although empirical support for it appears to be declining as new systems are investigated (e.g. Jeschke *et al.* (2012) *NeoBiotica* **14**, 1–20).

[3] In the Mediterranean where rabbits are native, Austin's 'ancient rogues' include the Iberian lynx, Spanish imperial eagle, red fox and badger; see Lees AC, Bell DJ (2008) *Mammal Review* **38**, 304–320. The quote, from *Wallaroo Times and Mining Journal* (Port Wallaroo), 20 December 1865, p. 5 references Jonathan Swift's (1710) 'A description of a city shower'.

[4] Fleming PJS *et al.* (2017) *The Rangeland Journal* **39**, 523–535.

[5] *The Argus* (Melbourne), February 25, 1867, p. 1.

[6] Mykytowycz R, Gambale S (1965) *CSIRO Wildlife Research* **10**, 111–123.

[7] *News* (Adelaide), 14 November 1935, p. 10.

[8] *Weekly Times* (Melbourne), 19 May 1877, p. 6; *Leader* (Melbourne), 10 December 1870, p. 6.

[9] Glen AS, Short J (2000) *Australian Zoologist* **31**, 432–442.

[10] *The Argus* (Melbourne), 3 August 1855, p. 5; *The Sydney Morning Herald*, 22 April 1859, p. 4.

[11] *The Sydney Mail and New South Wales Advertiser*, 10 October 1885, p. 769; Leopole AS, Wolfe TO (1970) *CSIRO Wildlife Research* **15**, 1–17; Lunney D (2001) *Rangeland Journal* **23**, 44–70.

[12] *The Riverina Grazier* (Hay), 22 July 1887, p. 2; *South Australian Chronicle and Weekly Mail* (Adelaide), 17 March 1877, p. 6; *Southern Argus* (Port Elliot), 1 July 1897, p. 3; Hofstede L, Dziminski MA (2017) *Australian Mammalogy* **39**, 227–237.

[13] In 1875–77 severe drought conditions extended across southern Queensland, South Australia and western New South Wales; *Gippsland Times*, 6 February 1875, p. 4; *Western Star and Roma Advertiser*, 1 January 1876, p. 3; *Colac Herald*, 25 February 1876, p. 3; *Queanbeyan Age*, 29 March 1876, p. 2.

[14] *The Kyogle Examiner*, 9 April 1926, p. 3.

[15] Brown managed Kallara Station through some of its most challenging years, particularly the rabbit plagues of the 1880s–90s. He was also apparently no friend to shearers: during the strikes of 1891 and 1894 he became immortalised in the lines: 'and we'll starve them to submission / said your great Kallara Brown.'

[16] *The Riverine Grazier* (Hay), 10 November 1883, p. 2; *The Sydney Morning Herald*, 14 August 1884, p. 4.

[17] E.g. *The Australasian Sketcher with Pen and Pencil* (Melbourne), 21 March 1889, p. 39; *Bathurst Free Press and Mining Journal*, 17 September 1889, p. 3; *The Sydney Mail and New South Wales Advertiser*, 1 September 1888, p. 437; *The Telegraph* (Brisbane), 27 September 1888, p. 2; *The Mercury* (Hobart), 21 May 1888, p. 2.

[18] *Sydney Morning Herald*, 14 December 1889, p. 7; 17 March 1888, p. 14; *Adelaide Observer*, 25 August 1888, p. 11; *South Australian Register* (Adelaide), 11 December 1888, p. 7; *The Australasian* (Melbourne), 26 May 1888, p. 1138.

[19] *The Richmond River Herald and Northern Districts Advertiser*, 21 March 1890, p. 4.

[20] *The Sydney Mail and New South Wales Advertiser*, 23 August 1890, p. 403.

[21] *Australian Town and Country Journal* (Sydney), 2 November 1889, p. 14; 30 November 1889, p. 13; *Riverine Grazier* (Hay), 29 November 1889, p. 1; *South Australian Chronicle* (Adelaide), 28 September 1889, p. 12.

[22] *South Australian Register* (Adelaide), 19 October 1888, p. 3; 3 September 1878, p. 6; *Queanbeyan Age*, 24 March 1914, p. 3; *The Brisbane Courier*, 11 October 1892, p. 6.

[23] *The Queenslander* (Brisbane), 22 October 1892, p. 1.

[24] I based this calculation on the stated loss of carrying capacity that occurred on stations in western New South Wales and, assuming that nine rabbits is equivalent to one dry sheep and a range in 1892 roughly bounded by the Eyre Peninsula, northern central South Australia, Charleville in south-west Queensland, the eastern Cobar peneplain region, the eastern Riverina and north-west Victoria (see Stodart E, Parer I (1988) Colonisation of Australia by the rabbit *Oryctolagus cuniculus* (L.). Canberra, Commonwealth Scientific and Industrial Research Organisation).

[25] Cooke BD (1982) *Australian Wildlife Research* **9**, 465–476.
[26] *The Advertiser* (Adelaide), 7 January 1892, p. 5.
[27] *South Australian Chronicle* (Adelaide), 9 January 1892, p. 10.
[28] *The Bendigo Independent*, 25 January 1904, p. 6; *The Sydney Morning Herald*, 21 November 1905, p. 4.
[29] *The West Australian* (Perth), 15 January 1901, p. 7.
[30] *The Sydney Morning Herald*, 23 December 1890, p. 3; 24 December 1890, p. 5; *Chronicle* (Adelaide), 19 September 1896, p. 41; *The Brisbane Courier*, 12 October 1892, p. 3; *The Riverine Grazier* (Hay), 25 October 1892, p. 2.
[31] Cooke BD (1982) *Australian Wildlife Research* **9**, 465–476.
[32] Accounts of the impacts of rabbits on shrubs can be found in: *The Sydney Mail and New South Wales Advertiser*, 9 July 1892, p. 66; *The Brisbane Courier*, 18 May 1892, p. 7.
[33] *The Queenslander* (Brisbane), 8 October 1892, p. 693; *Goulburn Herald*, 20 May 1892, p. 4; *The Brisbane Courier*, 18 May 1892, p. 7.
[34] *The Queenslander* (Brisbane), 5 August 1905, p. 31; This phenomenon was also seen at Kanowna, WA, in 1905 (*Kalgoorlie Western Argus*, 14 November 1905, p. 12).
[35] *The Sydney Morning Herald*, 9 April 1902, p. 6.
[36] Hamilton WD (1971) *Journal of Theoretical Biology* **31**, 295–311.
[37] Berdahl AM *et al.* (2018) *Philosophical Transactions of the Royal Society B: Biological Sciences* **373**, 20170009.
[38] Helbing D, Farkas I, Vicsek T (2000) *Nature* **407**, 487–490.
[39] *South Australian Register* (Adelaide), 18 September 1895, p. 5.
[40] *The West Australian* (Perth), 27 May 1898, p. 2.
[41] As immortalised in AB Paterson's 'Hay and Hell and Booligal' (1896).
[42] *The Riverine Grazier* (Hay), 21 October 1892, p. 3.
[43] *Evening News* (Sydney), 19 June 1902, p. 3.
[44] *The Riverine Grazier* (Hay), 21 October 1892, p. 3.
[45] *Kilmore Free Press*, 1 January 1903, p. 3.
[46] *The West Australian* (Perth), 27 May 1898, p. 2.
[47] *Evening News* (Sydney), 21 April 1896, p. 5.
[48] *Adelaide Observer*, 3 March 1900, p. 2.
[49] This interesting phrase was used by the journalist and politician Edward Davis Millen to describe country near Menindie (Menindee) in Millen ED (1899) 'Our western lands'. *The Sydney Morning Herald*, 25 November 1899, p. 5.

14
Niobe's ruin

Does Nature take revenge on man's audacity?
– The Brisbane Courier, *23 June 1900, p. 8*

ON DECEMBER 31, 1894, THE GOVERNMENT ASTRONOMER, HENRY Chamberlain Russell, was asked by a reporter from the *Daily Telegraph* about his views on the weather experienced over the past year across New South Wales. In a world where most people learnt of weather conditions across the colony from sporadic agricultural news reports or accounts of visitors to the country that read much like romantic travelogues, this was a most anticipated event, for Russell was a minor celebrity, a bastion of objectivity and synthesis in a disparate world. This year, however, Russell had little to report: rainfall over most of the colony had been 'satisfactory';[1] general rains had fallen in October, leading to flooding along the Murray,[2] and the 653 735 acres (264 557 ha) of wheat grown across the colony were expected to yield a healthy 11.96 bushels per acre (Fig. 14.1).[3] In short, there was nothing to suggest that anything other than favourable conditions might be on the horizon.

In 1900, just 5 years after Russell's interview, *The Brisbane Courier*[4] published what I think sits comfortably among the most compelling accounts of drought in Australian history. It was written by the great journalist and explorer Archibald Meston based on his experience in western Queensland, which was held fast the grip of the worst drought since European settlement, a 'malignant Titan' that stretched across the entire continent.[5] The calamity that he witnessed – vast sandy wastes, livestock in the thousands lying dead or trapped in an agony of thirst and famine – would have left many lost for words. But Meston was no ordinary traveller: for him, the passages that best described this Odyssean realm of death and shadows[6] were found in the works of Kipling, Swinburne, Will Alexander,[7] Shelley's 'Revolt of Islam' and 'Prometheus unbound', and above all Byron's 'Childe Harold's pilgrimage'. Reading his account, which is humbly titled 'The Western Drought', is as much a tour of melancholy, suffering and human fragility as it is of the drought-affected land itself.

> *... the sky became*
> *Stagnate with heat, so that each cloud and blast*
> *Languished and died; the thirsting air did claim*
> *All moisture, and a rotting vapor passed*
> *From the unburied dead, invisible and fast.*
>
> – *Percy Bysshe Shelley*, 'Revolt of Islam', Canto X

Fig. 14.1: Scenes of agricultural progress. Top left: harvesting a rich crop at Canning Downs Station on the Darling Downs, Qld in 1894, the year before the onset of the Federation Drought. National Library of Australia. Top right: the rain follows the plough: Green Hills Farm, Darling Downs, 1894. National Library of Australia. Bottom left: optimism ran high in the early days of western pastoralism as stations improved their infrastructure. Note the soil drift in the lee of the posts of these stockyards in western New South Wales, now filled with rabbit holes. Bottom right: bullock and horse teams arriving and departing from the Moree rail yard, ca. early 1900s. NRS-17420 State Rail Authority Archives Photographic Reference Print Collection (NRS-17420-2-39-[39]), NSW State Archives.

Perhaps Meston's most poignant insight was to describe the West as the 'Niobe of Australia', who sat 'in her voiceless woe, an empty urn within her withered hands, her dead children lying around her, far scattered over one vast, unbroken scene of desolation'.[8] Romantic allusions aside, the fact that the vitality of the inland, once brimming with human hubris, had been utterly shattered and left both physically and metaphorically barren gives you some indication of the immense and intractable crisis visited on the newly formed nation by the great drought that gripped the continent during the 1890s and early 1900s. The language Meston uses tells us that he could never have anticipated that a drought of such severity could ever strike Australia, let alone conceived of the scale of the agricultural and ecological destruction that it left in its wake.

It is often said that the calamitous state of the inland rangelands during the Federation Drought came as great shock because most people, even the weathered souls of the 'outback', had never witnessed such an event. I have reservations about this claim because antecedents did exist, namely the 1860s drought in South Australia and lesser but still severe droughts in 1870s and 1880s (Chapters 11, 13). Nevertheless, memory of these events was fading, and the Federation Drought undoubtedly ranks among the most extreme and protracted in Australian history and worse than anything since at least the 1840s. The eighth year, 1902, was the worst, especially in eastern Australia, when large areas received less than one-third of average rainfall, although rainfall deficiencies even extended into 1903 in central New South Wales and Queensland (Fig. 14.2).

Foreseeable or not, the drought, as we will see, proved devastating to Australian society and environments in ways that cannot be explained by rainfall alone. To understand why, we must draw on the domain of complexity theory and consider these impacts as a 'black swan' event or, in more formal language, a temporal conjunction of exceptional and unpredictable circumstances that led to radical change in both the physical system and human understanding of it. We have already met with the first of these exceptional events – the rabbit plagues of the 1880s–90s – but the parallel explosion in sheep and cattle numbers was an even more important factor in most areas, itself driven by unusual conditions in economic and labour markets. Let us find out what went wrong.

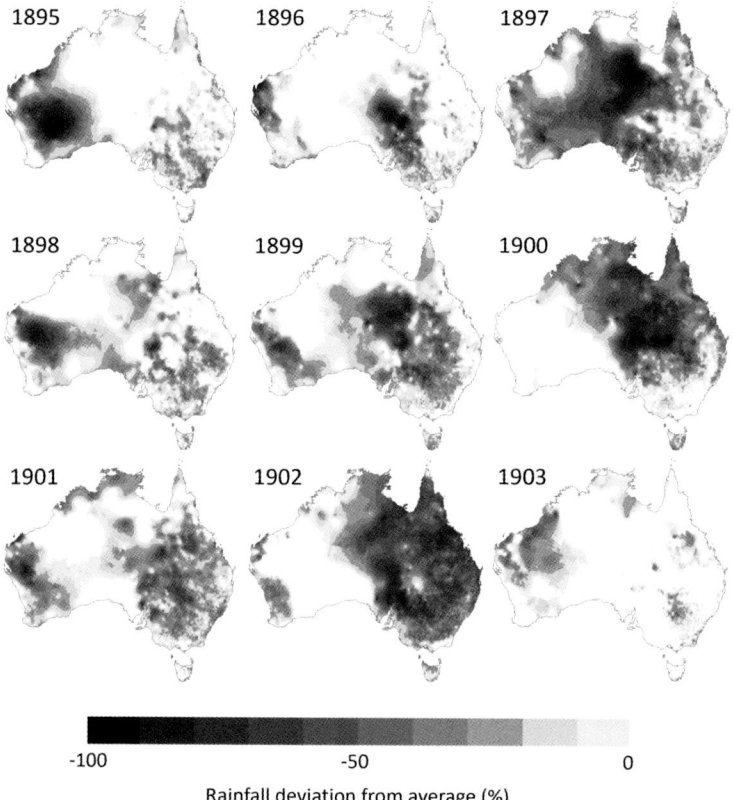

Fig. 14.2: Rainfall during the Federation Drought, with deviation from the long-term (1889–2022) average shown in grayscale. Areas with above average rainfall are shown in white. Data from more remote areas should be viewed with caution.

STRANDED

The period between 1885 and 1892 saw a massive run-up in sheep numbers across much of Australia. This was stimulated by the general profitability of the wool industry, a run of favourable years and expansion of pastoralism into semi-arid rangeland country, and probably at certain times to compensate for low wool prices caused by an increase in supply both domestically and from overseas producers,[9] international tariffs on wool imports,[10] and rising wages and disruptions caused by a union movement among shearers.[11] The increase over this period was staggering, with 30.2 and 11.0 million head added to already overburdened pastures in New South Wales and Queensland respectively. In Western Australia, where growth had been slightly slower, the high-water mark was reached in 1891 with some 2.5 million sheep, while in southern and eastern Australia it occurred in 1892–93, when 61.8 million animals were present in New South Wales (8 million west of the Darling), 21.7 million in Queensland, 13.2 million in Victoria, and some 765 000 in South Australia. Cattle numbers had also grown by 49% in New South Wales (14.3 to 21.3 million) and 45% in Queensland (4.3 to 6.2 million) over the same interval (Fig. 14.3). Optimism was high that the potential would continue to grow as station infrastructure, particularly fencing and water provision, was pursued with an almost religious zeal.[12]

The boom came abruptly to an end in 1891–93 when two critical factors intervened. The first was the collapse of the 1870s–80s economic boom and development of a deep financial depression, and the second the onset of drought conditions, which, as we have seen, spread across central and northern Australia into central and western New South Wales. In Western Australia, the great drought of 1890–92 struck the north-west pastoral zone from the Gascoyne and Murchison to the Pilbara, killing a quarter of the state's flock,[13] and also the Kimberley, where station hands were kept busy pulling starving cattle out of waterholes (Fig. 14.4). In South Australia sheep numbers declined by around 500 000 due to drought, rabbits and wild dog attacks.[14] Severe drought also struck some 700 000 km^2 of grazing country from the Barkly Tableland south to the Flinders Ranges and east to southern Cape York, killing hundreds of thousands of cattle and sheep, including core breeding animals.[15]

In the doomed lands of western New South Wales where rabbits were now 'king of the country',[16] pastures failed, and since squatters lacked any capacity to store feed on a sufficient scale to cover the shortfall, cattle and sheep began to join the millions of rabbits in death.[17] An 1892 parliamentary visit to Paddington Station, 100 km south-west of Cobar, to investigate the crisis was greeted by the skeletons of sheep, cattle, horses, kangaroos and emus whitening the ground. In one of the great examples of Australian understatement, the devastated country that they had passed through was considered 'not of a pleasant nature'.[18] Small wonder, then, that inducements were apparently required by local politicians before they would venture into the ravaged terrain, a situation decried by local station owners.[19]

Despite the carnage, in 1895 New South Wales, Queensland and South Australia were still home to a combined 80 million sheep and almost 10 million cattle (Fig. 14.3). Unfortunately, dry weather that had been incubating in central Gippsland during 1894 began to spread with a rapidity that stoked fears that a more general drought might be taking hold.[20] These

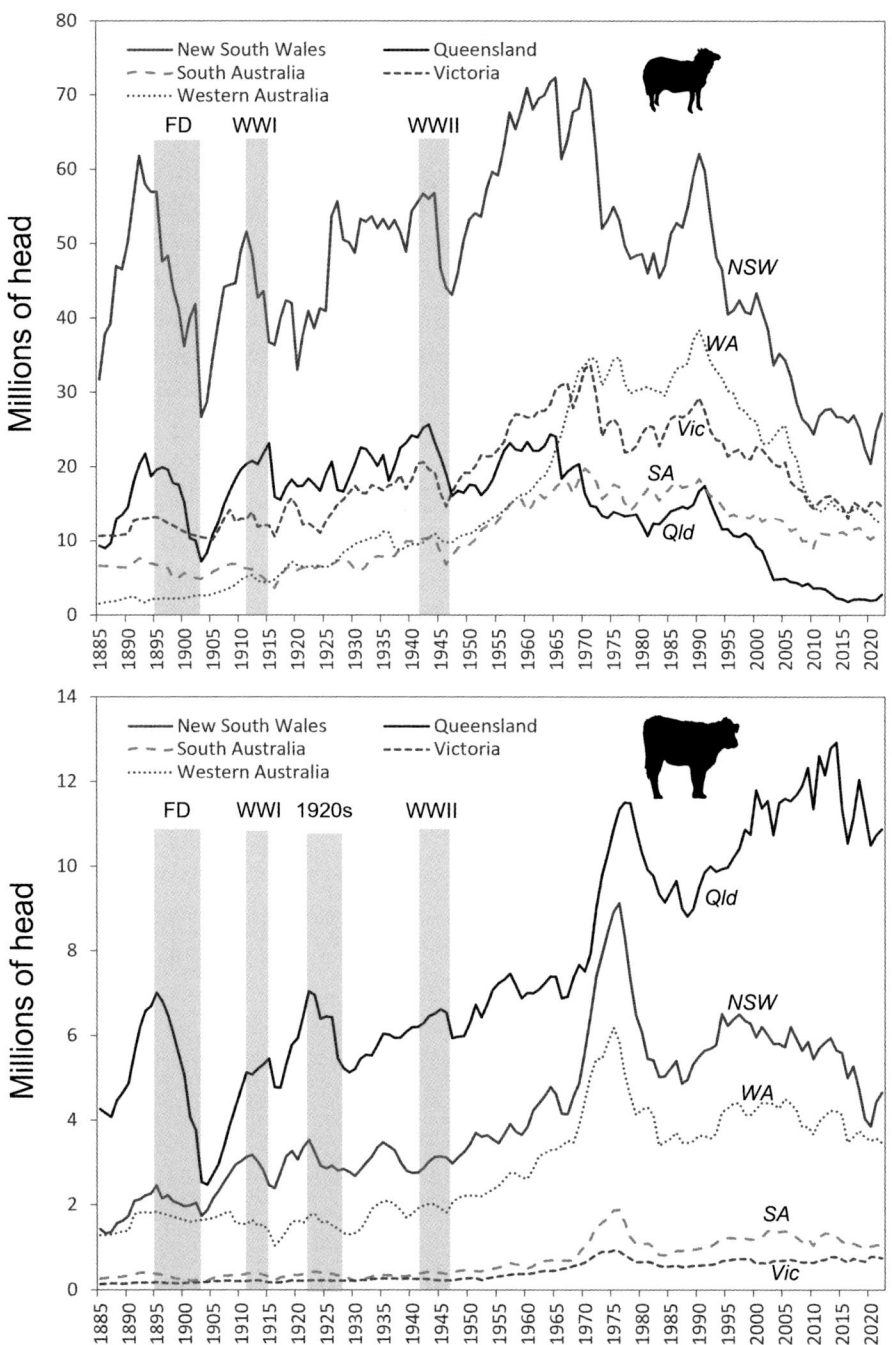

Fig. 14.3: Numbers of cattle and sheep in Australia, 1885–2022. Major droughts are shown in grey shading: FD = Federation Drought, WWI = World War I, WWII = World War II. Data from Meat and Livestock Australia Market Information.

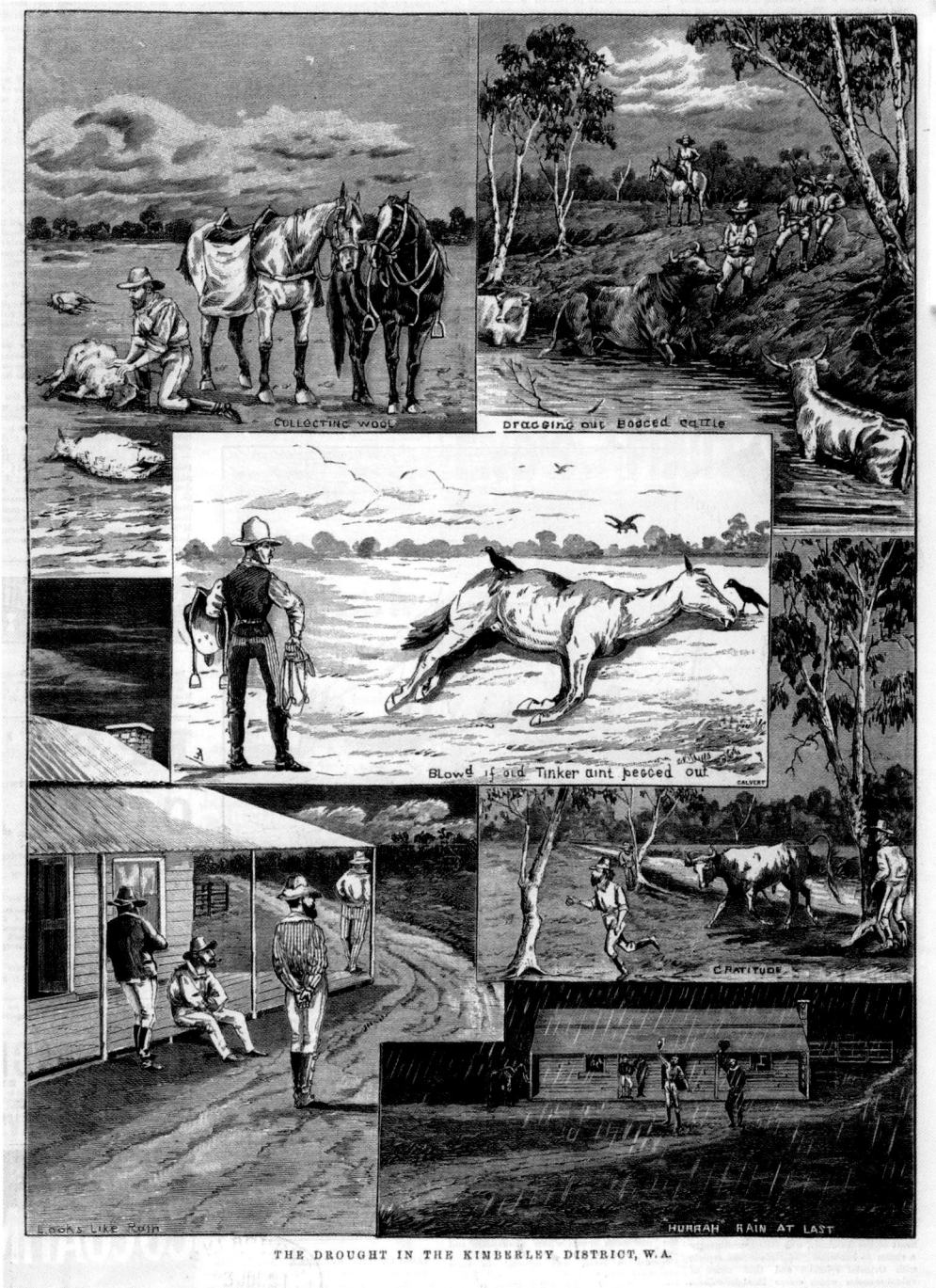

Fig. 14.4: Calvert, S. (1891) *The drought in the Kimberley District, W. A.* [print: wood engraving]. David Syme and Co., Melbourne. Source: State Library of Victoria, IAN 01-07-91 P.13, description in *Illustrated Australian News*, 1 July 1891.

worries proved well founded, for by the middle of 1895 water and pasture were running short across inland New South Wales, south-western Queensland, virtually all of Victoria and south-eastern South Australia.[21] The winter was harsh: extreme cold froze waterholes and creeks, burst pipes and brought low-level snow across the south-east of the continent, and by September tens of thousands of sheep were beginning to die from malnourishment and exposure.[22]

Fig. 14.5: By 1897 the railway network was well developed in the south-east, with major railheads in central New South Wales and Queensland serving the western inland. However, there were still vast regions of the remote interior that were completely dependent on the network of travelling stock routes for transportation. When these failed the consequences for inland stations were catastrophic. National Library of Australia, nla.obj-231828963. Inset: during the great drought livestock owners felt squeezed by duties placed on the railway freight of fodder. From The squeeze of the grazier. *Clarence and Richmond Examiner*, 2 December 1902.

The speed at which the drought took hold caused the rapid loss of pastures on overstocked properties. Indeed, although water deprivation is often perceived as the major challenge associated with drought, it is often food supplies that 'give out' first, particularly in hilly country, areas with cold climates, land adjacent to major river systems, or on land serviced by bores. Prior to the advent of mass livestock transportation this posed a diabolical problem on remote stations, but graziers situated closer to the railways or bullock and teamster routes could buy in fodder, albeit at often exorbitant prices. By the late 1890s rail transport was available to Hay, Condobolin, Bourke and Moree in New South Wales, to Charleville, Barcaldine and Hughenden in Queensland and to Oodnadatta in South Australia (Fig. 14.5), and during the drought these were major hubs for feed distribution to the outer districts (Fig. 14.1). During 1902 the traffic was enormous: between April and July some 10 663 tons of fodder sourced from Victoria passed through Corowa and thence onto the New South Wales rail network for the drought-stricken areas.[23] The imposition of duties on rail transport of fodder by the newly formed Federal Government proved highly contentious (Fig. 14.5), but at least many were able to secure feed this way.[24]

Fig. 14.6: Top left: MacFarlane, J. (1889). *A cruel death* [print: wood engraving]. State Library of Victoria, IAN 12-01-89 P.5. Top right: Melbourne: Alfred Martin Ebsworth. (1886). *A waterhole during the drought* [print: wood engraving]. State Library of Victoria, AS 24-08-86 P.132. Bottom left: *Carcases of cattle at the Gum Holes, Bowra Station, ca. 1900*. John Oxley Library, State Library of Queensland, negative number: 36333. Bottom right: Melbourne: Alfred Martin Ebsworth. (1886). *A reminiscence of the drought in central Australia* [print: wood engraving]. State Library of Victoria, AS 27-07-86 P.124. Such scenes were common across inland Australia during the Federation Drought.

For those in the vast tracts of the remote interior the choice was stark: either wait and hope for rain – a strategy doomed to fail during a decade-long drought – or take their chances on the 'long paddock' (public reserves for stock movement) and attempt to drive their starving animals to better feed. Most of these stock routes and roads followed traditional Aboriginal pathways, the routes of European explorers, or informal tracks along major watercourses.[25] It is a strange geographical oddity of inland Australia that the headwaters of the northernmost catchments of the Eyre Basin all lie in the tropical and subtropical rainfall belts and, despite receiving generally low annual rainfall (300–400 mm), often receive sufficiently heavy falls during early spring and summer to cause major flows down the Burke, Hamilton, Thomson, Barcoo and other northern rivers. During the 1870s–90s such events provided enough water for drovers to move enormous numbers of cattle into the fertile floodplain and channel country that awaited European settlement in the 'three rivers country' of south-western Queensland. But this is a true desert region, and these actions left hundreds of thousands of cattle dependent on rainfall hundreds of kilometres away for water, and on sufficient local rain to provide grass for movement between river systems. It was a house of cards doomed to collapse.

By 1900, after 5 years of drought, the situation in Queensland had evolved into the worst ever seen. Virtually every river west of the Great Dividing Range had been reduced to sparse chains of stagnant waterholes or had dried completely, and few if any stock routes had any feed. In the three rivers country hundreds, and even thousands, of cattle were bogged in individual holes, reaching in vain for the thin skin of stagnant water that retreated daily from reach. Multiple rows of carcases in every stage of decomposition lined the edges; on the Gregory River the dead were packed so closely that their remains formed stepping stones for others seeking water (Fig. 14.6).[26] Those still alive waited in dull despair for the end, forced to endure attacks by crows – those 'loathed ministers of pain and fear'[27] – which stood on their backs, pecking at eyes and flesh (Fig. 14.6).[28] Travellers to that country reported hearing the roaring of bogged cattle over vast distances,[29] a chorus of the damned in a land forgotten by God.[30]

Desperate squatters moved around the stock routes like players on a demented chess board (Fig. 14.7). Most pushed for greener pastures closer to the coast: in Queensland, the Gulf Country, the Barkly Tableland,

Fig. 14.7: Melbourne: David Syme and Co. (1893). *Travelling cattle in search of water* [photomechanical reproduction: halftone]. State Library of Victoria, IAN 01-02-93 P.1.

Adelaide and coastal areas serviced by the central and northern railway lines were major destinations, but many stock routes further inland were closed due to lack of feed and water.[31] Others drovers simply followed the rumour of feed to keep some animals alive, the more enterprising seeking out long-disused roads in the hopes that none had passed there before.[32] In one famous example, the drover Dave Coleman walked a mob of some 5000 sheep, which became known as 'The Israelites', across 5000 miles (ca. 8000 km) of Queensland, an indeed epic journey that took many years.[33] In Victoria, huge mobs of sheep arrived in Gippsland and the south-west, where conditions had been more favourable, while in New South Wales the hungry mouths were driven to the Snowy Mountains region and the north-east coast and hinterland.[34]

But vast numbers died in transit, being too weak to walk, encountering waterless stretches too wide to cross, or suffering from eye diseases caused by dust irritation that resulted in fever, starvation and death.[35] Indeed, even Sidney Kidman, one of the greatest pioneers of managing natural disasters in Australian history[36] (Box 14.1) is said to have lost every one of 2000 bullocks put on the road at Carcoory, near Birdsville, looking for feed and water, and an entire mob of 1000 breeding cows heading for Pandie Pandie in South Australia. In another famous disaster, the drover Jack Clarke was able to save only 70 bullocks out of 500 arranged for purchase by Kidman after a sandstorm and great heat struck the mob between Mungeranie and Cooper Creek, the eyes of the unfortunate beasts said to resemble balls of fire in the dust.[37]

In the arid wastes of New South Wales, South Australia and Queensland, the scenes in 1901 were so horrific that the landscape was compared unfavourably to Dante's Inferno.[38] Here, starvation was the great leveller: cattle as thin as rakes ate horse droppings in town streets,[39] sheep, cattle and horses consumed dead rabbits to get at the phosphorus-laced grain inside,[40] and mummified skeletons littered the ground as far as the eye could see, the carcases too desiccated to decompose.[41] In one bizarre scene, the Minister for Lands William Patrick (Paddy) Crick was greeted by a 'triumphal arch' festooned with the bones of dead sheep and cattle – victims of 'King Drought' – while touring drought-affected western New South Wales (Fig. 14.8).[42]

The Federation Drought caused what is almost certainly the greatest animal welfare tragedy ever to strike the pastoral regions of Australia. The collapse that occurred between 1892 and 1903, a decline of 50 million sheep and 5 million cattle in New South Wales and Queensland alone, has no parallel before or since. But it also caused great human suffering, from the impoverishment and starvation of settlers and Aboriginal people to the economic ruin and

---

Box 14.1: Sidney Kidman, king of risk

The massive scale of Australian droughts, which dwarfed even the largest stations, posed a nearly insurmountable problem for early pastoralists seeking prospects in the isolated reaches of central Australia. But necessity is the mother of invention, and Sidney Kidman, as a young man with considerable outback experience, set out in the 1880s to tackle the problem on a truly grandiose scale. Seeing an opportunity to raise cattle in Queensland and the Northern

Box 14.1: (Continued)

Territory to supply Adelaide, the eastern cities and even a growing refrigerated shipping market overseas, he and his brother Sackville began to buy stations in two great chains – one from Marree in South Australia into central and northern Northern Territory, the other north from Broken Hill and Marree to the channel country along Cooper Creek and the Diamantina and Georgina rivers. The task was made easier in the troubled 1890s as pastoralists succumbed to economic depression, rabbits and drought.

As the old saying goes, 'there is always a drought somewhere in Australia', but Kidman thought that by owning land across multiple climatic zones he would be able to weather the worst that nature could throw at him, an approach we today call *spatial resilience*. He was also a pioneer in grazing management, limiting stocking rates on his properties and using his network of station managers, drovers, Aboriginal people and many others to provide crucial information on conditions around his properties and opportunities for buying, selling and moving stock. His vast mobs were always on the move. By 1903 his empire sprawled over almost 100 000 km² of country, and it continued to grow until the time of his death in 1935 when he owned or controlled more than 60 stations from the northern tropics to the Flinders Ranges. Much of his success can be attributed to his multi-dimensional approach to managing climatic and economic risk, which was well ahead of its time.

It was not always clear sailing, however. During the late stages of the Federation Drought, his chain of 17 stations still had major gaps, and he lost as many as 70 000 cattle and more than half a million pounds of assets. During the continental-scale World War I (1912–15) drought he lost 85 000 cattle and 50 000 sheep, but the fact that this occurred over 50 stations stocked with vastly larger herds is now seen as vindication of his grand design.

 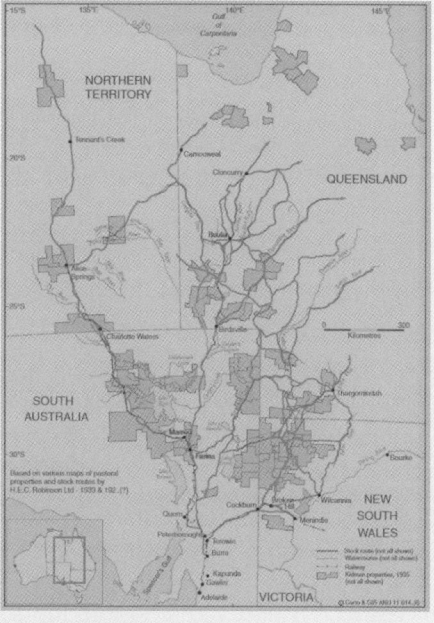

Left: Sidney Kidman in 1919. State Library of South Australia, BRG 57/5. Right: Sidney Kidman properties in 1935. Reproduced from Dobes L (2012) with permission from CSIRO Publishing.[36]

complete abandonment of stations as they became consumed by debt, rabbits, dust and sand. Many properties and even small towns were left stranded by the closure of the stock routes and cessation of coach runs, leaving people with few supplies apart from tea, sugar, flour, salt and beef the consistency of wood.[43] There was incredible poverty: in one sad tale a mallee labourer and his wife were forced to push their sick daughter in a perambulator *140 miles* (225 km) to Bendigo to seek medical treatment, being unable to afford the train fare.[44] There are countless such stories of misery, but, fortunately, also of hope – for out of the destruction grew the seed of generosity, with many people giving significant sums of money to establish drought relief funds and other charities for the needy across the country.

Fig. 14.8: An arch made of bones greets Minister for Lands William Crick on his tour of drought-affected country in western New South Wales in 1902. The image appeared in *The Sydney Mail and New South Wales Advertiser*, 31 May 1902.

We will revisit some of these themes in later chapters, but here I will leave you with a final image of the 'Great Drought'. In impoverished times people must go to great lengths to survive, and in western New South Wales and Queensland there were many men and few jobs. Those that were available were often gruesome, for although dead livestock were worth little, they weren't worth *nothing*. A trade sprung up in collecting the hides, horns and wool from carcases and men scavenged what they could where they could: one pair of men on Cooper Creek collected over 10 000 horns and 15 cwt (a long hundredweight, equal to 112 lb or 50.8 kg) of horse hair from dead cattle and horses.[45] Others followed mobs of sheep on the travelling stock routes, making £1 per day skinning sheep that dropped while walking, the carcases dragged into huge piles and burnt.[46]

Landless men and women also took to collecting fleeces from animals left dead on reserves and runs, which they would then sell to hawkers for a small sum of money. But, in a remarkable lack of charity, many were fined or sentenced to labour for not first seeking permission from the owner, the argument that the offenders saw little harm in selling wool from animals left dead on the ground falling on deaf ears. Wool collected from sheep carcases was known as 'dead wool', and I can think of no more fitting term.[47] About the only mercy was that sheep were often so desiccated that when they died they simply dried into a husk without going through the usual stages of decay.[48]

But let us end on a positive note. The systematic culling of sheep during the years of drought and depression led to an improvement in the quality of the remaining breeding sheep and, increasingly, a greater focus was placed on wool production from high yielding animals. The result of these agricultural advances was that in the aftermath of the Federation Drought wool production rose much more rapidly than sheep numbers (i.e. wool production per animal increased), which relieved pressure on the land.[49] And, although major losses of livestock did occur during later droughts (Fig. 14.3) and catastrophic events still plague grazing areas of Australia today, the worst excesses have been gradually mitigated by improvements in transportation networks, stocking management and animal welfare standards. Perhaps there is solace in that.

Endnotes

[1] *The Daily Telegraph* (Sydney), 1 January 1895, p. 5.

[2] *The Daily Telegraph* (Sydney), 16 October 1894, p. 5; 16 October 1894, p. 5.

[3] *The Corowa Free Press*, 5 January 1895, p. 7.

[4] Meston A (1900) 'The Western Drought.' *The Brisbane Courier*, 16 June 1900, p. 13.

[5] *The Yass Courier*, 23 September 1902, p. 2.

[6] Meston described his journey to western Queensland as descending 'with some Ulysses into a realm of death and shadows, where Silence had spread her dark pavilion on the waste Deep'. *The Brisbane Courier*, 16 June 1900, p. 13.

[7] Namely, Kipling's '*The Ballad of East and West*', Swinburne's '*Hymn to Proserpine*', and Will Alexander's '*Towards the Primeval Lightning Field*'.

[8] *The Queenslander* (Brisbane), 23 June 1900, p. 1191.

[9] *Weekly Times* (Melbourne), 17 October 1885, p. 3; *The Maitland Mercury and Hunter River General Advertiser*, 4 July 1885, p. 19; *The Week* (Brisbane), 3 July 1886, p. 13.

[10] E.g. *Morning Bulletin* (Rockhampton), 28 February 1888, p. 5.

[11] The economic factors associated with the boom in livestock numbers are discussed in Boehm EA (1971) *Prosperity and depression in Australia 1887–1897*. Clarendon Press, Oxford.

[12] *Empire* (Sydney), 30 June 1873, p. 4.

[13] Bolton GC (2008) *Land of vision and mirage: Western Australia since 1826*. University of Western Australia Press, Crawley, WA, p. 64; *The West Australian* (Perth), 17 June 1893, p. 6.

[14] *Adelaide Observer*, 13 August 1892, p. 31; 5 November 1892, p. 32; *Border Watch* (Mount Gambier), 7 May 1892, p. 4; *Evening Journal* (Adelaide), 28 May 1892, p. 6.

[15] *Australian Town and Country Journal* (Sydney), 26 November 1892, p. 14; *Evening Journal* (Adelaide), 28 May 1892, p. 6; *The Brisbane Courier*, 24 February 1893, p. 6; *Mackay Mercury*, 10 January 1893, p. 3; *Queensland Times, Ipswich Herald and General Advertiser*, 2 November 1893, p. 3; *Gympie Times and Mary River Mining Gazette*, 1 March 1894, p. 3; *The Western Champion and General Advertiser for the Central-Western Districts* (Barcaldine), 6 March 1894, p. 6.

[16] *The Daily Telegraph* (Sydney), 4 February 1892, p. 5.

[17] *The Sydney Morning Herald and New South Wales Advertiser*, 20 August 1892, p. 412; *Freeman's Journal* (Sydney), 12 March 1892, p. 15.

[18] *The Sydney Morning Herald*, 2 July 1892, p. 10.

[19] *The Daily Telegraph* (Sydney), 30 May 1892, p. 5.

[20] *The Snowy River Mail and Tambo and Croajingolong Gazette*, 12 January 1895, p. 3; *The Argus* (Melbourne), 8 February 1895, p. 6; *Cootamundra Herald*, 23 March 1895, p. 8; *Evening News* (Sydney), 13 March 1895, p. 5; *National Advocate* (Bathurst), 10 April 1895, p. 2.

[21] *The Sydney Mail and New South Wales Advertiser*, 4 May 1895, p. 890; *The Maitland Daily Mercury*, 20 May 1895, p. 4; *The Australasian* (Melbourne), 13 July 1895, p. 6; *Adelaide Observer*, 13 April 1895, p. 5.

[22] The extreme cold outbreaks that occurred in July 1895 probably rank as among the most severe to hit south-east Australia since European settlement; *The Express and Telegraph* (Adelaide), 22 July 1895, p. 2;

The Australasian (Melbourne), 27 July 1895, p. 34; *The Daily Telegraph* (Sydney), 10 July 1895, p. 5; *The Daily Northern Argus* (Rockhampton), 3 July 1895, p. 3; *Warwick Examiner and Times*, 25 September 1895, p. 3.

[23] *The Corowa Free Press*, 11 July 1902, p. 3.

[24] *The Daily Telegraph* (Sydney), 15 August 1902, p. 7; *Evening News* (Sydney), 26 May 1902, p. 4.

[25] Spooner PG, Firman M, Yalmambirra (2010) *The Rangeland Journal* **32**, 329–339.

[26] *Examiner* (Launceston), 5 July 1900, p. 8; *Yass Evening Tribune*, 14 January 1901, p. 4.

[27] *The Brisbane Courier*, 16 July 1900, p. 13. The quote is an adaptation of 'We are the ministers of pain, and fear' from Percy Bysshe Shelley's *Prometheus Unbound*.

[28] *The Shoalhaven Telegraph*, 1 May 1901, p. 12.

[29] *Gympie Times and Mary River Mining Gazette*, 22 November 1900, p. 4.

[30] *Yass Evening Tribune*, 14 January 1901, p. 4.

[31] *The Sydney Mail and New South Wales Advertiser*, 19 April 1902, p. 968.

[32] *The Sydney Stock and Station Journal*, 5 January 1900, p. 5.

[33] The precise details vary in different newspaper reports; see *The North Western Courier* (Narrabri), 25 July 1929, p. 9; *The North-Western Watchman* (Coonabarabran), 29 January 1942, p. 4; *The Richmond Herald and Northern Districts Advertiser*, 30 April 1909, p. 8.

[34] *The Western Port Times and Phillip Island and Bass Valley Advertiser*, 3 October 1902, p. 2; *The Narracoorte Herald*, 25 November 1902, p. 1; *The Daily Telegraph* (Sydney), 23 December 1901, p. 4; *The Brisbane Courier*, 7 June 1902, p. 14.

[35] *Western Mail* (Perth), 10 March 1900, p. 8.

[36] Dobes L (2012) *The Rangeland Journal* **34**, 1–15.

[37] Bowen J (2007) *Kidman: The forgotten king*. HarperCollins Publishers, NSW, pp. 110–113.

[38] *The Shoalhaven Telegraph*, 1 May 1901, p. 12.

[39] *Adelaide Observer*, 31 October 1896, p. 29.

[40] *Queensland Times, Ipswich Herald and General Advertiser*, 24 October 1896, p. 4; *The Gundagai Independent and Pastoral, Agricultural and Mining Advocate*, 1 April 1905, p. 2.

[41] *Yea Chronicle*, 5 June 1902, p. 4.

[42] *The Sydney Mail and New South Wales Advertiser*, 31 May 1902, p. 1368.

[43] *The Sydney Mail and New South Wales Advertiser*, 7 July 1900, p. 8.

[44] *The Bendigo Independent*, 23 December 1902, p. 2.

[45] *Bairnsdale Advertiser and Tambo and Omeo Chronicle*, 29 November 1900, p. 4.

[46] *The Armidale Chronicle*, 7 December 1898, p. 2.

[47] *Wagga Wagga Advertiser*, 20 June 1903, p. 2; *Yass Evening Tribune*, 11 September 1902, p. 3.

[48] *Chronicle* (Adelaide), 23 November 1944, p. 27.

[49] Boehm (1971), p. 94.

15
Crucified

The history of every nation is eventually written in the way it uses its soil.
— *Arthur Calwell,* The Herald *(Melbourne), 1944*[1]

PERHAPS NO GREATER IMAGE OF THE GREAT PAROXYSM OF SOIL EROSION that gripped eastern Australia in the mid-1940s, at height of the World War II drought, has ever been produced than Russell Drysdale's *Crucifixion* (1946) (Plate 14). The painting, which depicts a huddled ragged figure and the skeletal remains of trees in a nightmarish red and yellow eroded landscape, was inspired by his journey around western New South Wales in November 1944, when the region was in the grip of one of the driest periods in recorded history. Joining special reporter Keith Newman of *The Sydney Morning Herald* at the behest of Warwick Fairfax, he witnessed a landscape of massive sand drifts, dust storms, and the carcases of livestock and trees that had died in 'an agony of thirst'. Crucifixion, then, seemed a fitting way to describe not only the brutalisation of the land but the metaphorical erosion of 'human lives and human hearts'[2] caused by the ravages of drought and a European agricultural system unwilling to recognise the fundamental limits of the Australian environment.[3]

Drysdale's macabre sketches caused a considerable stir in urban circles and did much to inflame concerns that inland Australia was threatened by a dust bowl much like the one that had only recently struck the Great Plains of the United States.[4] The regular passage of major dust storms over coastal cities, which peaked during 1944–45, and record heatwaves (e.g. in 1939) reinforced these tensions, and there was genuine concern that the desert was rapidly moving east and encroaching on ever more productive country, a fate that had befallen myriad older civilisations.[5] Newspapers followed with sensational articles like 'How to save this continent', 'Half the country is in the air!', and 'Blood suckers in the cities',[6] and although sand dunes from the arid zone were not actually threatening to consume eastern Australia, as some of the more alarmist articles suggested, the problem of soil degradation and desertification was real enough.[7] The issue was finally pursued in earnest by key politicians like Arthur Calwell, whose epiphany came after witnessing a dust haze on a flight between Sydney and Melbourne so thick that it 'blotted out visibility' at 7000 feet (2134 m) altitude,[8] and, after much political jockeying, ameliorative programs to end soil erosion finally gained major support at the federal level.[9]

Australian politicians have a long and depressing history of reacting to major ecological crises long after they have become instantiated in the landscape. The terrible aeolian erosion (from the Greek word *Aiolos* or *Aeolus*, the god of the winds, and *aiolos*, which means 'changeful' or 'shifting') witnessed in the 1940s is no exception, for at that time it was not a new phenomenon but the endpoint of a series of degradational episodes that had begun more than a half-century earlier.[10] As we have seen, thanks to a series of droughts combined with overgrazing by growing numbers of sheep and rabbits, anthropogenic dust storms began to sweep across eastern Australia during the early to mid-1880s, and these seem to have intensified significantly during the acute drought year of 1888, affecting pastoral country from northern South Australia to the Riverina.[11] The tendency of sandy soils to drift following cropping and overgrazing was also becoming obvious in some areas, particularly in the sandhill country of the lower Murray region and along the Darling River near Wilcannia, where blinding sandstorms and drifts 3 feet (1 m) deep occurred in 1882.[12]

A strong case can be made that by the onset of the Federation Drought in the mid-1890s an Australian version of a 'dust bowl' had become firmly established across the semi-arid and arid south-eastern interior, and that this lasted until the mid-1940s, and, in central Australia, until the 1960s.[13] At the time the American scientist RG Bowman, who had experience with the US 'Dust Bowl', seems to have thought that erosional features of the drought-stricken country in south-eastern Australia were not strictly comparable to those seen in the US 'Dust Bowl',[14] and the scale, causes and soil types affected certainly had a uniquely antipodean flavour. But the aetiology of both disasters also had much in common: in each the destruction of fragile vegetation followed by the onset of multi-year drought conditions had caused the mobilisation and drift of sand and soil particles during colossal dust and sandstorms, which in turn led to the ruin and dislocation of agricultural people dependent on those systems for income. In this sense both events were the modern manifestation of the ancient problem of land degradation, but commensurate with the enormous scale of Western-style agriculture in the colonial new world.

LOSING GROUND

The dust bowl never ceases, never sleeps, never waits, never lets up, and is expanding all the time.
– The Daily Telegraph *(Sydney), 3 December 1946*

In Plate 15 I have reconstructed the spread of severe 'dust bowl' conditions in eastern and central Australia over this period, focusing on newspaper accounts of large drifts of sand and soil that blocked roads or railway lines or buried fences, dams, irrigation channels and even homesteads. The results of this work indicate that by the end of the Federation Drought in 1903 two main hotspots had emerged, the first of which was the belt of semi-arid cropping country from the West Coast district of South Australia to the Mallee and Wimmera districts of Victoria. Here the primary problem was the clearing of native vegetation, comprising mainly mallee eucalypts and a variety of shrubs and grasses, on sand hills and sandy loam soils.[15] Sand dunes in this region, which are of aeolian origin and formed under drier conditions in the past, have limited water-holding capacity and low cohesion when dry and are thus very susceptible to wind erosion.[16]

Inside Goyder's Line in South Australia the failure of crops on sandy substrates caused significant localised drift, but the problem was much worse in the drier agricultural Hundreds, which became a sea of 'sand and sorrow' in which the entire topsoil was blown away to the depth of the furrows, smothering farmhouses.[17] In the lower Murray and north-west Victoria 'Mallee snow storms' stripped the bark from gum trees, cut the earth from around fences and trees, and stripped sand off dunes, depositing it on roads, railways, and in the growing network of irrigation channels used to transport water from the Wimmera storages to farmers downstream (Fig. 15.1).[18] The magnitude of soil loss on the sand hills was such that trees were left suspended in the air, held up only by their roots (Fig. 15.2).

In contrast to the cropping belt, soil degradation in the semi-arid and arid rangelands of south-west Queensland, north-eastern South Australia and western New South Wales began with overgrazing and trampling of exposed soils by sheep and cattle, which in turn caused structural breakdown and loss of nutrients from the soil profile, erosion by wind and water, and salinisation. The degree to which different soil types were susceptible depended strongly on their physical and chemical properties and position in the landscape. In general, heavily textured alluvial grey cracking clay and black earth soils dominated by Mitchell grass, spear grasses, saltbush, bluebush and other grassland or shrubland species were either minimally impacted upon or were able to recover.[19] Fortunately, vast areas of such country exist across the inland plains and have always been among the most favoured by farmers and graziers.

But virtually any country dominated by sand or sandy loams was at risk of blowing away, and many of the starkest Australian 'dust bowl' landscapes ever witnessed occurred during the droughts of the 1920s and 1930s, when record dry conditions gripped the northern pastoral country of South Australia, south-west Queensland and north-west New South Wales (Plate 1). Gradually, the inland evolved into a great inland sea of shifting sand, the dormant dunes reawakening like great red siliceous slugs (Fig. 15.3). Driving sandstorms waged a constant battle against outposts like Innamincka and Farina[20] as well as stations throughout the arid zone; some clung to survival at the very margins of viability, while many others were abandoned to the ever-encroaching drift (Fig. 15.3).[21] In New South Wales, raging windstorms tore up the ground, causing damage to fences and dams on Sidney Kidman's stations estimated at £50 000.[22] Drifting sand and soil covered the dingo and rabbit-proof fences on the New South Wales–Queensland border, allowing an influx of dingos to prey on sheep that had not already died of starvation, thirst or suffocation.[23]

In flatter country enormous 'scalds' formed in sandy loams and sandy clay loams when an upper 2–4 inches (5–10 cm) deep A horizon of light surface soil was stripped from the land by wind and water, exposing a clay-based B horizon below.[24] When exposed to heavy raindrops that fragment large surface aggregates into smaller particles that clog and seal the soil pores, scalded areas develop a hard crust that is impervious to water.[25] Vast areas of these scalds can still be seen today on 'blown-out' country, since without intensive rehabilitation these are,

Of life there is nothing. It is a wilderness of abysmal loneliness.
– Captain Frank Hurley, on the condition of country around Broken Hill, 1928. Barrier Miner, *1 November 1928, p. 1*

Fig. 15.1: Drift in the mallee country of Victoria and South Australia caused much damage to infrastructure and ongoing costs to remove sand. Top left: drifts 10 feet deep on the Brownport Highway, south of Mildura. Top right: drifts were reportedly 50 feet high near this house in the Mildura district. Centre left: sand drift encroaching on a house in the Mildura–Euston region. This and the prior two photographs were published in *The Illustrated London News*, 27 January 1945, and caused quite a stir in Australia. Centre right: fence covered in drift near Walpeup, Vic. A second fence has been built on top of the first. State Library of Victoria, rwls/u487. Bottom: sand being removed from the Mittyack Railway Line, Victoria, after a storm. Report of Committee Appointed to Investigate Erosion in Victoria, 16 February 1938.

Fig. 15.2: Sand drift became a serious problem after native mallee vegetation had been cleared on sand hills and other sandy soils and drought killed the crops and pastures that replaced it. Top left: newspaper article from *The Mail* (Adelaide), 27 August 1932 showing tree roots exposed by drifting sand near Veitch, SA. Top right: soil loss near Veitch, SA. From: Australia's peril: Dangers of soil erosion. *Chronicle* (Adelaide), 21 December 1944. Bottom left: satellite image of typical mallee sandhill country near Veitch, SA, showing an uncleared patch of vegetation in the top left and cleared dunes below (Google Earth, 15 March 2023). Bottom right: mallee roots exposed to a depth of over 2 m, probably in the Euston–Mildura district. From: Australia's 'dust-bowl' problem. *The Illustrated London News*, 27 January 1945.

for all practical purposes, permanent features of the landscape (Plate 16).[26] Although clay-rich soils are not normally prone to drift, they can do so when denuded and pulverised by stock, and some of the worst wind erosion was experienced on the grey, brown and red clay (vertosol) country of the Riverina.[27]

Fig. 15.3: Top left: sandhills near William Creek, 60 km west of Lake Eyre, 1928. Dunes began to grow significantly in some areas during the 1920s and 1930s, due to successive severe droughts and grazing. Photo: Michael Terry, National Library of Australia, nla.obj-149197852. Top right: Lake Harry Station, 70 km south-east of Lake Eyre, ca. 1938. State Library of SA, B 77568/28. Centre left: sand deposited during a single sandstorm, Miranda Station, SA, 1919. Photo: Herbert Basedow, National Museum of Australia, 1985.0060.2799.001. Centre right: yards at Parachilna, SA, engulfed in drift sand, ca. 1938. The Parachilna Hotel is visible in the left background. State Library of SA, B 77568/20. Bottom left: drought at Bedourie, western Queensland, ca. 1929. Note the artesian bore on the right and the sand drift in the background. State Library of Queensland, IE297759. Bottom right: windswept country near Broken Hill, 1938. From *Walkabout*, 1 November 1938, National Library of Australia, nla.obj-566923190.

After the Federation Drought, when much of the initial soil degradation occurred,[28] a battle raged between recovery during wet years, which were few, and retraumatisation during dry years, which were many. The drift problem intensified in the north-western districts of Victoria and arable parts of South Australia, especially during 1914–15, mainly due to insufficient use of fallowing to ensure that subsequent crops have sufficient stored soil water to withstand drought during the growing season. On older country, this was attributed to the prevalence of a 'get rich quick' mentality among farmers, but on recently cleared land it was equally linked to the need to kill re-emerging mallee shoots by growing and then burning off crops over several successive years.[29] As is so often the case, the perennial need to service debt was usually at the root of the problem.[30]

Among the first serious efforts to deal with the problem was the creation of the South Australian *Sand Drift Act 1923*, which allowed government and councils to establish windbreak reserves and other remedial actions on drift-affected land, and to act against landowners responsible for sand drift on their properties. By 1933 wind erosion and sand drift in the mallee country was costing hundreds of thousands of pounds in channel- and road-clearing and train-derailment expenses, which the Victorian Sand Drift Committee blamed on indiscriminate clearing of vegetation, thoughtless cultivation of sandy areas and the lack of comparable legislation.[31] Under worsening drought in the mid- to late 1930s and early 1940s the situation could no longer be ignored and, after years of inaction and perceived propaganda,[32] numerous state-level initiatives were finally established to document the gravity of the problem,[33] establish recovery strategies,[34] and undertake scientific investigation and engagement.[35] Early research by Francis Ratcliffe, of the newly created Council for Scientific and Industrial Research, and later the establishment the *Journal of Soil Conservation Service of NSW* and other forums ensured that science would play and increasingly influential role in decision making.

THE DEVIL'S CAULDRON

I will show you fear in a handful of dust.

– TS Eliot, 'The Wasteland'

Although the long-term soil degradation that took place during the 'dust bowl' era would become one of the greatest challenges to agricultural and environmental sustainability in Australia, I think that it was, above all, the dust storms and great heat that occurred during this era that made life nearly unbearable for farmers and 'townies' alike. Now, both have always been a feature of the Australian inland – major dust storms and heatwaves have struck regularly and appear to have been on the increase over the past two decades – and there is much concern that both will become more frequent and severe in the future as the climate changes. But the sheer intensity, frequency and impact of dust storms during the 1890s–1940s was incredible, and several heatwaves among the worst ever experienced. I can but give you a brief account of some of the greatest of these events, hoping that little is lost in the retelling.

One thing that has struck me about the great dust storms of this era is that they were almost of transcendental significance to the souls of those who were forced to endure them. This often began in the hours preceding the storm itself: the air was very close, and the atmosphere was often described as taking on a strange hue, sometimes accompanied by dust being carried high aloft by northerly 'brickfielder' winds, engendering a sense of uneasiness in those venturing outside.[36] But then, as the winds swung to the south-west, a huge cloudbank would appear in the distance, a roiling wall of dust extending from horizon to horizon (Plate 17). These often towered hundreds of metres above the ground, sweeping over entire mountain ranges,[37] and even capable of downing planes.[38] The clouds of dust most often resembled huge plumes of smoke; one that approached Narrandera, NSW on 14 March 1900 was 'rolling and curling in a tremendous density, like the smoke of a gigantic prairie fire'.[39] So impressive was this storm that those present said that it was 'magnificent in sight', and that 'nothing approaching it in grandeur was ever witnessed here'. The colours were almost kaleidoscopic, shifting hue as different soil substrates were picked up by the wind, but closer to the arid centre storms invariably arrived with an eerie bright red glare or light.[40] I have only ever seen one such dust storm, west of the Pilliga Forest, NSW in 2019, and I can understand why they were often spoke of in biblical terms.

When the worst dust storms finally arrived, they plunged the land into an inky blackness; even in the middle of the day the sky was engulfed in an 'awe-inspiring' gloom as dark as night.[41] The dust was often so thick it was impossible to see across the street and headlights would penetrate the murk for only for a few feet, forcing the suspension of foot and motor traffic and lighting to be turned on even inside shops.[42] This total 'eclipse' of the sun could last for hours, before setting in a sky that looked as if on fire. At times it felt apocalyptic, imbued with a sense of impending calamity, and there are many reports, perhaps with some degree of embellishment, of women being terrified to the point of hysterics as the sun was blotted out. Thousands of birds could be seen driven on the wind before the storms, and horses, cattle and sheep stampeded, many ultimately collapsing to suffocate in the dust before being buried.[43] For those people caught in the barren wastes the consequences could be serious, with many becoming lost, including children. Most were found,[44] but not everyone: in 1919 one man was found dead from thirst and exhaustion 10 miles (16 km) from Wilcannia after becoming lost in a dust storm.[45]

In the wake of the storms dust coated everything, indoor and out, enough, it was said, to break the heart of housewives. Fine dust penetrated even the smallest nooks and crannies. I can give you no better account than this passage, written in 1924: 'A very fine dust began to fall, and so fine was the dust that it fell quietly and imperceptibly. The dust fell and crept into every wrinkle and crevice. It filled our eyes, ears, nostrils, and it crept into the rolls of our swags and pockets. It covered our hats and saddles, and it looked like flour sprinkled upon our hair, upon the horses' manes and tails, and behind the bullock's ears. The dust was everywhere. Nothing could keep it out. Even the water in the water-bag was covered with a fine scum from the dust, and our parched lips with a dusty slime. The heat was intense, the air suffocating, and the perspiration streamed down our faces in muddy beads.'[46] Even dinner wasn't safe: 'Now, mates, roll up quick and lively, or the soup'll be silted up afore yer gets to it'![47]

There were many great dust storms during the 'dust bowl' era, but I think that one surpassed all others in sheer *impossibility*, one that if passed down only in the oral record we might dismiss as a fanciful retelling of actual events. This storm had its birth in South Australia on the morning of 12 November 1902, the year of the 'Black Drought',[48] raising thick dust in Adelaide and the drought-stricken agricultural land to the east. By midday it was passing over Ballarat and

Melbourne, before reaching a peak during the afternoon across northern central Victoria. The storm was preceded by heat and hours of gale force winds, damaging houses and blowing down trees, before a great cloud bank arrived from the south-west.[49] As it reached its peak towns across the region were consumed by darkness, people fainted, and as light rain began to fall it did so as mud, infusing tanks with a light yellow-red hue.

So massive was the storm that it crossed southern New South Wales overnight, and Sydney in the morning of the 13th, where the dust was so thick that the harbour shoreline was obscured to boat traffic and a pall still visible 12 miles out to sea,[51] before finally petering out in northeast New South Wales and the Darling Downs late in the evening. Over the next day the dust crossed the Tasman Sea, with vessels reporting the 'unusual experience' of having encountered a dust storm at sea, before finally reaching the South Island of New Zealand.[52] The scale of the dust storm alone puts it among the greatest ever witnessed in Australia, and comparable to the 'Red Dawn' dust storm of September 2009 (Plate 17), but I think it is the strange accounts that came from the area between Wycheproof and Albury on the afternoon of 12 November that cement it as a truly incredible event.

Here, just before the arrival of the storm, lit matches seemed to resemble tiny jets of acetylene gas, as noticed by many people, and when the storm arrived, 'fireballs' were reported falling to the earth in showers of sparks. This was probably caused by the electrification of airborne dust and sand, a phenomenon reported in other dust storms,[53] and subsequent igniting of organic matter lifted from the soil, but whatever the cause they were significant enough to be blamed for igniting houses and even timber in the shaft of the New Barambogie Mine near Chiltern. These events, and a general red glow in the heavens caused many people to genuinely fear that the world was about to end, consumed by fire.[54] If any general conclusion can be drawn from these accounts, it is that during unprecedented climatic events complex phenomena can arise that are almost impossible to foresee, which, given that the earth's climate is rapidly in the process of transforming into a new and poorly understood state, does not fill one with confidence.

In summer, the emotional strain caused by dust storms was usually augmented by ferocious heat, with temperatures often peaking between 100°F (37.8°C) and 113°F (45.0°C) under northerly winds.[55] Before one great storm struck Hay, NSW on 26 February 1900, for example, scorching hot winds that had developed by 9 am turned into a blast furnace of 112°F (44.4°C).[56] But still hotter conditions were experienced during the 'dust bowl' years, for lacking any moisture and hence evaporative cooling the land was primed for heatwave conditions. Then, as today, heatwaves occurred in most years, but two events stand out. The first occurred

Like a mighty conflagration
Enveloping the nation!
And we cry in suffocation,
Oh, the dust–dust–dust!

Tom Black, 'The dust storm', *1909*[50]

> **Box 15.1: The 1895–96 heatwave**
>
> The heatwave that prevailed during October 1895–January 1896 ranks as the most devastating, protracted, continent-wide period of heat experienced since European settlement. In eastern Australia the fiercest temperatures set in immediately after the new year and persisted almost until the end of the month. Exceptionally hot temperatures were measured across New South Wales and southern Queensland; at Bourke, the temperature reached 40°C every day between 4 January and 25 January, peaking at 48.6°C. The nights were stifling, with Brewarrina recording an unbearable 42.8°C at midnight. Most temperatures recorded during this heatwave are biased by a degree or two above modern standard measurements, but the heat was extreme by any standard, and the worst ever experienced in the colony.
>
> Hundreds of people died from sunstroke and heat 'apoplexy', and virtually all inland towns, especially in northern New South Wales, suffered considerable loss of life. The elderly and very young were worst affected; in Goulburn 12 infants are said to have died from the heat, and another 10 in Albury. In Bourke dozens of people passed away, some collapsing outdoors, others expiring despite receiving medical attention, and such was the suffering that there was some panic among the residents, with many fleeing to cooler regions by train. During the heatwave the death rate in Sydney doubled, and there was great demand for graves and funeral trains. The heat was so great that horses and other livestock were struck dead across the inland, an event which only occurs under the most severe of conditions.
>
> The two greatest heatwaves of the 'dust bowl' era, 1895–96 and 1939, now rank among the most lethal natural disasters in Australian history, and there is concern that under climate change rising temperatures are again posing a growing threat to human health.
>
>
>
> *The Bird O' Freedom* (Sydney), 25 January 1896.

in October 1895–January 1896, a terrible event that killed 435 people (Box 15.1; Fig. 15.4). The second took place in January 1939, when 420 people were killed, and ranks as the most severe in New South Wales history. Many records were set which remain unbroken today, including the highest temperature every recorded in New South Wales, 50.0°C at Wilcannia on 11 January 1939.

Fig. 15.4: Top: the 1895–96 heatwave gave the word 'inferno' a whole new meaning. *Melbourne Punch*, 30 January 1896. Bottom: record temperatures were experienced across eastern Australia during the 1939 heatwave. *News* (Adelaide), 16 January 1939.

At Narrabri the temperature reached 40°C or more for 11 straight days (7–17 January 1939), 5 days of which reached 45°C or more, with a peak of 47.5°C. To put this in perspective, there have only been 8 days of 45°C or more since 1939 in Narrabri – 2 in 1940, 1 in 1943, 3 in 1952 and 1 each in 2014 and 2017.[57] Interestingly, the notable gap in very hot temperatures from the 1950s until the 2000s reflects a broader mid-century decline in maximum temperatures observed in central and eastern New South Wales during this period, probably due to the sustained increase in rainfall.[58] When I was a university student, the 1939 heatwave was still legendary across the inland, but today this memory is fading, lost in the dusts of time, because few people who witnessed it remain alive. Furthermore, with improvements in technology and health care, and a reduction in severe heatwaves during the wetter decades of the 1970s–90s, the past century has seen heat-related deaths trending downwards, and although events like the January–February 2009 heatwave caused similar overall mortality, the rate per capita was vastly higher during the 'dust bowl' era.[59]

END OF AN ERA

In eastern Australia, the long dry period that had prevailed since the 1890s gradually began to relinquish its grip in 1947, and then emphatically with the onset of extremely wet conditions in 1950 and 1955–56 (Fig. 12.1). This is not to say, however, that landscape degradation did not continue, nor that the mistakes of the past were not repeated. As I have mentioned, central Australia experienced a terrible drought from the late 1950s until 1965 (Fig. 12.1; Plate 1), and with a near-doubling of cattle numbers since 1946, pasture deterioration and soil erosion again emerged as serious challenges.[60] When this drought shifted into western New South Wales and Queensland during the mid-1960s (Plate 2) at a time when farmers were struggling under an increasing debt burden and low commodity prices, the health of the grazing country, and especially the mulga lands, again declined, due to overgrazing and pressure on woody fodder plants. This in turn led to extensive wind and water erosion and then invasion by woody weeds during subsequent wet years, particularly in the 1970s. The return of severe drought during the 2000s and 2010s has shown that there can be no permanent victory over soil and ecosystem degradation, and you might be forgiven for wondering if any gains have been made at all (Plate 16).

Nevertheless, the response to soil erosion and pasture decline during the 1890s–1940s ushered in a 'golden age' of research and conservation, in which dust storms began to decline, broad-scale soil reclamation projects and education programs were introduced, animal welfare standards improved, and sophisticated climatic, ecological and agricultural models that greatly increased the ability of land managers to plan for extreme drought events were developed. The events of the 1960s reinforced the need for a deeper understanding of pasture maintenance and carrying capacities in highly stochastic semi-arid ecosystems, and formed a crucial platform in the emergence of Australia as a global leader in rangeland research and management.[61] Although undoubtedly aided by improved climatic conditions, reduced rabbit populations and technological improvements, these gains could not have occurred without a real shift in thinking and determination against ongoing economic, social and adversity, and deserve much admiration.

Endnotes

[1] *The Herald* (Melbourne), 18 December 1944, p. 7.

[2] *Molong Express and Western District Advertiser*, 16 February 1945, p. 2.

[3] One of Newman's many reports on the drought was published in *The Sydney Morning Herald*, 16 December 1944, p. 5. Themes in Drysdale's painting are discussed in Klepac L (1996) *Russell Drysdale*. Murdoch Books, North Sydney.

4 *The Canberra Times*, 17 November 1944, p. 2.

5 *The Daily Telegraph* (Sydney), 17 September 1946, p. 8; *Warwick Daily News*, 5 November 1936, p. 8.

6 Idriess I (1944) *The Sydney Morning Herald*, 2 December, p. 9; 'Martingale' (1944) *Western Mail* (Perth), 30 November, p. 45; *Molong Express and Western District Advertiser*, 16 February 1945, p. 2.

7 Ratcliffe FN (1937) 'Further observations on soil erosion and sand drift, with special reference to south-western Queensland'. Pamphlet No. 70, Commonwealth of Australia, Council for Scientific and Industrial Research.

8 *The Herald* (Melbourne), 18 December 1944, p. 7.

9 *Westralian Worker* (Perth), 1 February 1946, p. 7.

10 McKeon GM et al. (2004) *Pasture degradation and recovery in Australia's rangelands: Learning from history.* Queensland Department of Natural Resources, Mines and Energy.

11 *The Herald* (Freemantle), 25 March 1882, p. 2; *The Daily Telegraph* (Sydney), 1 May 1886, p. 9; *The Australasian* (Melbourne), 28 June 1902, p. 1507; *The Port Augusta Dispatch, Newcastle and Flinders Chronicle*, 6 January 1888, p. 3; 17 January 1888, p. 3; *The Express and Telegraph* (Adelaide), 27 January 1888, p. 3; *The Australian Star* (Sydney), 13 November 1888, p. 6; 24 December 1888, p. 6.

12 *South Australian Register* (Adelaide), 12 June 1886, p. 6; *The Herald* (Freemantle), 25 March 1882, p. 2.

13 Cattle SR (2016) *Aeolian Research* **21**, 1–20.

14 *The Land* (Sydney), 2 May 1941, p. 2.

15 Johns GG, Tongway DJ, Pickup G (1984) Land and water processes. In *Management of Australia's rangelands*. (Eds GN Harrington, AD Wilson and MD Young) p. 35. CSIRO, Melbourne.

16 Beadle NCW (1948) *The vegetation and pastures of western New South Wales: With special reference to soil erosion.* Thomas Henry Tennant, Sydney, p. 55.

17 *Adelaide Observer*, 26 February 1898, p. 2; *Chronicle* (Adelaide), 25 January 1898, p. 20; *Evening News* (Sydney), 23 August 1898, p. 5; *The Port Augusta Dispatch, Newcastle and Flinders Chronicle*, 26 August 1898, p. 3.

18 *Adelaide Observer*, 19 June 1897, p 3; *The Horsham Times*, 19 December 1902, p. 4; *The Ballarat Star*, 20 January 1903, p. 1.

19 Johns GG, Tongway DJ, Pickup G (1984) Land and water processes. In *Management of Australia's rangelands*. (Eds GN Harrington, AD Wilson and MD Young) pp. 25–40. CSIRO, Melbourne.

20 *The Advertiser* (Adelaide), 9 June 1936, p. 16; *The South Eastern Times* (Millicent, SA), 19 June 1928, p. 1.

21 *The Register* (Adelaide), 19 February 1926, p. 14; *The Advertiser* (Adelaide), 1 March 1929, p. 22; *Recorder* (Port Pirie), 14 October 1931, p. 3; *Australian Christian Commonwealth* (SA), 13 December 1929, p. 7.

22 *News* (Adelaide), 15 February 1929, p. 8; *The Queenslander* (Brisbane), 12 May 1900, p. 899; *The Daily Telegraph* (Sydney), 15 October 1901, p. 5; *The Express and Telegraph* (Adelaide), 14 November 1902, p. 2; *Border Watch* (Mount Gambier), 21 May 1902, p. 1; *The Sydney Morning Herald*, 5 October 1904, p. 6; *Dubbo Dispatch and Wellington Independent*, 9 March 1900, p. 4.

23 *Sydney Mail*, 30 June 1920, p. 36; *The Daily Telegraph* (Sydney), 14 May 1921, p. 13; *Observer* (Adelaide), 26 November 1927, p. 7.

24 *The Queanbeyan Observer*, 10 October 1902, p. 2.

25 Johns GG, Tongway DJ, Pickup G (1984) Land and water processes. In *Management of Australia's rangelands*. (Eds GN Harrington, AD Wilson and MD Young) pp. 25–40. CSIRO, Melbourne, p. 31; Warren JF (1965) *Australian Geographer* **9**, 282–292.

26 *South Australian Register* (Adelaide), 8 July 1897, p. 5.

27 Beadle (1948), pp. 55–56.

28 New South Wales. Royal Commission to Inquire into the Condition of the Crown Tenants (1901) *Western Division of New South Wales Royal Commission to Inquire into the Condition of the Crown Tenants.* Government Printer, Sydney.

29 Drift into irrigation channels and across roads and railway lines was a major problem; *Western Champion* (Parkes), 4 February 1915, p. 13; *The Ballarat Courier*, 30 January 1915, p. 10; *Minyip Guardian and Sheep Hills Advocate*, 6 July 1915, p. 2.

30 *Weekly Times* (Melbourne), 6 May 1916, p. 44.
31 Sand Drift Committee, State Rivers and Water Supply Commission, Melbourne, Victoria (1933) 'Report on Sand Drift Problem in Mallee Areas'. H. J. Green, Government Printer, Melbourne.
32 *Macleay Argus*, 11 February 1936, p. 3.
33 Victoria. The Soil Erosion Committee, Department of Lands and Survey, Melbourne (1938) 'Report of Committee Appointed to Investigate Erosion in Victoria'. H. J. Green, Government Printer, Melbourne; *Daily Mercury* (Mackay), 22 March 1947, p. 6; *The Dalby Herald*, 31 May 1938, p. 3; *Macleay Argus* (Kempsey), 11 February 1936, p. 3.
34 *The Sydney Morning Herald*, 23 April 1936, p. 11.
35 *Newcastle Morning Herald and Miner's Advocate*, 11 November 1933, p. 8; *The Henty Observer and Culcairn Shire Register*, 2 July 1937, p. 3; *Manilla Express*, 20 November 1934, p. 4; *The Herald* (Melbourne), 18 December 1944, p. 7.
36 *The Horsham Times*, 1 March 1898, p. 3.
37 *The Advertiser* (Adelaide), 20 March 1900, p. 6; *Wellington Times*, 3 January 1924, p. 4.
38 *The Albany Advertiser*, 8 February 1945, p. 5.
39 *The Daily Telegraph* (Sydney), 15 March 1900, p. 6; *The Corowa Free Press*, 30 December 1941, p. 3; *Balonne Beacon* (St. George), 7 February 1914, p. 3.
40 *The Riverine Grazier* (Hay), 22 January 1915, p. 2; *The Scrutineer and Berrima District Press*, 15 April 1896, p. 5.
41 *The Scrutineer and Berrima District Press*, 15 April 1896, p. 5.
42 *The Riverine Grazier* (Hay), 27 February 1900, p. 2; 21 November 1944, p. 2.
43 *Tweed Daily* (Murwillumbah), 22 January 1915, p. 3; *The Herald* (Melbourne), 13 November 1902, p. 2; *Hamilton Spectator*, 5 January 1909, p. 4; *Darling Downs Gazette* (Qld), 22 January 1915, p. 4.
44 *Shepparton Advertiser*, 13 June 1929, p. 2; *The Daily Telegraph* (Sydney), 5 January 1899, p. 3; *Maryborough Chronicle, Wide Bay and Burnett Advertiser*, 28 December 1937, p. 7; *Burra Record*, 3 March 1926, p. 3.
45 *The Daily Mail* (Brisbane), 26 December 1919, p. 3.
46 *The Sydney Morning Herald*, 24 May 1924, p. 13.
47 *Traralgon Record*, 28 April 1899, p. 3.
48 *Quambatook Times*, 28 January 1914, p. 3; *The Forbes Advocate*, 29 November 1929, p. 5.
49 *The Argus* (Melbourne), 14 November 1902, p. 6.
50 *Port Lincoln, Tumby and West Coast Recorder*, 26 May 1909, p. 2.
51 *The Argus* (Melbourne), 14 November 1902, p. 6.
52 *The Age* (Melbourne), 17 November 1902, p. 5.
53 Zhang H, Zhou Y-H (2020) *Nature Communications* **11**, 5072; Esposito F et al. (2016) *Geophysical Research Letters* **43**, 5501–5508.
54 *Australian Town and Country Journal* (Sydney), 19 November 1902, p. 12; *The Bendigo Independent*, 14 November 1902, p. 3; *Queensland Times, Ipswich Herald and General Advertiser*, 22 November 1902, p. 7; *Evening News* (Sydney), 15 November 1902, p. 6.
55 *The Riverine Grazier* (Hay), 21 November 1944, p. 2; *Australian Town and Country Journal* (Sydney), 1 March 1902, p. 15; *The Charleville Times*, 8 November 1940, p. 5; *Daily Advertiser* (Wagga Wagga), 15 January 1938, p. 6; *The Forbes Advocate*, 9 January 1942, p. 4; *Sunraysia Daily* (Mildura), 18 December 1934, p. 5.
56 *The Riverine Grazier* (Hay), 27 February 1900, p. 2.
57 Based on data from Narrabri Bowling Club, Narrabri West Post Office, and Narrabri Airport AWS.
58 Nicholls N, Della-Marta P, Collins D (2004) *Australian Meteorological Magazine* **53**, 263–268.
59 Coates L et al. (2014) *Environmental Science and Policy* **42**, 33–44.
60 Condon RW, Newman JC, Cunningham GM (1969) *Journal of the Soil Conservation Service of New South Wales* **25**, 47–92; Chippendale GM (1963) *Journal of the Australian Institute of Agricultural Science* **29**, 84–89.
61 McKeon et al. (2004).

16
The desert of dry bones

The drought seems to have driven almost everything out of the country.
– Lawrence A Wells, surveyor of the Elder Exploring Expedition, 1891[1]

LIFE ON THE EDGE

In 1902 the scientist and explorer John Walter Gregory coined the phrase 'the dead heart' to compare the modern arid climate of the Lake Eyre Basin to the more fertile conditions that once supported herds of extinct megafauna there in the distant past.[2] It was a pithy description of the land he encountered, one of 'appalling scarcity of water' in which the bones of animals, both ancient and modern, littered the ground.[3] The name stuck, and through countless repetition in books and the popular media since, it has become deeply inculcated in the Australian psyche. I, like many an Australian child, grew up with a vision of the Red Centre as a vast wilderness of lost explorers, of mighty ergs of shifting sand and barren stony plains, and, above all, of desolation.[4]

Ecologists, of course, take issue with the concept of a 'dead heart', and justifiably so, for Australian deserts are not waterless or lifeless wastes – animal life is diverse, waterholes, soaks and springs dot the landscape and, apart from a few very restricted areas, rainfall is usually sufficient to allow drought-hardy vegetation to stabilise sand and thus prevent sand dune mobility.[5] In fact, the Australian arid zone is, by global standards, a relatively 'wet' desert.[6] Only around 10–15% of the continent averages less than 200 mm of rain annually, and even the very driest areas near Lake Eyre have a mean annual rainfall of 100–125 mm,[7] vastly more than the hyper-arid climates of the Atacama, Sahara, Namib, Gobi, Arabian and Mojave-Sonoran deserts, which receive less than 5–50 mm.[8] The closest analogue is probably the Kalahari desert, which is also comparatively well vegetated.[9]

Nevertheless, Australian deserts have been, and remain, brutal landscapes that seem perfectly calibrated to produce pain. For they are not a dead heart, but an arrhythmic one that pulses with the diastole and systole of death and renewal, as irregular and unpredictable as the rain. Here, where footholds are tenuous, drought is synonymous with desiccation and hunger, a 'great thirst land' in which death is often the most prevalent feature of the landscape.[10] There are many tales to tell of those who lived through such times, and those that did not.

> *[I']t is most unfortunate that the popular idea of a desert is very different from the scientific one, and usually takes the form of a wind-tortured Sahara of drifting sand ...*
> – HH Finlayson, *1935,* The Red Centre

A WRETCHED COUNTRY

I was not sorry to leave the wretched place, which we left as dry as the surrounding void.

– Ernest Giles, October 1873, near Warburton, WA[11]

We pick up our story in the early 1870s, when the arid zone was gradually being dissected like a great hide laid over the red sand to dry. Chastened by the harshness and poverty of the semi-arid pastoral districts, hopeful explorers were trying their luck in the vast unexplored regions of the central and western deserts, seeking undiscovered pastures or some inland drainage basin where water might make an unlikely appearance. In the west, pastoralists were pushing north along the narrowing coastal stock route to the Pilbara, where the dunes of the Great Sandy Desert reach the sea, and in Central Australia the construction of the Overland Telegraph Line, following in the tracks of Stuart, was completed in 1872. Cattle and sheep moving up from South Australia would soon be met by overlanders from Queensland *en route* to the vast Mitchell grass plains of the Barkly Tablelands.

Beyond these frontier regions we know relatively little about rainfall patterns before the mid-1880s, although extreme droughts likely occurred in the 1830s–40s and in 1864–66. James Foley, in *Droughts in Australia* (1957), indicates that late 1860s were extremely dry in the Pilbara and both western Queensland and New South Wales, and most deserts were probably also affected around this time.[12] After this, however, explorers began to encounter more favourable conditions. Ernest Giles reported good rock holes and springs during his expeditions through the MacDonnell Ranges in 1872 and over the Ayers and Musgrave ranges in 1873–74,[13] while William C Gosse, who travelled through the country north of Uluṟu to the Reynolds Ranges in 1873, and John Forrest, who passed through the Tomkinson, Mann and Musgrave ranges in 1874, both found the country well grassed, waterholes well supplied, and even some springs and flowing creeks.[14] They were lucky, for in later decades traversing these arid regions on horseback became difficult or impossible.[15]

But here droughts are never banished for long. An unfortunate party, led by the Lutheran Pastors Hermann Kempe and Wilhelm Schwarz to establish a mission on the Finke River at Hermannsburg, struck terrible conditions in 1875–76, and after battling sand, heat and thirst, the disintegration of one of their wagons and the loss of 1100 sheep, they struggled into Dalhousie Springs, where they were trapped until early 1877.[16] Rainfall was low across northern South Australia in 1880 and 1881 – Dalhousie Springs received only 3.28 inches (83 mm) of rain in these 2 years[17] – and again during late 1883 to early 1884, which caused a decline in wool production and loss of lambs across the northern pastoral regions. In 1883, William Whitfield Mills found the area around the Everard, Musgrave and Tomkinson ranges to be stricken by drought, with many native wells dry and wildlife scarce[18], and 3 years later the explorer David Lindsay found the eastern MacDonnell Ranges to be gripped by a 'great drought', with 'permanent' waterholes, along with the Todd River and Giles Creek, dry.[19]

Despite this, conditions were favourable enough of the time to underpin a speculative boom in large-scale cattle and sheep operations along the Overland Telegraph Line and across subtropical Northern Territory, and to facilitate ongoing pastoral expansion into the

Murchison, Gascoyne and Pilbara regions of Western Australia. The first cattle stations in central Australia, Owen Springs and Undoolya, were established in 1872 and stocked by 1873 and 1876 respectively,[20] and within 10 years many more had sprung up in north-western MacDonnell Ranges and along the Overland Telegraph. The years 1878–87 marked a crucial turning point in the commencement of the region's 'painful but permanent exploitation',[21] for by 1885 nineteen stations were also operating in the Victoria River and Barkly Tableland country.[22] The scale of these operations was simply colossal: the largest, Barrow Creek Pastoral Co., held a 25-year lease over 20 000 square miles of land (12.8 million acres) near Barrow Creek and took possession of 5000 cattle in 1883.[23] By this time there were already some 20 000 head in the southern Territory, and 22 000 more *en route* to stations further north.[24] By 1888 the pastoral districts of the Territory were carrying around 200 000 cattle and 100 000 sheep, a disproportionate number of which were heavily concentrated in the more marginal southern districts.

THE LAND HALF WON

> *This does not look like a country subject to drought; there is too much grass, which all appears to spring from strong roots.*
> *– Letter from Barrow Creek, Northern Territory, published in* The Queenslander *(Brisbane), 1872*

The pain began in the northern pastoral districts of Western Australia, where dry conditions that began in 1890 had, by the end of 1891, grown into the worst drought in decades, if not the history of the colony (Plate 2). Extreme conditions prevailed from the Kimberley region to Geraldton in the south, but the Murchison Region suffered the most, where overstocking greatly exacerbated the dry conditions, resulting in severe stock losses and land degradation (Fig. 14.4).[25] East of Geraldton sheep were dying as they were being shorn, and travelling horse and bullock teams died before reaching Millewa, less than 100 km to the west.[26] Stations from the Upper Irwin and Gascoyne to the Fortescue River lost thousands to tens of thousands of sheep,[27] and native species were no more fortunate, with kangaroos lying dead near waterholes and patches of dead or dying mulga, gum trees, spinifex and mallee scattered across the landscape (Fig. 16.1).[28]

Rainfall was better in 1893, but dry conditions soon returned to the Gascoyne and Pilbara. Stations south of Onslow received only 7–8 inches of rain (175–200 mm) in the 3 years ending March 1897, about one-quarter of the long-term average,[29] stations were abandoned in the pastoral country between the De Grey and Murchison rivers, and up to 100 miles (160 km) inland was rendered 'almost uninhabitable'.[30] Central and northern Australia fared little better, particularly in 1892–93 when the 'magnificently grassed' plains and 'permanent water' taken

> *The drought of 1890-1 will long be remembered with horror ...*
> *– HGB Mason,* The West Australian *(Perth), 25 November 1891*

Fig. 16.1: Left: Annean Station, south-west of Meekatharra, Western Australia, ca. early 1892. This photograph, taken by Frederick Elliott, medical officer and photographer of the Elder Scientific Exploring Expedition, illustrates the severe conditions that prevailed across the Murchison region during the 'Great Drought' of 1891–92. State Library of South Australia, B 499/88. Right: the Elder Scientific Exploration Expedition at Queen Victoria Spring, Great Victoria Desert, WA, 23 September 1891. National Library of Australia, nla.obj-895314314.

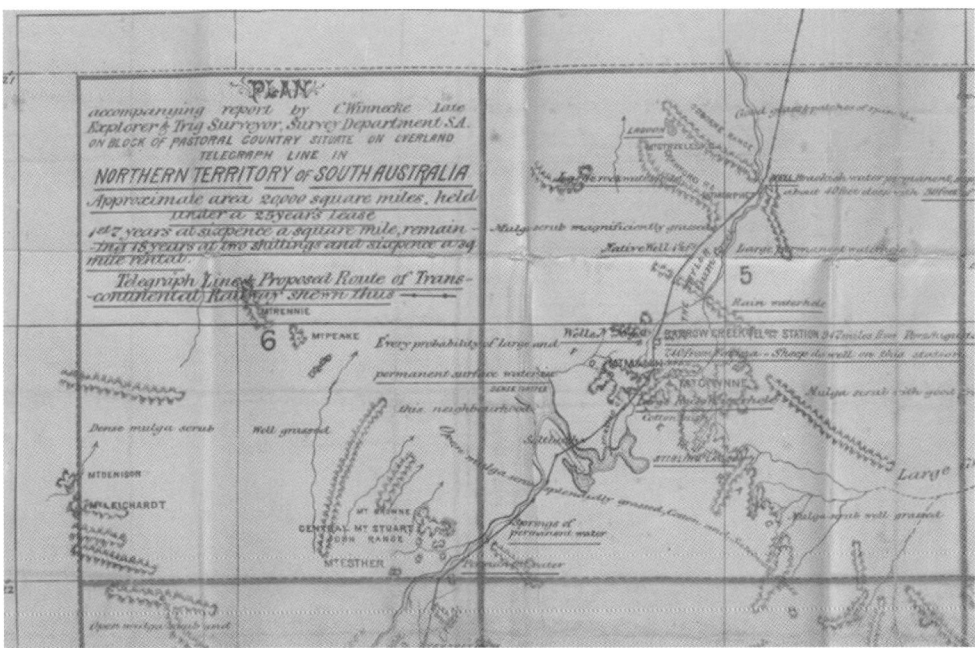

Fig. 16.2: Partial plan of country around Barrow Creek, ca. 1882. The map was based on a survey conducted by Charles Winnecke; the full map and accompanying report are provided in 'Report on pastoral country about Barrow's Creek Telegraph Station, Overland Telegraph, N. T., 20 000 square miles, 25 years' leases'. Frearson and Brother, King William Street, 1882. National Library of Australia, nla.obj-27895014.

up on the promise of early survey plans vanished (Fig. 16.2). Perhaps one-third of the cattle herds on the Barkly Tablelands died, including 5000 head on Alexandria Downs alone, and thousands were lost in the northern pastoral districts of South Australia.[31]

Droughts of varying severity then struck periodically until 1903, a great anhydrous ratchet that inexorably drove cattle operations encumbered by poor roads, high labour costs and

> **Box 16.1: Drought in arid zone ecosystems**
>
> During the most severe Australian droughts it is not uncommon to see patches of trees and shrubs dying from lack of moisture, and widespread dieback events in the south-east and south-west of the continent over the past two decades have received much public and scientific interest. Plant communities from the subtropics to montane ecosystems have been damaged, and these ecosystems may be threatened if climate change allows insufficient time for recovery.
>
> Interestingly, despite being populated by many species that are extremely tolerant of moisture stress, arid and semi-arid vegetation can also die during exceptionally dry periods. This is not a new phenomenon, since extensive dieback of trees and shrubs was observed in these environments as early as the 1840s and in 1891–1902, pre-dating the introduction of livestock. My colleagues and I have shown that during the Federation Drought much of the continent suffered vegetation dieback and decline of native animals. But probably the most extensive episodes in Central Australia occurred in the 1920s–40s, where photographic evidence shows entire woodlands of dead trees, and again during the great drought of 1958–65, when rangeland vegetation was significantly degraded.
>
> More recently, the severe drought of 2019–20 killed large areas of spinifex, mulga, desert oak and many other species that can be seen in the region around Uluṟu–Kata Tjuṯa National Park. These species are of great cultural significance to the Aṉangu people and critical to the biodiversity of these desert ecosystems, and research is being conducted into the longer-term implications of such events.
>
>
>
> Left: a woodland of *Eucalyptus coolabah* killed by drought on the Birdsville Mail Route, ca. 1930s. National Library of Australia, nla.obj-142575527. Right: desert oak killed by drought in 2019–2020, with Uluṟu in the background. Photo: Robert Godfree.

Aboriginal predation of stock ever closer to financial ruin.[32] It was a hazardous, depressing and above all unprofitable life, and 1890–94 saw the sale, closure or abandonment of many stations between Oodnadatta and the southern Northern Territory, and a sharp contraction in the area that contained stock.[33] For some, the losses were tremendous: the Barrow Creek Pastoral Co., lost tens of thousands of pounds' worth of investments after first giving up 75% of its leased land and, shortly after, disposing of its entire cattle herd.[34] Even divine providence was in short supply, for the Lutherans at Hermannsburg were forced to largely abandon the mission after the 1892–93 drought had left wells brackish, orchards and gardens unproductive and much of the local population suffering from illness.[35]

At the same time, the great 19th-century era of Australian exploration was sputtering to a close, as many of the last desert crossings took on an Odyssean quality, like the Elder Exploring Expedition's epic march across the Great Victoria and Great Sandy Deserts to Queen Victoria Spring in June 1891 (Fig. 16.1)[36], Carnegie's 1896–97 expedition between Coolgardie and Hall's Creek, and the tragic 1899–1900 Border Exploration Syndicate Expedition, with its tales of intrigue and murder.[37] The more mundane simply struggled through vast areas of drought-killed spinifex, mulga, mallee and wildlife, consumed by the search for water (Box 16.1).[38] Finally, the sense of optimism that the struggle for water might one day be vanquished from the centre of the continent was lost forever as visions of inland seas and abundant water evaporated not only in the imagination but also in front of the very eyes of the settlers who had taken up land on the promise of greater permanence.

THE MOLOCH OF CIVILISATION

He needs must goe, the Devil drives.
– Charles Cotton, 'Scarronides; or, Virgile travestie', *1664*[39]

For Aboriginal people of the arid zone who had witnessed countless droughts, the departure of rains in 1891 must have invoked traditional movement patterns aimed at preserving water and food resources as they had in prior centuries. However, these were now increasingly being disrupted by European exploration or prospecting expeditions seeking water in natural or 'native' rock holes and wells, and as these were vital drought refuges the invaders were often met with diversionary tactics or even outright hostility by local people. One such incident took place near Mount Webb, WA when Carnegie's party convinced two Aboriginal people to reveal the location of their camp, and then spent the afternoon and night digging out a small well located nearby to water their parched horses. Aboriginal men who had returned from hunting watched the proceedings with 'intense annoyance' and threatened the party by brandishing their spears. Horses can drink more than 10 gallons (46 L) and camels 25 gallons (114 L) per day or more,[40] and the fury of Aboriginal people watching carefully preserved water plundered for stock can only be imagined (Fig. 16.3).

Closer to the Overland Telegraph and the expanding pastoral districts, however, a far more pressing issue had emerged, for the trickle of European settlers that had gradually encroached from districts to the south and east, had, during the 1870s–80s, grown into a flood of men and hooves. At first motes in the red landscape, the herds soon had grown into the tens of thousands, and when dry times returned, they trampled the ground, stripped the ground of cover, and silted the waterholes so crucial to survival.[42] Interestingly, drought was viewed by at least some Aboriginal people as a beacon of hope in these dark times; in one anecdote, a man named Gillet is said to have blamed the drought on God, who was playing a joke on

Walk, white fellow, walk.
– Anangu man during attack on the Giles Expedition Party, Officer Creek, SA, 1872[41]

the white squatters for building dams and bores, which by obviating the need for rain, had put Him out of work.[43]

Others took a more direct approach to dewater the European invaders. Walter Baldwin Spencer tells in 1901 the story of a Kaytetye (= Kaititya) man at Barrow Creek, 'head man of the water or rain people' (head of the rain totem) who 'did not care very much for the white man' and, having complete control over rainfall in that part of the country, was 'not going to allow any more rain to fall until the drought had killed off the white men and their cattle so that the blacks could have all the country to themselves again'.[44] And, after drought conditions worsened in 1902 stations in the vicinity of Hanson and Lander Creeks, west of Barrow Creek, were indeed abandoned.

The cultural impact of the drought, however, was ultimately asymmetric for, as game and water disappeared, many Aboriginal people struggled to simply survive. Some shifted to hunting rabbits, which now occurred across southern districts in plague numbers (Fig. 16.4),[45]

Fig. 16.3: In August 1896 David Carnegie and his team became desperately short of water in the Gibson Desert ca. 500 km north-east of Kalgoorlie. Facing death, they captured an Aboriginal man who took them to Empress Springs, shown here, where they finally found water more than 25 feet underground. From Carnegie DW (1898) *Spinifex and sand*. C. Arthur Pearson, London.

but others resorted to stealing from stores or killing of cattle and sheep, either as reprisal for incursions onto their land or directly for food. At Barrow Creek competition with stock for water supplies may have led to an outbreak of cattle killing in 1891, while in 1895–96 Alyawara (= Alyawarre) people reportedly killed 600 cattle that had wandered into the hills where they were camped.[46] This 'black drought' unsurprisingly led to conflict and violence with station owners.[47] One of the worst incidents occurred in the East Kimberley in 1893, when 23 Miriwoong men were shot after an altercation with police and Aboriginal assistants sent to investigate the killing of bullocks and two horses, 'victims sacrificed to the ruthless Moloch of Australian civilisation'.[48]

Inevitably, with the country destitute of plants, fish and game, desperate and starving Aboriginal people were forced to disperse to missions and stations for food during 1901–02,

> *My word, God he cobbon saucy long a whitefellow.*
>
> – *The drought.* The Spectator *(Perth), 8 June 1901*

Fig. 16.4: As native animals declined due to competition from cattle and predation by cats and foxes, Aboriginal people were forced to rapidly alter their diet and livelihoods. Top: hunting rabbits in the southern Musgrave Ranges. National Museum of Australia, 1985.0060.1270. Bottom: Indigenous man shepherding goats and sheep near Oodnadatta, SA. Note the dying woody vegetation in the foreground. Both photographs by Herbert Basedow, 1903. National Museum of Australia, 1985.0060.2144.

particularly in far northern South Australia and along the Overland Telegraph.[49] Upwards of 50 people were receiving food at Tennant Creek, and there are reports of several people dying of malnutrition while travelling to, or at, Barrow Creek.[50] Others were collected from scattered locations and placed on stations, settlements or new reserves to facilitate the dispersal of rations, or arrived at government depots.[51] On cattle and sheep stations able-bodied people learned new skills with astonishing speed and took up positions as stockmen, farm hands and housekeepers (Fig. 16.4) – as early as 1892 a majority of station employees in parts of central Australia were Aboriginal[52] – but were then trapped in a state of perpetual Tantalean torment, reliant on the very system that had deprived them of their traditional livelihood.

> *One white man with his herds can render impossible the health and happiness of two hundred aborigines.*
>
> – *Charles Duguid (1963)* No dying race. *Rigby, Adelaide*

In these remote areas the pastoralist remained 'a law unto himself'[53] and while some treated Aboriginal people with compassion, at best the circumstances were penurious: salaries were meagre and sometimes paid in clothes, boots, mosquito nets, basic foods and tobacco, or a combination of money and rations.[54] Frequently, food rations were insufficient to avoid starvation, and able-bodied people still relied on wild food to supplement their diets. During drought, the old and young suffered most: Charles Duguid tells of one incident in 1935 in which he saw emaciated children and women clamouring for entrails of a bullock that had been killed.[55] At worst, treatment was so depraved as to earn station managers nicknames like 'Brutal Brockman'.[56] In the longer term, however, conflict with white men over women proved an even bigger problem, for it fuelled severe social destabilisation. According to the famous welfare worker and anthropologist Daisy Bates, the destruction of ancient social and sexual taboos after contact with 'low whites' was especially devastating to Aboriginal society.[57]

Unfortunately, the work provided on stations often proved as ephemeral as the rain. Faced with immense losses of stock, stations had little choice but to reduce labour and charitable assistance, forcing Aboriginal people to rely entirely on government relief, walk to nearby stations or towns, or even move back to remaining waterholes and channels to gather whatever food remained.[58] Missions were also sometimes unable to produce food: the Kaparlgoo Mission on the South Alligator River, for example, suffered severely during the 1901 dry season, when the lagoon on which it depended completely dried out. As the vegetable supply ran short, able-bodied Aboriginal people were forced to travel long distances to hunt.[59] Other missions similarly affected include the Mapoon Mission in Cape York and, as we have seen, the Hermannsburg Mission in the MacDonnell Ranges.[60]

Over the next three decades ever more people were drawn off Country and into the orbit of government depots, missions, and cattle and sheep stations. The next protracted drought occurred in 1911–15 (the 'WWI Drought', Plate 2), and again large numbers of Aboriginal people with little or no prior contact with white people were forced from the interior and onto government provisions.[61] The economic cost was growing; tens of thousands of pounds were now being spent annually in Western Australia alone, which included the purchase and establishment of the Moola Bulla native settlement scheme in 1910 near Halls Creek, WA, and thousands of pounds of flour and sugar along with clothing and blankets and other goods were being distributed in northern South Australia alone.[62] Disease heightened the misery; in 1919 Herbert Basedow encountered 'tribes' in northern South Australia suffering from syphilis and declining rapidly in number, which contributed to the widely held belief that Aboriginal people were at that time a dying race.[63]

The wet years of 1920–21 offered some reprieve, but drier weather returned in in 1922–26, and a drought of record severity struck in 1927–29. Beneath rainless skies and burning summer heat cattle broke down trees for food and pawed the spinifex from the ground, before finally dying, the pyres of burning carcases sending up plumes of smoke for months at a time.[64] Gradually, whole landscapes were transformed into waterless hellscapes[65], and life for Aboriginal people once again descended into a battle for survival against starvation, high infant mortality, disease and social decline. According to the journalist and early conservationist Arthur Groom, this problem was exacerbated by the splitting of once-balanced tribal units into smaller, vulnerable groups due to settlement and drought,[66] but also in no small part to the extended incarceration of many people in local gaols.[67]

Perhaps no event better encapsulates this nexus of evils than the infamous massacre that took place on and around Coniston Station, 230 km north-west of Alice Springs, in August and September 1928. Prior to 1924 Aboriginal people of the remote Tanami Desert, including the Warlpiri, had successfully remained on Country, but after 2 years of severe drought and facing famine, most decided to disperse and seek food – a prudent decision since some who stayed ultimately perished. But many made for better watered country near the Hanson and Lander rivers, which was occupied by cattle stations, and here tensions rose, especially after a spate of cattle killings.[68] Matters came to a head when dingo trapper Fred Brooks, who had apparently provided food until his supplies ran low, was killed by Warlpiri men for failing to provide the promised 'tucker' of flour, tea, sugar and tobacco for taking a woman Marungali (Marungardi) to do washing, and then for the night.[69] The punitive expeditions undertaken by mounted police, in which at least 31 Warlpiri, Anmatyerre and Kaytetye people, and likely very many more, were massacred, are known to this day by the Walpiri people as 'the killing times'.[70]

AILURU PULKA MULAPA

A very long time ago, in a time of great hunger ...
– Nganyintja Ilyatjari, Pitjantjatjara woman[71]

The Anangu word for drought, ailuru, traditionally refers to a period dry enough to make plant food scarce and life for people difficult.[72] When I flew into Pitjantjatjara country to visit Uluru-Kata Tjuta National Park in June 2022, the mark of ailuru was everywhere, from the leafless desert oak reaching upwards from the red sand to the blackened rings of spinifex branded like some malignant pox into the landscape. The drought that had killed these iconic plants had been short but very severe – 2019 had been the driest year on record, with just 27 mm of rain falling, and 2020 was also much drier than normal[73] – and I had come to map the dead vegetation and investigate the future role of climatic extremes on central Australian ecosystems (Box 16.1). But it soon became clear that the Anangu people could recall much bigger droughts in the past, the ailuru pulka, which were so harsh that their ancestors were faced with the very real prospect of dying from famine or thirst, and a few, called ailuru pulka mulapa, that even forced people from their land.[74]

There is evidence that this occurred during 1891–1902, when SG Hübbe, RT Maurice and WR Murray encountered people living in a state of semi-starvation in Anangu country due to a lack of game and plant foods.⁷⁵ The next major drought occurred during 1911–15, which is notable for being one of the few occasions in which permanent territorial shifts due to water shortages and an overall climate drying trend has been carefully documented. During this great drought Pitjantjatjara people were forced into the eastern Musgrave Ranges, which had received patchy thunderstorms, which in turn caused the displacement of the traditional Yankunytjatjara more than 200 km to the south, towards the limit of their territory. Although a few Yankunytjatjara returned under white protection at the Moorilyanna Sheep Station, the Ernabella (Pukatja) area subsequently remained a major summer watering place for the Pitjantjatjara.⁷⁶

Anthropological evidence suggests that during ordinary dry times different Aboriginal groups would fall back to central permanent water supplies, which reinforced rigid territorial boundaries, but in the most extreme droughts peripheral refuges could be shared by more than one group of people, and in desperate circumstances some were forced to decide between staying in Country and hoping for rain, or trespassing, which was extremely hazardous. This was apparently a common situation, for on the periphery of the western deserts terms such as ngatari was used to describe such strangers.⁷⁷ The situation that unfolded in the eastern Musgrave Ranges in 1914–15 may have been precipitated by a similar dynamic, although it was probably facilitated by the declining population size of Yankunytjatjara in the east due to disease and migration to European settlements.⁷⁸

An even more severe period of nearly continuous rainfall deficits then occurred between 1922 and 1938, with major droughts occurring in 1922–29 and 1931–38 (Plate 1). This period coincides with the A̱nangu memory of two successive droughts, or of one very long one, that affected northern South Australia around this time.⁷⁹ The 1922–29 drought caused famine among Aboriginal people across central Australia; conditions were described as being so dry that native 'banana greens' or doubah (*Marsdenia australis*) died or were eaten by animals, parakeelya (*Calandrinia balonensis*), spinifex and summer grasses set no seed, kangaroos migrated away, and no witjuti grubs could be found on the roots of acacias. In the Petermann Ranges large areas of the hardy desert oak (*Allocasuarina decaisneana*) were even killed.⁸⁰

In 1928–29 these conditions caused abject starvation and malnutrition among the western Kukatja-Luritja and Yumu (= Jumu) people near Lake Amadeus and Lake Bennett, manifesting as a form of scurvy with soft puffed flesh and swollen belly and feet. While some children died and other people were left crippled, many fortunately walked to Hermannsburg Mission, where they received medical attention, and it is estimated that 50% of children under the age of 15 might have died if they had received no aid.⁸¹ Probably referring to the same event, Pastor Freidrich Wilhelm stated that many arrived like 'human scarecrows', blinded from the dust and heat and led by young women who could still see. Sadly, not everyone made the arduous journey successfully: one woman who attempted to walk from Haast Bluff, 90 waterless miles

> *It wasn't so much the shadow of death – it was death itself ...*
> – Pastor Freidrich Wilhelm, Pastor at Hermannsburg Mission⁸⁴

Fig. 16.5: This photograph, which shows a group portrait of Aboriginal women and children at the camp of NWR Dumas, Musgrave Ranges, SA, ca. 1938, illustrates the severity of conditions experienced towards the end of the 1920s–30s drought period. Note the bare country and dead shrubs in the background. State Library of South Australia, B 77568/79.

away (145 km), perished, along with one of her two children, a boy named Tuntelkarina. At the height of the drought 17 people were being buried at the mission each month.[82] Later reports suggested that ultimately hundreds of people caught in the Haast Bluff and Yuendumu districts died trying to reach civilisation.[83]

Worse was to come. By 1936 an unparalleled drought was taking hold across central and northern Australia, and in the following year Patrol Officer TGH Strehlow reported that most Arrernte, Kukatja-Luritja and Pintupi had deserted their country. The Petermann Ranges were still permanently inhabited by perhaps 200 Pitjantjatjara people, but others had drifted to dogger (dingo-hunting) camps, stations like Erldunda and Lynda Vale (also called Lyndavale, 230 km south-west of Alice Springs), and the newly established Ernabella Mission (Fig. 16.5).[85] But in 1939 an expedition that set out to investigate the plight of people in this remote region encountered only five men, eight women and 12 children. Under drought conditions large gatherings of people often broke into progressively smaller and more scattered groups to access declining food resources, provided water was available,[86] and on this occasion boys of 16 were encountered who had not undergone their first initiation rite, which normally occurred at age 12, because it had not been possible to gather enough men together for this and other important ceremonies.[87]

But when an older man named Palinkana refused to accompany the expedition south from Docker River (Kaltukatjara), it became apparent that a much more tragic series of events had also taken place. In 1938, people had crossed from the Petermann to the Mann and Tomkinson

> *... at a time when everything had dried out with the drought, there is no meat and the people become too weak to hunt. Some just sit down and died together, but the stronger ones go to try and find meat for their families. The old people die, but the younger folk manage to live on and care for their children. The young sons go out and climb the rocky hillsides, far from the camp, and if they have a good day, they bring home their catch and share it with their families. The parents climb kurrajong trees, searching for seed pods, taking what they find back to camp where they clean and grind the seed, and share it with their sons.*
>
> – Nganyintja Ilyatjari[90]

ranges looking for food, but shallow and intermediate waters in the district were fast drying up, and when the last had gone all hope had vanished. The stronger people moved towards another permanent water spring, but the sick, elderly and mothers with small children had to remain and ultimately died a cruel death. This account was confirmed by a dogger at Mount Cavanagh Station, who witnessed the death and starvation of people west of the Mann Ranges after going south looking for better conditions, the expedition's guide, Tjuintjara, and many senior men and women since.[88] In 1939 it remained a 'desert of dry bones' to which Palinkana would not return.[89]

When the Pitjantjatjara returned after the drought broke in 1939, they found that part of their country had died, too. Gone were many small- to mid-sized mammals so crucial to culture, including the numbat (Anangu name = Walputi), western quoll (Partjata), desert bandicoot (Walilya, Makura), bilby (Ninu, Marura, Tjalku) and burrowing bettong (Mitika), while rabbits and feral cats recovered quickly, probably exacerbating the decline.[91] For many native species the arrival of the fox in the 1920s appears to have been decisive; populations devastated by the long drought and left without ground cover must have had little chance when faced with this added predation pressure.[92] Today, Pitjantjatjara are heavily involved in plans to reintroduce these and other locally extinct species back into the wild.

While the loss of life during the 1920s and 1930s was exceptional, other disasters have occurred even more recently in the most remote parts of central and western Australia. Wally Dowling, the famous 'barefooted drover', claimed, in a series of articles written in the 1950s, that he frequently encountered starving Aboriginal people while crossing the Canning Stock Route, and that on many trips he found people dead around wells.[93] I have found this difficult to verify, although Dowling drove stock across the Canning from ca. 1942 to 1955, and severe droughts struck the region in 1943–46 and 1948–50.[94] In the late 1940s a group of more than 30 people who had been trapped at Lake Mackay died after heading south to a waterhole only to find it dry, giving some credence to Dowling's account, as does the fact that that the great drought of 1951–54 (Plate 2), which was probably the worst ever to strike tropical Northern Territory and Western Australia, definitely caused malnourishment of 'nomadic natives' on the northern Canning.[95] It has been estimated that 10–25% of desert 'tribal' groups of perhaps 450 people may have perished every two generations; if even close to correct,[96] this indicates that in pre-European times drought was the most dangerous hazard faced by central Australian societies.

Endnotes

[1] Lindsay D (1893) *Journal of the Elder Scientific Exploring Expedition, 1891-2*. C. E. Bristow, Government Printer, Adelaide, p. 175.

[2] From: *The Age* (Melbourne), 22 February 1902, p. 13; for full account of Gregory's expedition see Gregory JW (1906) *The dead heart of Australia: A journey around Lake Eyre in the summer of 1901-1902, with some account of the Lake Eyre basin and the flowing wells of central Australia*. John Murray, London.

[3] *The Argus* (Melbourne), 24 January 1902, p. 6; *The Age* (Melbourne), 22 February 1902, p. 13.

[4] From the North African Arabic word *'irj*, an erg is a large area of wind-swept sand with virtually no vegetation cover.

[5] In Australia, partially active dunes are mostly restricted to parts of the Simpson, Strzelecki and Great Sandy Deserts; dune behaviour and composition are discussed in Hesse PP, Simpson RL (2006) *Geomorphology* **81**, 276–291 and Hesse P (2011) *Geomorphology* **134**, 309–325.

[6] The concept of central Australian deserts as being 'wet' occurs in Hesse P (2011) *Geomorphology* **134**, 309–325.

[7] Based on average rainfall maps provided by the Australian Bureau of Meteorology (http://www.bom.gov.au/).

[8] The ratio of annual precipitation (AP) to the potential evapotranspiration (PET) is often used to quantify aridity (aridity index = AP/PET); hyper-arid climates are usually defined as having an aridity index of 0.03–0.05 or less. Australia currently has no hyper-arid areas; see Larkin *et al.* (2020) *Scientific Reports* **10**, 6653.

[9] The similarity between dunes in the Kalahari and Central Australia is noted in Goudie A (1970) *South African Geographical Journal* **52**, 93–101.

[10] Cf. Maurice RT (2022) *Across the Great Thirst Land*. Hesperian Press, Carlisle, WA; Cormac McCarthy, *Blood meridian*, Picador, London.

[11] Giles E (1889) *Australia twice traversed: The romance of exploration, being a narrative compiled from the journals of five exploring expeditions into and through central South Australia, and Western Australia, from 1872 to 1876*. Sampson Low, Marston, Searle & Rivington, London. Despite the title Giles uses the term 'wretched' to refer to various aspects of desert life and travel no less than 46 times.

[12] Foley JC (1957) *Droughts in Australia: Review of records from earliest years of settlement to 1955*. Commonwealth of Australia Bureau of Meteorology Bulletin No. 43, reproduced by Department of Defence Production, Central Drawing Office, Maribyrnong, WA.

[13] Giles (1889). Giles found the Gibson Desert region less hospitable, although not as dry as that encountered by explorers during the 1890s.

[14] Gosse WC (1874) *W. C. Gosse's explorations, 1873. Report and diary of Mr. W. C. Gosse's central and western exploring expedition, 1873*. Government Printer, Adelaide, 1874; Forrest J (1998) *Exploration in Australia*. Friends of the State Library of South Australia, Adelaide.

[15] Finlayson HH (1935) *The Red Centre. Man and beast in the heart of Australia*. Angus and Robertson, Sydney.

[16] Groom A (1952) *I saw a strange land*. Second edition. Angus and Robertson, Sydney, pp. 23–33.

[17] *South Australian Register* (Adelaide), 30 April 1883, p. 6.

[18] *Port Augusta Dispatch*, 5 January 1885, p. 1; *South Australian Register* (Adelaide), 1 January 1884, p. 1.

[19] *South Australian Register* (Adelaide), 30 November 1886, p. 6.

[20] McLaren G, Cooper W (2001) *Distance, drought and dispossession: A history of the Northern Territory pastoral industry*. Northern Territory University Press, Darwin, p. 10.

[21] Duncan R (1967) *The Northern Territory pastoral industry 1863–1910*. Melbourne University Press, Carlton, Victoria, p. 35.

[22] Data from Table 2 in Duncan (1967).

[23] *The Australasian* (Melbourne), 17 February 1883, p. 26. At the time the block of land held by the Barrow Creek Pastoral Company was the largest in Australia; *Border Watch* (Mount Gambier), 24 February 1883, p. 4.

[24] *South Australian Register* (Adelaide), 8 August 1883, p. 7.

[25] *Eastern Districts Chronicle* (York), 28 November 1891, p. 4; *The West Australian* (Perth), 25 November 1891, p. 2; *Western Mail* (Perth), 4 July 1891, p. 11; 5 December 1891, p. 14; *The Australian Advertiser* (Albany), 5 October 1891, p. 3; *Victorian Express* (Geraldton), 2 October 1891, p. 3.

[26] *The Australian Advertiser* (Albany), 5 October 1891, p. 3; *Victorian Express* (Geraldton), 2 October 1891, p. 3.

[27] *Eastern Districts Chronicle* (York), 7 August 1897, p. 3.

[28] Lindsay (1893), pp. 154–188; *Western Mail* (Perth), 5 December 1891, p. 14.

[29] *The West Australian* (Perth), 12 May 1897, p. 5; *The Inquirer and Commercial News* (Perth), 2 April 1897, p. 16.

[30] *The Inquirer and Commercial News* (Perth), 2 April 1897, p. 16; 28 May 1897, p. 6.

[31] *The Telegraph* (Brisbane), 12 January 1893, p. 2; *Queensland Times, Ipswich Herald and General Advertiser*, 2 November 1893, p. 3; *The Telegraph* (Brisbane), 12 January 1893, p. 4; *Northern Territory Times and Gazette* (Darwin), 17 November 1893, p. 3; 24 March 1893, p. 3; Duncan (1967), p. 50.

[32] *Adelaide Observer*, 27 April 1889, p. 31; McLaren G, Cooper W (2001) *Distance, drought and dispossession: A history of the Northern Territory pastoral industry*. Northern Territory University Press, Darwin, p. 36.

[33] E.g. Mount Burrell, Owen Springs, Eringa, Macumba and Dalhousie Springs; see Duncan (1967).

[34] *Adelaide Observer*, 30 August 1890, p. 21; 9 September 1893, p. 5; *The Register* (Adelaide), 5 January 1917, p. 9; 26 September 1928, p. 6.

[35] *Adelaide Observer*, 23 June 1894, p. 43; *South Australian Register* (Adelaide), 23 January 1893, p. 6. The drought here was broken in early 1894 with a fall of 8.5 inches (216 mm), see *Adelaide Observer*, 24 February 1894, p. 12.

[36] Lindsay (1893). In a remarkable feat of endurance, Lindsay's camels crossed 536 miles in 35 days with only 8 gallons (36 L) of water each, in one case going more than 2 weeks without drinking.

[37] Hill HW (2009) *Desert, drought, and death: The border exploration syndicate expedition to the Rawlinson Range 1899-1900*. (Eds M Chambers, PJ Bridge). M. G. Creasy and Hesperian Press, Carlisle, WA; Carnegie DW (2004) *Spinifex and sand: A narrative of five years' pioneering and exploration in Western Australia*. Sydney University Press, Sydney.

[38] Lindsay (1893), p. 175; Hubbe SG (2018) *Samuel Grau Hübbe and the South Australia to Western Australia Stock Route Expedition*. (Ed. AG Peake). Hesperian Press, Carlisle, WA; Maurice RT (1904) *Extracts from journals of explorations: Fowler's Bay to Rawlinson Ranges and Fowler's Bay to Cambridge Gulf. With Plans*. C. E. Bristow, Government Printer, Adelaide, p. 7.

[39] This quote appears in 'Our letter box'. *The Western Australian Goldfields Courier*, 12 October 1895, p. 4, as 'Needs must when the devil drives'.

[40] These are typical estimates, but during hot and dry expeditions consumption could be much higher; see Giles E (1889) *Australia twice traversed: The romance of exploration, being a narrative compiled from the journals of five exploring expeditions into and through central South Australia, and Western Australia, from 1872 to 1876*. Sampson Low, Marston, Searle & Rivington, London.

[41] Giles (1889), pp. 175–180. Officer Creek flows southward out of the Musgrave Ranges in South Australia.

[42] Sadly retold by Wendy Gordon Sprigg, who wrote 'O great white rulers of the world / We only ask of you the right to live. / This land / Was ours before your alien tread pressed / The sand / In all its reaches we were free to roam.' *The Herald* (Melbourne), 17 December 1928, p. 1.

[43] *The Spectator* (Perth), 8 June 1901, p. 3.

[44] Spencer W (2013) *Walter Baldwin Spencer's Diary from the Spencer and Gillen Expedition 1901–1902*. (Ed. J Gibson, transcribed by H. Milton), pp. 62–63.

[45] *The Register* (Adelaide), 2 December 1901, p. 5.

[46] Koch G (1993) *Kaytetye country: An Aboriginal history of the Barrow Creek area*. (Eds G Koch, H Koch). Institute for Aboriginal Development Publications, Alice Springs; Hagen R, Rowell M (1978) *A claim to areas of traditional land by the Alyawarra and Kaititja*, Central Lands Council, Alice Springs; cited in Layton R (1989) *Uluru: An Aboriginal history of Ayers Rock*. Aboriginal Studies Press, Canberra.

[47] *The Register* (Adelaide), 16 June 1902, p. 9; *The Western Australian Goldfields Courier* (Coolgardie), 12 October 1895, p. 4; *Coolgardie Miner*, 16 February 1895, p. 4; *Western Mail* (Perth), 26 December 1891, p. 13.

[48] *Nor'West Times and Northern Advocate* (Roebourne), 30 September 1893, p. 2; *The W.A. Record* (Perth), 5 October 1893, p. 8.

[49] South Australia (1901) 'Report of the Protector of Aborigines for the year ended June 30, 1901'. C. E. Bristow, Government Printer, Adelaide; *Examiner* (Launceston), 10 April 1902, p. 5; *The Sydney Mail and New South Wales Advertiser*, 5 November 1902, p. 1190.

[50] *Adelaide Observer*, 24 May 1902, p. 38; *The Advertiser* (Adelaide), 16 May 1902, p. 6; *Examiner* (Launceston), 10 April 1902, p. 5.

[51] *The Telegraph* (Brisbane), 17 September 1903, p. 3; *Coolgardie Miner*, 15 April 1898, p. 5.

[52] Duncan (1967), p. 73.

[53] *Sunday Times* (Perth), 4 January 1903, p. 13.

[54] Duncan (1967), p. 73; South Australia (1901); *Western Mail* (Perth), 6 December 1902, p. 9.

[55] Duguid C (1963) *No dying race*. Rigby, Adelaide, p. 138; Love JRB (1915) *The Aborigines: Their present condition as seen in northern South Australia, the Northern Territory, North-West Australia and Western Queensland*. Arbuckle, Waddell & Fawckner, Melbourne.

[56] *West Australian Sunday Times* (Perth), 8 December 1901, p. 12; 6 October 1901, p. 14.

[57] Bates D (1973) *The passing of the Aborigines: A lifetime spent among the natives of Australia*. Pocket Books, a division of Simon and Schuster, Inc., New York, pp. 64–65.

[58] *Toowoomba Chronicle and Darling Downs General Advertiser*, 2 May 1901, p. 6; *Evening News* (Sydney), 21 February 1903, p. 4; *Australian Town and Country Journal* (Sydney), 6 September 1902, p. 23; *The Brisbane Courier*, 10 June 1903, p. 5.

[59] *The Advertiser* (Adelaide), 16 April 1902, p. 6. This event, and a more detailed history of the mission, is provided in Freier P (2008) *Journal of Northern Territory History* **19**, 17–32.

[60] Roth WE (1903) *Annual report of the Northern Protector of Aboriginals for 1902*. George Arthur Vaughan, Government Printer, Brisbane.

[61] *The West Australian* (Perth), 5 May 1913, p. 5; 6 May 1913, p. 6; *Chronicle* (Adelaide), 8 July 1911, p. 42.

[62] *The Register* (Adelaide), 1 November 1911, p. 6; *The West Australian* (Perth), 6 May 1913, p. 6; *Sunday Times* (Perth), 18 January 1914, p. 11. A history of Moola Bulla is provided in Rumley H, Toussaint S (1990) *Aboriginal History* **14**, 80–103.

[63] *The Register* (Adelaide), 27 November 1919, p. 7.

[64] Lockwood D (1950) *The Mail* (Adelaide), 25 March, p. 7.

[65] *News* (Adelaide), 28 October 1930, p. 5; *The Mail* (Adelaide), 22 September 1928, p. 15; *Morning Bulletin* (Rockhampton), 2 February 1929, p. 11.

[66] Groom A (1952) *I saw a strange land*. Second Edition. Angus and Robertson, Sydney, pp. 37–38.

[67] South Australia (1930) 'Report of the Chief Protector of Aboriginals for the year ended June 30, 1929'. Harrison Weir, Government Printer, Adelaide; *The Advertiser* (Adelaide), 13 September 1928, p. 13.

[68] Meggitt MJ (1965) *Desert people: A study of the Walbiri Aborigines of Central Australia*. The University of Chicago Press, Chicago and London, pp. 23–34.

[69] *The Advertiser* (Adelaide), 13 September 1928, p. 13; *Northern Standard* (Darwin), 9 November 1928, p. 1.

[70] Wray-McCann R (1986) *The Canberra Times*, 31 March, p. 14.

[71] Hilliard WM (1968) *The people in between: The Pitjantjatjara people of Ernabella*. Hodder and Stoughton, London, p. 200.

[72] Goddard C (1922) *Pitjantjatjara/Yankunytjatjara to English dictionary*. Second Edition. Institute for Aboriginal Development, Alice Springs.

[73] Rainfall data from Yulara, a small resort town 4 km north of the Park boundary; data obtained from Australian Government Bureau of Meteorology, Station No. 15635.

[74] Goddard (1992). Some Anangu use the phrase 'ailuru pulka mulapa' to emphasise the severity of a really bad drought.

[75] Mr. R. T. Maurice's Expedition North of Fowler's Bay. *Proceedings of the Royal Geographic Society of Australasia: South Australian Branch*, vol. 5, Session 1901–1902, pp. 33–39. Vardon & Pritchard, Adelaide; Maurice RT (1904) *Extracts from journals of explorations: Fowler's Bay to Rawlinson Ranges and Fowler's Bay to Cambridge Gulf. With Plans*. C. E. Bristow, Government Printer, Adelaide, p. 29; *The Sydney Mail and New South Wales Advertiser*, 5 November 1902, p. 1190; *Chronicle* (Adelaide), 14 March 1896, p. 15.

[76] Tindale NB (1974) *Aboriginal tribes of Australia: Their terrain, environmental controls, distribution, limits, and proper names*. University of California Press, Berkeley, pp. 69–70; Tindale NB (1972) Chapter 6: The Pitjandjara. In *Hunters and gatherers today: A socioeconomic study of eleven such cultures in the Twentieth*

77 Tindale (1974), pp. 65–79.
78 Yankatjatjara people may have departed the Musgraves and moved east towards the Central Australia (Overland) Railway in 1929; see Hilliard (1968) but c.f., Tindale (1972). The process may have occurred in several phases.
79 Baker LM (1991) 'Kanpi Biological Survey 1991. Anangu Information. Anangu Pitjantjatjara Council and South Australian National Parks and Wildlife Service Joint Biological Survey of the Anangu Pitjantjatjara Lands'. Unpublished report cited in Robinson AC *et al.* (Eds) (2003) 'A Biological Survey of the Anangu Pitjantjatjara Lands, South Australia, 1991-2001'. Department for Environment and Heritage, South Australia.
80 Mackay D (1929) *Geographical Journal* **73**, 258–264; Tindale (1974).
81 Tindale (1974).
82 Lockwood D (1950) *The Mail* (Adelaide), 25 March, p. 7.
83 *Barrier Miner* (Broken Hill), 10 April 1952, p. 1.
84 Lockwood D (1950) *The Mail* (Adelaide), 25 March, p. 7.
85 The Parliament of the Commonwealth of Australia (1938) 'Report on the Administration of the Northern Territory for the year ended 30th June, 1937'. L.F. Johnston, Commonwealth Government Printer, Canberra; Layton R (1989) *Uluru: An Aboriginal history of Ayers Rock*. Aboriginal Studies Press, Canberra; *Northern Standard* (Darwin), 1 May 1936, p. 6.
86 Meggitt MJ (1968) *Desert people: A study of the Walbiri Aborigines of Central Australia*. The University of Chicago Press, Chicago, pp. 49–51.
87 Duguid (1963); *The Advertiser* (Adelaide), 11 August 1939, p. 24; c.f., page 44 in Albrecht FW (1965) *Journeying with Missionary F. W. Albrecht in 1939*. Pages 37–47 in *Lutheran Almanac: Yearbook of the United Evangelical Lutheran Church in Australia*, 1965. Lutheran Book Depot, Adelaide.
88 Duguid (1963), pp. 58, 71 and elsewhere; Albrecht (1965); Robinson AC *et al.* (Eds) (2003).
89 Albrecht (1965).
90 Hilliard (1968), p. 200.
91 Robinson AC *et al.* (Eds) (2003).
92 Finlayson HH (1961). *Records of the South Australian Museum* **14**, 141–191.
93 Hill E (1953) *Australian Geographical Walkabout Magazine*, vol. 19, March 1953; 'People' 3 April 1957, cited in Cane S (1990) Desert demography: A case study of pre-contact aboriginal densities in the western desert of Australia. In *Hunter-gatherer demography: Past and present*. (Eds B Meehan and N White) pp. 149–159. Oceania Monograph 39, University of Sydney, Sydney.
94 One possible incident occurred in 1949: *The Daily News* (Perth), 3 November 1949, p. 3. Dowling appears to have operated on the Canning Stock Route from at least 1942 (*Western Mail* (Perth), 14 September 1950, p. 14) until 1955 (*The Daily News* (Perth), 24 September 1955, p. 3).
95 Cane (1990); *Sunday Times* (Perth), 14 June 1953, p. 7.
96 Cane (1990).

Part 5

Life in the Anthropocene

17
River of tears

The rivers and the lakes are ours. We sow the seed, and plant the trees. We fertilize the earth by overflowing it. We stop, direct, and turn the rivers: in short, by our hands we endeavor, by our various operations in this world, to make, as it were, another nature.

– *Cicero*, De Natura Deorum[1]

PLANET OF THE HUMANS[2]

In May 2000 atmospheric chemist Paul J Crutzen and biologist Eugene F Stoermer proposed the term 'Anthropocene' to describe the current geological epoch in which human activity has come to dominate the global environment.[3] They dated the beginning of this epoch to the late 18th century when anthropogenic emissions of carbon dioxide and methane were becoming detectable in the atmosphere, and since then human dominion over global carbon, nitrogen and sulphur cycles, biodiversity, landscapes and even global net primary productivity has grown unabated. While the date of this transition is controversial, one of the more interesting inclusions in this list of globally transformative activities were the practices of dam building, river diversion and extraction of fresh water for use by humans, the latter of which by 2002 had reached 50% of global accessible freshwater resources.[4]

Coincidentally, the year 2000 also marked the end of a half-century interval of relatively high rainfall across eastern Australia (Plate 12), dominated by extremely wet years in the mid-1950s, mid-1970s and late 1990s (Fig. 12.1). Throughout this period much of the knowledge accumulated in the dark times of the 1890s–1940s had been lost to the public consciousness, but this all changed in 2001 when one of the most extreme and protracted droughts since European settlement took hold over what was now a much more heavily modified and exploited continent. As the rainless years once again stretched on for what seemed an eternity, old wounds were re-opened and new ones formed, and Australians, groaning under punishing water shortages and extremes of heat and fire, awakened to a new life in the Anthropocene.

The 2001–09 Millennium Drought and the 'Big Dry' of 2017–19 (Plates 1 and 2), visited terrible destruction on terrestrial ecosystems and dryland farming regions over much of the continent. Forests and native vegetation suffered extensive dieback, native animals declined, massive dust storms swept over the denuded landscape, livestock were lost, agricultural productivity fell and communities suffered economic decline.[5] Growing awareness of the implications of extreme events and climate change for the nation led to community outrage over government inaction, yet Australia doubled-down on its commitment to becoming a fossil-fuel-exporting superpower. We will consider these topics in later chapters, but here we begin by investigating something that I think is unique to the droughts of the recent Anthropocene – their devastating impacts on water, irrigation, industry, and the health of Australia's over-extracted and degraded inland river systems.

THE RIVER PUT TO WORK

In a river the water you touch is the last of what has passed and the first of what is to come. It is the same with the present time.

– Leonardo da Vinci[6]

'The River Murray, judged by its length of channel and area of catchment, should be one of the great streams of the world ...'.[7] So begins the 1902 Interstate Royal Commission on the River Murray, which was tasked with investigating the practicality of a general system of water conservation and distribution across the Murray–Darling Basin (MDB). When viewed on a global scale, Australia's great inland river system is indeed something of an anomaly: despite draining over a million square kilometres of country and having a maximum continuous length of 3672 km,[8] it has a typical annual natural discharge of only around 11 000–14 000 GL per year.[9] This is a less than one-tenth of the volume of water carried by most of the Earth's other great river systems – compared with the Columbia River, near where I did my PhD work in Oregon, which discharges a staggering 230 000 GL per year despite have a 38% smaller catchment, it is little more than a trickle.[10] The Rio Grande of North America and the Orange River of southern Africa are perhaps the best hydrological analogues and, interestingly, have a similar history of water resource development.[11]

This scarcity, however, brings great value, for the rivers of the MDB are the veins of the landscape that bring lifeblood to its waterless heart,[12] flooding thousands of riverine floodplains and wetlands, that, as water dries from the surrounding plains, stand out viridescently in the taupe landscape (Plate 19). Across the millennia Aboriginal people from the Ngarrindjeri of the Coorong and lower River Murray to the Bigambul, Kooma and Mandandanji of the Balonne survived on their rich assemblages of aquatic plants, birds, fish and reptiles, travelling along a great network of roads created by the foot traffic along the riverbanks. The rivers themselves, however, remained untamed, swelled by massive floods that periodically swept down from the highlands, then drained by droughts that reduced them to little more than shallow pools.

After European settlement, however, hungry eyes began to see the potential of the MDB's great rivers through an entirely different lens. As we have seen, during the early decades of agricultural expansion inland rivers were first used as corridors for moving stock into the fertile inland plains, and by the 1850s runs had occupied virtually the entire river frontages west to the Darling. The huge sheep and cattle stations and small communities that formed along the Darling and lower Murray now saw these rivers transform into inland highways, with scores of paddle steamers plying a trade that serviced remote communities across the west.[13] In a typical year snowmelt and winter rains supported a navigation season between Blanchetown on the Riverland of South Australia and the upper reaches at Albury of May to December, while the Darling River was typically navigable for half the year and the Murrumbidgee River to Hay only in months of high flow.

Until the 1880s water from the main rivers was used primarily for rural domestic and sanitary uses, stock watering, and some support for the gold industry, and while most towns and stations had some means of pumping water the quantities were small.[14] But when access to water meant the difference between economic success and failure, the rivers were far too valuable to remain unshackled for long. The tide began to turn with royal commissions in New South Wales (1884) and Victoria (1886) that considered the practicality of developing systems for water conservation and distribution to alleviate the impacts of droughts and floods, particularly in the southern MDB. These generated much interest in the development of new irrigation schemes, and by 1887 mean annual diversions (removal of water from the river) had reached more than 100 GL per year, virtually all in the Victorian reaches of the Murray. By 1900 the construction of major works on the Goulburn, Loddon and Broken rivers and establishment of numerous 'irrigation colonies' along the Murray, annual diversions had grown to more than 600 GL, ca. 90% of which were in Victoria (Fig. 17.1). By 1902 major works were also being undertaken on rivers across New South Wales.[15]

The extraction of around 5% of total annual discharge of the Murray may not seem important, but the 1902 royal commission revealed that people on the lower Murray were already concerned that use of water upstream for irrigation had rendered lower stretches of

Fig. 17.1: The dawn of a new era – construction of the Goulburn Weir in Victoria, ca. 1888–89. Completed in 1891 as the first major irrigation diversion structure built in Australia, the Goulburn Weir was considered an engineering marvel at the time. In 1902 the storage of the weir was estimated at ca. 19 GL although it is now stated to be 25.5 GL. State Library of Victoria, IE193206.

Fig. 17.2: Walking on the dry bed of the Murray River at Moorook, SA in 1915. Recent studies show that water extractions and diversions were large enough by this time to significantly reduce the flow of the river during drought. Note, however, the healthy crowns of the trees in the left background. State Library of South Australia, PRG 1258/2/644.

the river unnavigable during the Federation Drought. Larger boats with draughts exceeding 4 feet (0.3 m) were most affected, and for a river transport industry locked in a pitched battle with an expanding railway network the impact was catastrophic. Station owners around Lake Alexandrina, near the mouth of the Murray, also complained that inflows had been so reduced that the lake's water had become as salty as the sea and that fresh water for livestock had become unprocurable. Although saltwater intrusion here had also apparently occurred in the 1850s and probably the mid-1880s,[16] in 1900–02 flow was so low that saltwater fish were reportedly caught as far as 30 miles (48 km) upriver between Mannum and Murray Bridge.[17] In April 1901 the Murray even stopped flowing near Morgan, SA.

There was reluctance on the part of the commissioners to conclude that irrigation had anything to do with this situation, but recent research has raised the possibility that the April 1901 zero-flow event may have been caused by large irrigators pumping upstream at Mildura and Renmark at the time.[18] Either way, the number of diversions and smaller irrigation licences continued to steadily rise, and subsequent zero flow events (Swan Hill 1914–15, probably Morgan 1915, and Swan Hill in 1923) had an unambiguous anthropogenic signature (Fig. 17.2).[19] At the time, the contributing role of irrigation was widely understood, but in an apparent case of *shifting baseline syndrome* (Box 17.1), many modern ecologists have come to believe that under natural conditions the Murray River normally ceased flowing during periods of drought. In fact, there is scant evidence for this, at least above Mildura, although historical reports from 1839 and September 1850 suggest that it may have occurred during exceptionally dry conditions.[20]

Box 17.1: Shifting baseline syndrome

One of the great challenges to conserving global biodiversity is that major environmental changes often take place over timeframes that are long relative to the human lifespan. Thus, while many people notice over the course of their lives that the biodiversity or natural value of the place in which they live has deteriorated, each successive generation comes to accept an increasingly degraded state as a new baseline on which to measure subsequent change. This phenomenon, first conceptualised by the marine biologist Daniel Pauly in 1995, has become known as 'shifting baseline syndrome' or 'environmental generational amnesia'.

The problem is that each successive generation comes to see the increasingly impoverished state of the natural world as 'normal', and so the threshold for what is considered an acceptable amount of environmental deterioration is continually lowered. This is by no means a new problem, for, contrary to popular belief, animal assemblages, especially megafauna, were extremely impoverished in most parts of the world before the arrival of Europeans as a result of Indigenous human exploitation coupled with climatic change. However, over the past two centuries, and particularly since the 1950s, the pace of environmental decline has increased, and virtually everyone now live in places that are unimaginably different from anything resembling a 'natural' condition.

The changes that have taken place in north-west New South Wales in just two generations is a case in point. The two images below show a typical section of the Namoi River floodplain ca. 10 km east of Wee Waa in the mid-1950s (left; aerial photograph) and 2023 (right; Google Earth satellite image). In the mid-1970s most of the region still contained native floodplain grassland and woodland plant communities and was used to raise sheep and cattle. Within my lifetime these have been cleared to grow cotton and other crops, often leaving only a strip of semi-natural vegetation along the Namoi River. The river itself has undergone significant hydrological change and biodiversity loss due to the construction of dams upstream and water extraction for irrigation. Raising community awareness of environmental history and concepts such as rewilding may be a partial solution to this progressive problem.

The pace of change continued to accelerate. By 1922 MDB-wide diversions had reached ca. 2000 GL per year,[21] and over the coming two decades a series of locks and weirs were constructed along the Murray, each of which raised the base flow river height (head) by several metres.[22] These lifted groundwater levels and forced salty water to flow from saline aquifers into the river downstream, among many other hydrological impacts.[23] Other modifications

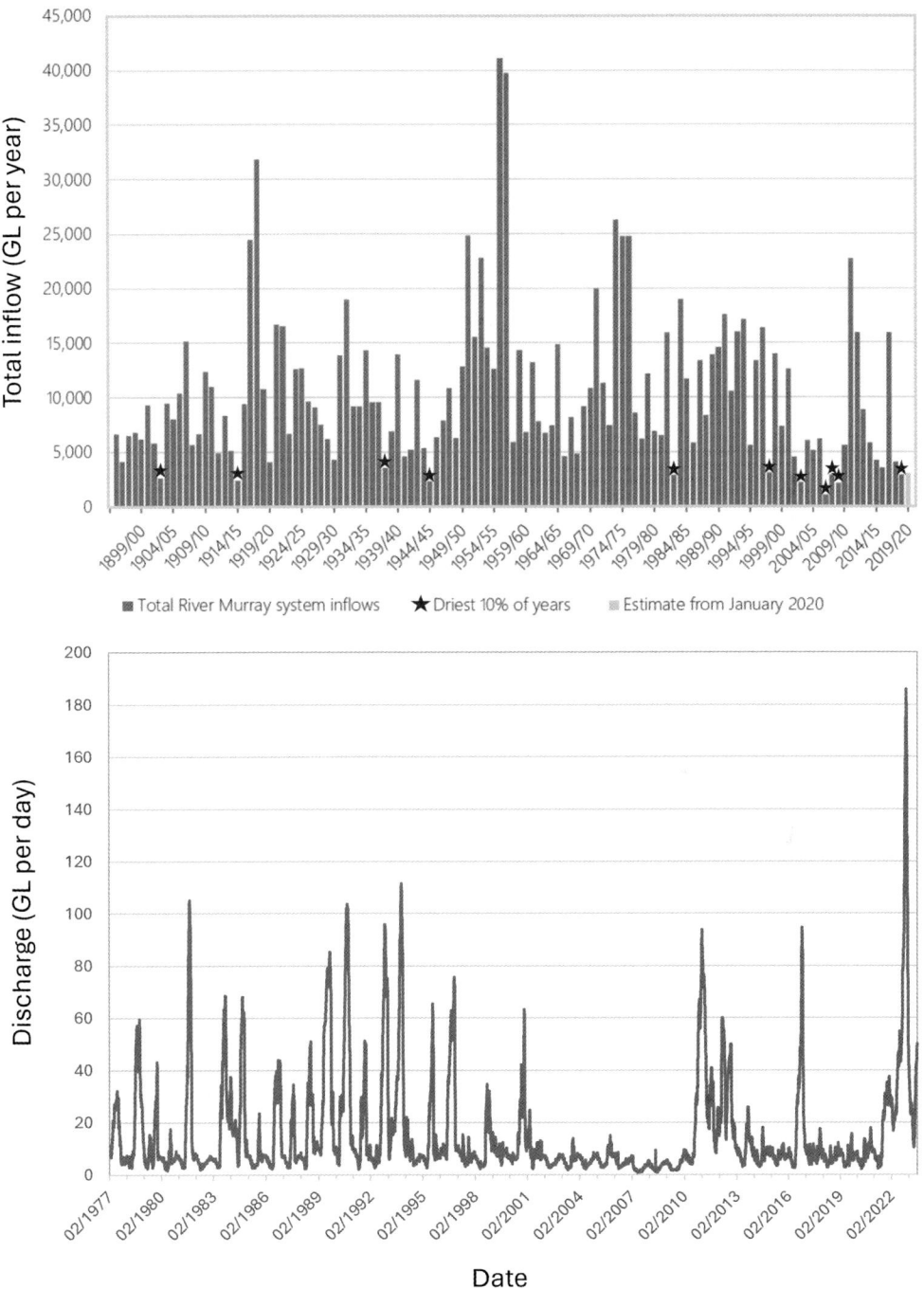

Figure 17.3: Top: modelled total inflows to the River Murray system, 1895–2020. Interim Inspector-General of Murray–Darling Basin Water Resources 2020, Impact of lower inflows on state shares under the Murray–Darling Basin Agreement, Canberra, data provided by MDBA. Bottom: River Murray discharge at the SA border, 1977–2023. Note the exceptionally low discharge during the Millennium Drought and recovery in 2010–12, and the massive flows in 2022–23. Data obtained from Water Data South Australia, site A4261001.

including the raising of levees, construction of canals, installation of regulators, construction of the Kiewa and Snowy Mountains hydroelectric schemes, installation of the Murray Mouth (Goolwa) barrages and construction of the massive Hume Dam above Albury in 1934. Major reservoirs and other water storage systems were constructed on virtually all major rivers in New South Wales, such that by the 1970s diversions had exceeded 8000 GL.[24] On the Murray these caused a shift in the flow regime from a natural winter–spring peak to a regulated summer–autumn peak, a decline in maximum but increase in minimum flows, and a reduction in the frequency and duration of extensive floods that connect the Murray River to its anabranches.[25]

Inevitably, the combined effect of these hydrological changes, as well as the local shifts in groundwater depth caused by vegetation clearing and irrigation, locked sensitive riparian plant and animal communities into an inevitable trajectory of degradation. At Barmah Forest the reduction in peak flood depths caused river red gum to invade low-lying floodplain grasslands, the area of which declined by more than half between 1930 and 1980.[26] Worse still, forests on more elevated floodplains that depended on the largest floods for survival had become what ecologists colourfully call 'zombie ecosystems' – having the 'living dead' quality of being alive but doomed to gradually die.[27] At the 17 700 ha Chowilla Floodplain, a Murray River Icon and Ramsar site that contains the largest remaining area of natural riverine forest in South Australia,[28] dieback became obvious in the 1960s, but strong inflows, deep floods and 'flushing flows' in 1956 and then again in 1974–76 (Fig. 17.3) all provided a stay of execution of around 12 years by replenishing fresh water and leaching salt from the soil profile.[29]

After the mid-1970s, however, a drying trend set in, first across the southern MDB in 1976–77, and then the east in 1979–80. The dry conditions intensified throughout 1981, when the mouth of the river closed, and drought conditions culminated in 1982–83 (Plate 2). As salt began to accumulate in elevated floodplain areas due to a lack of deep floods, trees again began to experience osmotic drought and die.[30] By 1990, 17 999 ha of vegetation along the length of the Murray was found to be severely degraded, primarily by saline groundwater (53% of the total area), drowning and waterlogging caused by raised river levels (37%) and drought or water stress (6.2%). Overall, black box was less healthy than river red gum, and the condition of floodplain forest and woodlands declined with distance downriver.[31] It was against this backdrop that the Millennium Drought began.

LAST STAND

In 1997–2000, when most of the continent was very wet, rain had dried up across the upper Murray watershed in the Australian Alps and north-central Victoria. Just five catchments – the Upper Murray, Kiewa, Mitta Mitta, Ovens and Goulburn-Broken – contribute more than 40% of the MDB's total water, and so the dry weather had a disproportionate impact on MDB inflows, which, during 1997–98, were in the driest 10% of years and the lowest since 1982–83 (Fig. 17.3). Contributions from both northern and southern catchments were better in 1999 and 2000, although far too little to generate deep, prolonged floods along the Murray. But the trouble really started when dry weather returned in 2001 and the continent was plunged into the severe drought of 2002–03.

Fig. 17.4: Top: dieback of river red gum at Chowilla Floodplain during the Millennium Drought (Photo: Robert Godfree). Middle: severe drought stress in a patch of young river red gum on elevated terrain in Barmah Forest (Photo: Lyndsey Vivian). Bottom: the author wading through a dense stand of *Juncus ingens* on Steamer Plain, Barmah Forest, 2011. Due to changes in flood regime *Juncus ingens* has replaced moira grass in many wetland habitats in Barmah Forest (Photo: Lyndsey Vivian).

River Murray system inflows for the year to May 2003 fell to 2220 GL, the fourth lowest since 1890, and 2-year inflows (2001/02 and 2002/03) amounted to just 6590 GL, the lowest on record (Fig. 17.3). By the middle of 2003 water storages had fallen to 24% of full active capacity, and irrigation allocations in Victoria and New South Wales were reduced to the lowest level since 1979.[32] At the South Australia–Victoria border, flow in the Murray River fell to under 3 GL per day in June 2001, marking the beginning of a 9-year period of exceptionally low flows in which discharge would stay below 10 GL per day for most of the time (Fig. 17.3).[33] This is less than half the volume required to begin inundating iconic floodplains along the Murray (e.g. more than 30 GL/day at Chowilla, Hattah Lakes and Gunbower Forest), and so most were effectively deprived of watering events for almost a decade.[34] Even at the Barmah Forest, where the 'Barmah Choke' limits bank capacity to only 10 600 ML (10.6 GL per day), floodplains and wetlands were left high and dry for years at a time.

When media reports of dieback in iconic stands of river red gum along South Australian reaches of the lower Murray began to trickle in it was obvious that ecosystem health was beginning to seriously suffer. Although locals had noticed stress among trees here as early as mid-2001, the problem greatly intensified in late 2002, and by 2003 patches of dead and dying trees were becoming conspicuous along the river (Fig. 17.4). One of the earliest to draw attention to the problem was Tony Sharley, manager of Banrock Station near Kingston on Murray, which had recently spent more than $1 million to restore 168 ha of degraded floodplain just upstream of the famed Overland Corner. No major flood had occurred here for 6 years, and trees three centuries old were withering and turning brown, including one riverbank canoe tree.[35] Increasingly, river red gum dieback was being seen as an indicator of broader biodiversity decline and landscape-level environmental change, with over-extraction for irrigation largely to blame. Even the Chief Executive of the Murray–Darling Basin Commission, Don Blackmore, thought that the system was 'on the cusp of catastrophe'.[36]

> *The cause of this is too much water being taken out of the Murray ... We can only bring this situation to people's attention and hope they respond.*
>
> *– Tony Sharley, Banrock Station, 2003*[37]

Evidence of the extent of the dieback soon emerged. A study conducted in 2002 revealed that more than half of floodplain trees between Wentworth, NSW and Renmark, SA were suffering from water stress, and the next year another revealed that around 80% of river red gum trees on South Australian floodplains were stressed, 20–30% severely so.[38] As the drought ground on the situation worsened: in 2004 the percentage of stressed trees between Wentworth and Renmark increased to 78%, and by 2006 70% of river red gum forests in the Victorian Murray River floodplain were also stressed.[39] By 2009, detailed vegetation modelling of five Living Murray Icon Sites (River Murray Channel, Chowilla floodplain and Lindsay-Wallpolla Islands, Hattah Lakes, Gunbower-Koondrook-Perricoota Forests and Barmah-Millewa Forest) revealed that 79% of forest and woodlands were moderately to severely degraded. Downstream floodplains like Chowilla and Hattah Lakes were most affected, but even in the Barmah-Millewa Forest only 21% of river red gum forests were considered healthy (Fig. 17.4). By 2009, around 40 000 ha of forest and woodlands that were in good condition in 2003 had become stressed.[40] Thousands upon thousands of trees now stood dead, the rows of rampikes lining the floodplains like the exposed vertebrate of some serpentine creature washed up and buried by the river itself.

KIDNEY FAILURE

While these were perhaps the most vivid manifestation of the drought, the entire system, visible or not, was undergoing rapid change. Flood-dependent plants and animals began to disappear, replaced by species adapted to drier or saltier conditions. Low-lying areas in Barmah Forest, for example, lost a significant portion of *Pseudoraphis spinescens* (moira grass), a stoloniferous perennial species that grows into large floating mats during floods. By 2014 only 51 ha of the thickest patches remained out of hundreds that existed in pre-European times, and the loss of this crucial keystone wetland species began to attract serious concern.[41] Worse still, research conducted by Lyndsey Vivian, whom I worked with as part of CSIRO's Water for a Healthy Country Flagship, showed that prospects for recovery were severely limited under the modern flow regime, in which floods were too shallow and brief to drown the 3–4 m tall stands of *Juncus ingens* (giant rush) that had gradually encroached over the former moira grass plains (Fig. 17.4).[42]

Other permanent wetlands were 'disconnected' from the Murray by falling river levels. Downstream of Lock 1 near Blanchetown, SA, for example, wetlands that been permanently inundated by the construction of the Murray Mouth barrages were left dry by 2008–09, leading to the loss of floating and submerged plant species that live in aquatic environments and their replacement by terrestrial species.[43] Others, however, were actively cut off as part of water conservation measures primarily intended to maintain water for human use. More than 30 managed wetlands were disconnected from the main river channel above Lock 1 by closing regulators and filling inlet channels to protect the water supply for Adelaide, which, in 2007, was placed on Level 3 water restrictions, and to reduce the risk that nutrient-rich or saline water might re-enter the Murray.[44] Since most of these wetlands are shallow and suffer high evaporative losses these measures reportedly saved 133 GL of water, and it was hoped that by progressively re-wetting the more vulnerable systems damage might be avoided.[45] Nevertheless, the drying of these wetlands and onset of visible signs of ecological stress, such as the slow death of river red gum in the Katfish (Katarapko-Eckerts Creek) Demonstration Reach, near Loxton, SA resulted in calls for their reconnection to the river.[46]

> *Rivers die from the bottom up.*
>
> – *Australian Conservation Foundation, 2010*[47]

These problems, however, were dwarfed by those unfolding in the 'kidneys' of the Murray – the Lower Lakes (Alexandrina and Albert), Murray Mouth and the coastal lagoon system of the Coorong.[48] This region rose to national prominence with the release of the iconic 1976 movie *Storm boy*, based on the 1964 novel by Colin Thiele, and in 1985 the Coorong, Lake Alexandrina and Albert Wetland were listed under the Ramsar Convention on Wetlands due to international significance as habitat for migratory waterbirds and outstanding biodiversity value. Quite apart from the captivating tale of life, death and redemption of *Storm boy*, one cannot fail to be struck by the wild beauty of and abundance within the Coorong in the 1970s, its great flocks of waterbirds, fierce seas and shifting sands all speaking of an environment still not completely bent to human will.

But at this time the health of the Coorong was being kept afloat by years of high rainfall, and when prolonged drought conditions returned, its ecosystems too faced collapse. Even under natural conditions, about half of total MDB flows were lost to ecosystems, evaporation and groundwater before reaching the Murray Mouth, and water diversions and storage had increased the sensitivity of these terminal ecosystems to low rainfall over upstream catchments. Ongoing dredging was required to keep the mouth open between October 2002 and 2010, at a reported cost of $40 million, and by the mid-2000s median annual flow at the barrages had fallen to 3075 GL, just 29% of the 10 764 GL estimated under natural conditions.[49] Daily flows at the mouth now ceased 40% of the time, and starved of fresh water, the Murray Mouth closed and the water level below Lock 1 fell below sea level for the first time in recorded history.[50] By 2009, Lake Alexandrina fell to ca. 1 m below sea level and the shores of Lakes Alexandrina and Albert retreated by hundreds of metres, exposing acid sulphate soils (Plate 20).[51] Salinity rose precipitously, reaching 6000 to 8000 $\mu S\ cm^{-1}$ across Lake Alexandrina, 10 times that of fresh drinking water, more than 20 000 $\mu S\ cm^{-1}$ in Lake Albert, and more than 180 000 $\mu S\ cm^{-1}$ in the southern Coorong.[52]

The result was a series of cascading changes that ended in breakdown of the system. Fine river sediments were replaced near the river mouth by marine sands, which, combined with a reduction in tidal influence, resulted in a decline in the food and habitat available for migratory wading birds. In the southern Coorong, tuberous sea tassel (*Ruppia tuberosa*) virtually disappeared due to low water levels and hypersaline conditions more than five times that of seawater, in turn affecting other wildlife, including waterbirds that forage on sea tassel turions (vegetative buds).[53] Populations of two small freshwater fish, the southern pygmy perch (*Nannoperca australis*) and the endangered Murray hardyhead (*Craterocephalus fluviatilis*) declined, and the Yarra pygmy perch (*Nannoperca obscura*) was apparently extirpated from Lake Alexandrina. This in turn affected piscivorous (fish-eating) waterbirds such as pelicans and grebes.[54] Many shifted to the northern Coorong, which had fresher water, while banded stilts (*Cladorhynchus leucocephalus*) flocked to the southern reaches to feed on brine shrimp that had proliferated in the salty water.

Perhaps most tragic was the plight of freshwater turtles, numbers of which had been declining since the 1970s.[55] In 2008, news broke that populations of two species, the southern river turtle (*Emydura macquarii*) and the Eastern long-necked turtle (*Chelodina longicollis*), were being killed in the Lower Lakes and Murray Mouth by infestations of Australian tubeworms (*Ficopomatus enigmaticus*). Australian tubeworms are salt-tolerant, colony-forming animals that secrete clusters or reefs of calcareous tubes in favourable habitat. This included the shells of turtles, which sank from the weight or gradually died of starvation due to decreased mobility. It is hard to forget the images of school children treating tubeworm-infested turtles, a process that apparently involved soaking them in fresh water for 24 h and then scraping off the deposits with spoons.[56]

A SYSTEMIC PROBLEM

During the initial stages of the drought there were hopes that disasters of this kind might be restricted to the Murray, especially since the more northern catchments of the Balonne, Gwydir, Namoi and Macquarie had suffered less severe rainfall deficiencies than those further south, particularly in 2004. But this optimism evaporated as it became apparent that all major river systems from southern Queensland to Victoria had been impounded and over-extracted for decades, and that here too fundamental ecosystem changes were taking place. Among the worst affected were the beautiful necklace of wetlands that form where the rivers slow and braid across the inland plains – once great barriers to Oxley, Sturt and Mitchell in their search for the inland sea, now reduced to a state of silent dessication.

> *Why cannot our Government Departments listen to the voice of nature?*
> – *RA Hudson, 1948*[57]

In the far north, the plight of the Ramsar-listed Narran Lakes, known to the Yuwaalaraay/Euahlay people as Dharriwaa, became the centre of a major war over water extraction from the Lower Balonne River, which rises as the Condamine River in the Darling Downs and becomes a network of smaller watercourses (the Narran, Culgoa, Bokhara, Ballandool, Briarie and Birrie rivers), waterholes and floodplains downstream of St George, Qld. At the centre of the controversy was the use of 'water harvesting', in which levee banks and other diversions are used to capture water from rivers and floodplains and then store it in large impoundments for later agricultural use. Although this had been practised on the Lower Balonne for decades on a small scale, during the 1980s landowners realised that these water licences, which had no volumetric restrictions, could be used to transform their properties from dryland grazing to irrigation-based cotton enterprises.

The first and most famous practitioner was Cubbie Station, on the Culgoa River, but during the 'wild west' of the 1990s many massive engineering projects to harvest 'off-river' water were constructed across the Lower Balonne. Total storage capacity exploded to well over 1000 GL, most of it in vast shallow dams, and by 2008 a staggering 55% of total available surface water for the entire catchment was being utilised. In some years, more than 500 GL were diverted for agricultural use.[58] The results were predictable: flows in the Culgoa and Bokhara rivers declined by half or more and floods on the Narran River, and particularly those with daily discharges of 50–100 GL per day became much less frequent. After 2001 floods virtually disappeared, and by 2005–06 terminal flood-dependent ecosystems of the Narran Lakes had become severely drought-stressed, with surveys revealing that up to half of river red gum, river cooba (*Acacia stenophylla*) and coolabah were dead.[59] Numbers and breeding of waterbirds across the Narran Ecosystem, once home to vast numbers of colonial species such as straw-necked ibis, glossy ibis and Australian white ibis, crashed. Breeding of these species is tightly correlated with floodplain and wetland inundation and since no floods of sufficient size occurred between 2000 and 2006, waterbirds deserted the area.[60]

Similar changes were taking place across virtually all inland Murray–Darling waterways. Wetlands and floodplains on the Gwydir River, downstream of the 1364 GL capacity Copeton Dam, were suffering from 'artificial drought' due to very high water extraction (41%) to supply, among other things, 85 000 ha of irrigated cotton in the western catchment; thousands of coolabah trees died.[61] The famed Macquarie Marshes, downstream of Burrendong Dam, suffered from water extraction and even water theft; thousands of hectares of river red gum forest, reed beds and other wetland vegetation lay dead or dying, frog and reptile populations collapsed, and by 2006 waterbirds, which once numbered in the hundreds of thousands, had effectively disappeared.[62]

Further south, river red gums died across the 15 000 ha Booligal Wetlands due to altered seasonality, frequency and duration of flows in the Lachlan, particularly downstream of the Wyangala Dam[63], while on Lowbidgee floodplain, half of which had already been damaged by dams, diversions and development and the loss of 60% of the pre-regulation water supply, bird numbers declined and large tracts of floodplain red gum were stressed or dead.[64] On the Darling, trees and lignum died on floodplains downstream of Wilcannia[65]; red gum stands on the lower Wimmera River and its terminal endorheic lakes were stressed or killed by increased salinity and lack of flooding, and major tributaries of the Murray including like the Broken and Campapse rivers suffered declining fish numbers.[66]

This makes depressing reading, but it illustrates the scale of riverine ecosystem decline that occurred during the Millennium Drought. With a few exceptions, these were far more severe than any observed during prior droughts, including 1895–1903 and the 1930s–40s. The crucial difference, of course, was the massive, basin-wide expansion of water diversions, storage and extraction for agriculture that had reduced downstream water availability and the frequency of deep rejuvenating floods, and even before 2000 the signs of strain were everywhere in declining bird and fish populations, dieback of tree populations, and the spread of invasive species.[67] The extreme and protracted dry conditions experienced during the Millennium Drought simply accelerated the decline of many ecosystems that modern regulated hydrological regimes had long ago rendered little more than historical artefacts.

Endnotes

[1] Marcus Tullius Cicero (1877) *Cicero's Tusculan Disputations; also, Treatises on the Nature of the Gods, and on The Commonwealth.* Literally translated, chiefly by C. D. Yonge. Harper and Brothers, New York.

[2] This title is taken from Jeff Gibbs's controversial (2009) environment documentary.

[3] Crutzen PJ, Stoermer EF (2000) *The International Geosphere–Biosphere Programme (IGBP) Global Change Newsletter* **41**, 17–18.

[4] Crutzen PJ (2002) *Nature* **415**, 23, doi:10.1038/415023a.

[5] Hoy A (2003) *The Bulletin*, volume 121, no. 7, 18 February; Australian Broadcasting Corporation (2010) 'Meekatharra properties desperate for drought aid'. *ABC News*, 23 March 2010; Byrnes M (2003) 'Australia drought puts kangaroo war in cross-hairs'. *Reuters News*, 25 May 2003; Razak I (2008) 'Drought threatens rare SA flora and fauna'. Australian Broadcasting Corporation Transcripts. *ABC*, 3 March 2008.

6 This quote, taken from Leonardo da Vinci's *Codex Trivulzianus*, in Italian is *'L'acqua che tocchi dei fiumi e l'ultima di quella che andò e la prima di quella che viene. Così il tempo presente.'*

7 Davis J, Murray, Burchell FN (1902) 'Interstate Royal Commission on the River Murray, representing the states of New South Wales, Victoria, and South Australia: Report of the commissioners. With minutes of evidence, appendices, and plans'. Sands & McDougall, Melbourne.

8 From the Condamine River to Murray Mouth.

9 Estimates of mean annual discharge of the Murray–Darling Basin vary but generally fall within this range; see Simpson HJ, Cane MA (1993) *Water Resources Research* **29**, 3671–3680; CSIRO (2008) 'Water availability in the Murray. A report to the Australian Government from the CSIRO Murray-Darling Basin Sustainable Yields Project'; Williams J (2017) *Journal & Proceedings of the Royal Society of New South Wales* **150**, 68–92. The annual discharge at Morgan was determined in the 1902 Interstate Royal Commission (p. 12) on the River Murray to be 12 884 GL in 1900, which was considered a mean flow year.

10 Hamilton P (1990) *Progress in Oceanography* **25**, 113–156.

11 Blythe TL, Schmidt JC (2018) *Water Resources Research* **54**, 1212–1236; Lange G-M, Mungatana E, Hassan R (2007) *Ecological Economics* **61**, 660–670. The Rio Grande is specifically mentioned in the 1902 Interstate Royal Commission on the River Murray.

12 From 'A river is water in its loveliest form; rivers have life, sound, movement and an infinity of variation, rivers are veins of the earth through which the life blood returns to the heart', by Haig-Brown RL (1974) *A river never sleeps*. Crown Publishers, New York.

13 In 1902 there were 90 registered steamers and about the same number of barges (Minutes of Evidence in Davis *et al*. (1902), p. 31).

14 Davis *et al*. (1902), p. 15.

15 Davis *et al*. (1902).

16 According to Alfred Catt, M.P., Chairman of Committees in the House of Assembly, South Australia, the lakes (Alexandrina and Albert); Davis *et al*. (1902), p. 11 of Minutes.

17 Davis *et al*. (1902) p. 34 and Minutes pp. 17–40; *The Advertiser* (Adelaide), 29 May 1902, p. 6.

18 One view is that under natural conditions the upper reaches of the Murray River were perennial but that flow ceased to the sea during the most extreme periods of drought, including the Federation Drought. Mallen-Cooper M, Zampatti BP (2018) *Ecohydrology* **11**, e1965.

19 Mallen-Cooper, Zampatti (2018) *Ecohydrology* **11**, e1965.

20 Mallen-Cooper, Zampatti (2018) *Ecohydrology* **11**, e1965. In September 1850 the Murray was reportedly 'practically dry' from Tocumwal to Moama, with 'no stream flowing at all'; see Davis *et al*. (1902), Minutes of Evidence.

21 Close A (1990) The impact of man on the natural flow regime. In *The Murray*. (Eds N Mackay and D Eastburn) pp. 61–74. Murray-Darling Basin Commission, Canberra.

22 There are now 14 lock-and-weir constructions on the Murray River, the last of which was completed in 1939.

23 Impacts discussed in Jolly ID, Walker GR, Thorburn PJ (1993) *Journal of Hydrology* **150**, 589–614; Thomas GA, Jakeman AJ (1985) *Land Use Policy* **2**, 87–102; Slavich PG *et al*. (1999) *Agricultural Water Management* **39**, 245–264; Walker KF, Thoms MC, Sheldon F (1992) In *River Conservation and Management*. (Eds PJ Boon, PA Calow, GE Petts) pp. 270–293. Wiley, Chichester.

24 Williams J (2017) *Journal & Proceedings of the Royal Society of New South Wales* **150**, 68–92.

25 Maheshwari BL, Walker KF, McMahon TA (1995) *Regulated Rivers: Research and Management* **10**, 15–38; Mac Nally R *et al*. (2011) *Water Resources Research* **47**, W00G05, doi:10.1029/2011WR010383.

26 Chesterfield EA (1986) *Australian Forestry* **49**, 4–15.

27 Zombie afficionados will immediately see the problems with this metaphor.

28 Cale B (2009) 'Literature review of the current and historic flooding regime and required hydrological regime of ecological assets'. A Report for the South Australian Murray-Darling Basin Natural Resources Management Board, Murray Bridge, South Australia; Overton I, Jolly I (2004) 'Integrated studies of floodplain vegetation health, saline groundwater and flooding on the Chowilla Floodplain South Australia'. CSIRO Land and Water Technical Report 20/04, May 2004.

[29] Slavich *et al.* (1999).

[30] Decadal-scale changes in salinity of the Murray are discussed in Hart B *et al.* (2020) *Water* **12**, 1829.

[31] Margules and Partners Pty Ltd, P. and J. Smith Ecological Consultants, Department of Conservation, Forests and Lands, Victoria (1990) 'River Murray riparian vegetation study: Prepared for the Murray-Darling Basin Commission.' Canberra, Murray-Darling Basin Commission (henceforth MDBC).

[32] MDBC (2003) 'Murray-Darling Basin Commission Annual Report 2002-2003'. MDBC Publication No. 11/04, MDBC, Canberra, ACT 2601.

[33] Calculated daily discharge data from Water Data SA site A4261001 indicate that the peak discharge observed between May 2001 and August 2010 occurred on 12 November 2005 with a volume of 15.1 GL per day.

[34] Overbank floodplain inundation commences at Chowilla at between 30 and 35 GL/day, Hattah Lakes at 37 GL/day, and Gunbower at 30 GL/day. Inundation of most red gum and black box woodlands and forests require much higher flows.

[35] Debelle P (2003) *The Sydney Morning Herald*, 1 March, p. 13; Littlely B (2003) *The Advertiser* (Adelaide), 8 January, p. 14.

[36] Fyfe M (2003) *The Age* (Melbourne), 6 June, p. 8.

[37] Cited in: Littlely B (2003) *The Advertiser* (Adelaide), 8 January, p. 14.

[38] MDBC (2003) 'Preliminary investigations into observed River Red Gum decline along the River Murray below Euston'. Technical Report 03/03. MDBC, Canberra, Australia.

[39] MDBC (2005) 'Survey of River Red Gum and Black Box health along the River Murray in New South Wales, Victoria and South Australia – 2004. MDBC, Canberra, Australia; Cunningham SC *et al.* (2009) *Ecosystems* **12**, 207–219.

[40] Cunningham SC *et al.* (2009) 'Mapping the condition of river red gum and black box stands in The Living Murray Icon sites. A milestone report to the Murray-Darling Basin Authority as part of contract MD1114'. Murray-Darling Basin Authority, Canberra.

[41] Vivian LM, Ward KA, Marshall DJ, Godfree RC (2015) *Australian Journal of Botany* **63**, 526–540; Chesterfield (1986).

[42] Vivian LM, Ward K, Godfree RC (2014) *Journal of Applied Ecology* **51**, 1292–1303.

[43] Nicol JM (2010) 'Vegetation monitoring of River Murray Wetlands downstream of Lock 1'. South Australian Research and Development Institute (Aquatic Sciences), Adelaide, SARDI Publication No. F2009/000416-1.

[44] ABC News (2007) 'New water restrictions hit Adelaide'. Australian Broadcasting Corporation, 1 January 2007; Government of South Australia (2007) 'River Murray Act 2003 Annual Report 2006-2007'. Prepared for the South Australian Parliament by the Minister for the River Murray; Durant R, Nielsen D, Reid CJ (2009) 'Monitoring of temporarily disconnected River Murray Wetlands Final Report'. Report prepared for South Australian Murray-Darling Basin Natural Resources Management Board.

[45] Jenkin C, Hyde B (2009) *The Advertiser* (Adelaide), 6 August.

[46] Wallace TA (2019) 'Eckerts-Katarapko floodplain tree condition survey data April 2015-March 2019'. Report produced by Riverwater Life Pty Ltd for the Department for Water and Environment, South Australian Government; Jenkin C (2008) *The Advertiser* (Adelaide), 1 August; Roberts J (2006) *The Australian* (Sydney), 7 November.

[47] Australian Conservation Foundation (2010) 'Submission to the Senate Rural Affairs and Transport Standing Committee Inquiry into the management of the Murray-Darling Basin'.

[48] Collectively now known as the Lower Lakes, Coorong and Murray Mouth (LLCMM) Icon Site. The 'kidney' metaphor, which refers to the role that the Lower Lakes and Murray Mouth play in flushing salt (as well as sediment and nutrients) from the system, is admittedly imperfect.

[49] Kingsford *et al.* (2011) *Marine and Freshwater Research* **62**, 255–265; references therein.

[50] Ye Q *et al.* (2014) 'Ecological responses to flooding in the lower River Murray following an extended drought. Synthesis report of the Murray Flood Ecology Project'. Goyder Institute for Water Research Technical Report Series No. 14/6.

[51] Banerjee O et al. (2013) *Ecological Economics* **91**, 19–27.
[52] Salinity and lake level data are available from the Water Data South Australia website at https://water.data.sa.gov.au/. Representative sites include A4261133, A4261158, A4260574, A4261155 and A4261165.
[53] Rogers DJ, Paton DC (2009) 'Changes in the distribution and abundance of *Ruppia tuberosa* in the Coorong'. CSIRO, Canberra; Paton DC, Bailey CP (2013) 'Annual monitoring of *Ruppia tuberosa* in the Coorong region of South Australia, July 2012'. University of Adelaide, Adelaide.
[54] Wedderburn S, Bailey C, Barnes T (2021) 'Condition monitoring of threatened fish populations in Lake Alexandrina and Lake Albert'. Report to the Murray–Darling Basin Authority and the South Australian Department for Environment and Water; Morris S (2008) *The Australian Financial Review*, 11 October 2008; Haxton N (2009) *Australian Broadcasting Corporation Transcripts*, 22 July 2009.
[55] Chessman BC (2011) *Wildlife Research* **38**, 664–671; Van Dyke et al. (2019) *Scientific Reports* **9**, 1998. *Emydura macquarii* is listed as Vulnerable in South Australia and Critically Endangered in Victoria.
[56] Australian Broadcasting Corporation News (2008) 'Students mount rescue campaign to save Murray turtles. A Milang primary school school scrapes the remnants of tubeworms off a turtle. Turtles rescued from South Australia's Lower Lakes'. *ABC News*, 24 December 2008.
[57] *The Land* (Sydney), 3 September 1948, p. 16.
[58] CSIRO (2008) 'Water availability in the Condamine-Balonne. A report to the Australian Government from the CSIRO Murray-Darling Basin Sustainable Yields Project'. CSIRO, Australia; Perrin A (2005) *The Courier-Mail*, 5 November 2005; Commonwealth of Australia (2006) 'Water policy initiatives. Final report. Standing Committee on Rural and Regional Affairs and Transport'. Senate Printing Unit, Department of the Senate, Parliament House, Canberra.
[59] MDBC (2008) 'The Narran Ecosystem Project: The response of a terminal wetland system to variable wetting and drying'. Final report to the Murray-Darling Basin Commission, MDBC Publication No. 40/08; CSIRO (2008).
[60] MDBC (2008).
[61] CSIRO (2007) 'Water availability in the Gwydir. A report to the Australian Government from the CSIRO Murray-Darling Basin Sustainable Yields Project'. CSIRO, Australia; *The Sydney Morning Herald*, 3 February 2007; Saintilan N, Ling J, Maher M (2010) In *Ecosystem response modelling in the Murray-Darling Basin*. (Eds N Saintilan, I Overton) pp. 55–66. CSIRO Publishing, Collingwood, Victoria.
[62] Woodford J (2004) *The Sydney Morning Herald*, 6 November 2004; Davies A (2006) *The Sydney Morning Herald*, 27 March 2006; Mitchell S (2007) *The Australian*, 29 October 2007; Clarke S (2007) *Australian Broadcasting Corporation Transcripts*, 25 June 2007; Morris S (2010) *The Australian Financial Review*, 14 January 2010.
[63] Armstrong JL, Kingsford RT, Jenkins KM (2009) 'The effect of regulating the Lachlan River on the Booligal Wetlands – the floodplain red gum swamps'. School of Earth & Environmental Sciences, University of New South Wales, NSW; Anonymous (2009) 'Red gum decline blamed on river management'. *ABC Regional News*, 26 February 2009; McLeod S (2009) *AM – ABC Sydney*, 26 February 2009.
[64] Kingsford RT, Thomas RF (2004) *Environmental Management* **34**, 383–396; Australian Broadcasting Corporation (2009) 'Yanga National Park still magnificent: Councillor'. *ABC News*, 5 January 2009.
[65] Wahlquist A (2007) *The Australian*, 28 April 2007.
[66] Fyfe M (2003) *The Age*, 21 April 2003.
[67] The decline of waterbirds has been particularly severe. See Kingsford RT et al. (2020) *Scientific Data* **7**, 172; Kingsford RT, Porter JL (2009) *Wildlife Research* **36**, 29–40; Casben L (2019) 'Waterbird population has fallen by as much as 90 per cent in Australia's east, shows 37-year study'. *ABC News*, 19 November 2019.

18
The great divide

This is what we think of the plan.

*– Griffith irrigator on the burning of copies of
the proposed Basin Plan, October 2010*[1]

THE ENVIRONMENTAL CRISIS CAUSED BY THE MILLENNIUM DROUGHT, WHICH gradually spread from an embryonic nucleus in the lower Murray to the entire Murray–Darling Basin (MDB), was matched in space and time by reputational damage that eventually threatened to engulf the public institutions and systems of governance that had failed to foresee or prevent it. The political magnitude of the problem cannot be overstated: environmentalists argued that water must be returned to over-extracted rivers, irrigators feared for their future, policy makers scrambled to make sense of a highly complex problem decades in the making, and much of the public began to see the disaster through the lens of climate change. The situation demanded action.

The first serious effort by governments to tackle the growing over-allocation crisis took place in June 2004, when First Ministers of New South Wales, Victoria, South Australia, the Australian Capital Territory and the Commonwealth Government signed the *Intergovernmental Agreement on Addressing Water Overallocation and Achieving Environmental Objectives in the Murray–Darling Basin*, a $500 million plan to return 500 GL of water to the Living Murray Icon sites; the Living Murray Business Plan was activated on 1 April 2005.[2] This got off to a slow start, however, as the states failed to implement the necessary water savings and as of 30 June 2006, as the drought was deepening, no water had been recovered for the environment.[3] By 2007, with an election looming, the Coalition government led by John Howard was under increasing pressure to address declining river health across the MDB (Fig. 18.1), and, stating that the 'old way' of managing the MDB had 'reached its use-by date', proposed a $10 billion National Plan for Water Security that aimed to recover water via improved irrigation efficiencies and water buybacks.[4]

What then unfolded was a truly watershed event: the passing of the *Water Act 2007* by the Australian Parliament and the subsequent mobilisation of a massive, cross-government plan to control and manage the entire water supply of one the world's largest river catchments – an ambition that was, to my knowledge, without precedent anywhere in the world. The Act established the Murray–Darling Basin Authority (MDBA), which was tasked with developing a plan for the MDB (the Basin Plan) that would focus on placing limits on the use of water by industries and communities and establishing mechanisms for reclaiming water for the environment. These decisions would be made by policy makers using sound application of what might best be termed applied or 'utilitarian' science. Fortunately, scientific interest in the plight of the drought-stricken Murray–Darling was on the rise, and although better modelling of sustainable diversion limits for MDB water resources was still required, the rapidly growing body of ecological, hydrological and economic data now provided a way forward.

From the outset, however, the development of the Basin Plan was politically flawed. For legal reasons the *Water Act 2007* was primarily interpreted as a piece of environmental legislation, and the entire process was centrally controlled and lacked the emphasis on negotiation and consensus typical of former water governance models.[5] This alienated irrigation communities, and when the MDBA released its 'Guide to the proposed Basin Plan' in October 2010, which recommended returning 3000–4000 GL of water to the environment (down from an initial maximum of 7600 GL per year) and major cuts to water allocations,[6] the response was immediate and angry (Fig. 18.2). On October 13 a thousand people sandwiched into an MDBA meeting at the Riverina town of Deniliquin, where a copy of the basin guide was set on fire.[7] The next day, between four and five thousand attended a volatile meeting at the Griffith Yoogali Club, where boxes of the guide were consumed in a bonfire, and a resident reportedly flung a costume horse's head at an MDBA representative.[8]

Overt public outrage on this scale is seldom seen in Australia, and recriminations began almost immediately. Faced with what appeared to be intractable differences between intense political pressure to consider economic and social costs on affected communities and legal advice that, under the Act, the MDBA must not compromise on environmental water needs, the chairman Mike Taylor resigned, ultimately to be replaced by former NSW Water Minister Craig Knowles.[9] In December 2012, following 2 years of political machinations, threats of legal action, reviews and a more formal public consultation process that included socio-economic concerns, an updated Basin Plan that mandated a return of 2750 GL to the environment was signed into law in December 2012.

But not everyone was happy: the Australian Greens Party echoed widespread concerns by stating that this volume was 'not adequate to achieve the majority of targets required by the Water Act for a healthy working river',[10] nor that it suitably addressed the likely impact of climate change on future run-off and stream flow across the MDB. Indeed, extensive rainfall-run-off modelling conducted by CSIRO, based on climate change projections from 15 global

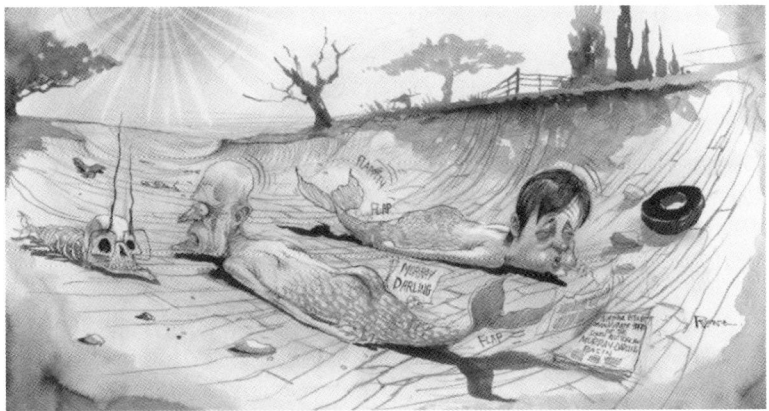

Fig. 18.1: *Flappity flap*, 2008. Reproduced with kind permission from David Rowe.

Fig. 18.2: The release of the 'Guide to the proposed Basin Plan' by the Murray–Darling Basin Authority caused considerable public anger over plans to reclaim water for the environment at the expense of irrigators and local communities, with food production a major point of contention. Photo: Lyndsey Vivian.

climate models under a medium global warming scenario, indicated that by around 2030, mean annual run-off across the MDB would decline by 9% relative to around 1990, but with much greater losses – in the order of 20–30% – also possible.[11] Such changes would put the delicate balance between water consumption and environmental requirements so painfully thrashed out during 2012–13 in jeopardy.

Ironically, by the time the proposed Basin Plan was signed into law very different climatic circumstances prevailed than those immediately after the introduction of the *Water Act 2007*. The 2009–10 El Niño that had ravaged the MDB peaked in December 2009, and by July 2010 La Niña conditions had returned. The ensuing 2010–11 La Niña, which was so strong that the additional rain over Australia, northern South America and South-east Asia caused the global mean sea level to fall by 5 mm,[12] broke down in May 2011, but was then followed by a second, weaker event in July 2011–January 2012. After very heavy rain fell across Victorian catchments in September 2010, flow in the Murray at the South Australian border steadily rose from less than 10 000 ML per day to a peak of just over 85 000 ML per day in mid-February 2011, the highest value since December 1993. Further peaks occurred in August 2011, April 2012 and October 2012, again after heavy rain across the upper catchments (Fig. 17.3). The Millennium Drought had finally broken, and a nation collectively gasped in relief.

RESILIENCE AND RECOVERY

The heavy rain and floods of 2010–12 gave scientists and land managers a once-in-a-lifetime opportunity to document the recovery of riverine ecosystems from a drought that caused impacts on a scale not seen for hundreds, or possibly thousands, of years. There was interest in the concept of *ecological resilience*, which we may define (at the risk of grossly oversimplifying a very complex area of research) as the broad ability of systems to regain their basic structure, processes and function, or remain mostly unchanged, when exposed to disturbance or stressors.[13] For floodplains, this includes the ability to oscillate between wet and dry phases, or states, without significant degradation.[14] In lay terms, we are interested in knowing whether ecosystems and species in the MDB were able to persist through the Millennium Drought or recover once flooding returned.

Thankfully, there now exists a vast literature on the degree of resilience in the MDB, and the overall answer is one familiar to ecologists but frustrating to others – simply put: it depends. Some species and populations recovered very little, or not at all, such as the Murray hardyhead, southern pygmy perch and the Yarra pygmy perch in the lower Murray[15] and populations of black box growing at elevations that require flows larger than those seen in 2010–12 for inundation.[16] In contrast, benthic invertebrates in the Murray Mouth and Coorong increased in abundance, although with a greater time lag than that observed during less severe periods of drought, while waterbirds abundance across the MDB increased, albeit temporarily.[17]

Many species that did recover often did so patchily. River red gum populations growing close to the main river or inundated during the floods survived or recovered better than those that remained dry,[18] and submergent aquatic plants that were killed in the lower lakes and Murray Mouth recovered in some areas and not others.[19] Murray cod, golden perch and silver perch generally increased relative to low-flow years but a major hypoxic blackwater event killed many native fish species, crayfish, shrimp and yabbies in long stretches of the river while leaving carp little affected.[20] At a broader scale, conditions for recovery were better in the southern MDB than in northern New South Wales and southern Queensland, which suffered from severe drought in 2013–15 and again in 2017–19. If this was not complex enough, scientists also disagree about whether the Millennium Drought exacerbated degradation of ecosystems that have long been in incremental decline or whether the observed changes are more indicative of a general pattern of fluctuating stability but with certain 'winners' and 'losers'.[21]

As we will see in the following chapter, controversy continued over the Basin Plan as well, particularly around oversight of water use in the northern MDB and the efficacy of buyback schemes. Further degradation, including extensive fish kills, again took place during the 2017–19 drought, and recent analyses of environmental watering regimes across the MDB indicated that wetlands have not received the water necessary to meet statutory conservation requirements.[22] More optimistically, provision of environmental water to target ecosystems in different parts of the MDB has had positive impacts on salinity, fish breeding, vegetation recovery, tree health and waterbird breeding, and knowledge of the most efficient way to deliver a given volume of environmental water to multiple ecosystems has also improved. The take-home message, I think, is that is extremely difficult to improve the ecological condition of a huge and heavily extracted water catchment that extends across multiple jurisdictions. As an even more water-restricted future looms under climate change, the fate of the MDB, in its fractious venosity, remains a story yet written.

THE VOICE OF THE RIVER

> ... water has a right to be recognised as an ecological entity, a being and a spirit and must be treated accordingly.
> – MLDRIN Echuca Declaration 2007

If, like me, you found wading through this mire of politics and conflict a tough slog, reflecting on some deeper philosophical dimensions to the Australian response to the water crisis caused by the Millennium Drought might be welcome relief. Now, at its core, the political tension that ignited during 2010–12 reflects the fact that the development of the Basin Plan represents a significant challenge to traditional constructions of water entitlements as being a state-controlled asset or an inherent right of irrigators. The fact that we can trace these constructs to the persistent influence of Australia's frontier past and assumed agricultural right to tame the environment[23] emphasises the degree to which they have been fundamentally shaped by powerful historical political and social forces,[24] and so they must be open to reconsideration if these societal values change. Indeed, other more marginalised perspectives, such as Aboriginal rejection of governmental claims to sovereignty over their traditional lands and rivers, challenge not only these collective beliefs but also the institutions that underpin them.[25]

While the Basin Plan did address the issue of water entitlements, its development involved two other important changes. First, the structural mode of decision making partially shifted from one of top-down centralised planning to one of 'decentralised regionalism' intended to enhance local deliberation and 'buy-in', although such systems have their own potential pitfalls.[26] Second, and I think even more importantly, science and technocratic rationalism were now placed at the core of environmental management across the MDB. This is not a new phenomenon in Australia,[27] and while it is now so ubiquitous that it goes virtually unnoticed by most of the population, I think that few people would dispute that the coupling of scientific objectivity with community concern represents an improvement over the almost exclusively extraction-based approach of prior years.

Nevertheless, I personally have also found it increasingly difficult to escape the conclusion that the relentless doubling-down on regulation, management and the rationalisation of the entire system in terms of optimising the use of water 'resources' has ultimately left the Murray–Darling perilously close to what Bill McKibben warned of in *The end of nature* (1989) – a post-natural, anthropogenic world in which there is nothing left but ourselves and landscapes that we have created.[28] I suspect that this may seem like hyperbole, but when I first visited Barmah Forest, a Living Murray Icon site, I was shocked to see the degree to which the Murray was regulated by dams, weirs and regulators. To me the riverscape resembled a planned garden;[29] the beauty of the unrestrained river, 'water in its loveliest form',[30] now lost in a channel used mainly for moving water downstream for human use.

Early critical theorists warned that technological rationalism could indeed lead to such outcomes, and, inevitably, the loss of human individuality, freedom and capacity to interact with the environment.[31] I cannot pretend to be a proponent of critical theory or postmodern reasoning, and whether some form of broad cultural *mors ontologica* has beset individuals or communities in the MDB with the loss of unregulated places should perhaps be best left to philosophers.[32] But these concepts do alert us to the fact that past ideas, problem framing and embedded institutional arrangements place significant constraints on the flexibility of new political or social structures to introduce alternative solutions to evolving environmental problems. These 'sticking points' or 'path dependencies'[33] can cause us to overlook deficiencies inherent in legacy modes of thinking and continue down trajectories that are maladapted to future challenges.

> *Enclosure, not wilderness, has become the major given of our lives.*
>
> – A Carl Bredahl Jnr[34]

Alternative models based not on command–control and technocratic structures but instead on appreciation of the inherent autonomy and self-determination of nature do exist. These ideas have a long history of development – the famous transcendentalist book *Walden* by Henry David Thoreau, for example, was published in 1854 – but recent decades have seen an explosion in the related domain of ecological *rewilding*, which essentially involves turning domesticated or degraded landscapes back over to natural processes.[35] Removing barriers or dams to return riverscapes to a more natural, self-regulated and connected state is among the most ambitious of these ideas,[36] but it is by no means a niche movement: the European Union's Biodiversity Strategy for 2030, for example, aims to restore 25 000 km of rivers to a free-flowing state primarily through the removal of obsolete barriers, and more than 200 barriers were removed across 17 countries in 2021 alone.[37] Similar efforts have been ongoing in the United States, where the removal of 57 dams in 2021 returned 2131 miles (3430 km) to a free-flowing state.[38]

> *Here was no man's garden, but the unhandselled globe.*
>
> – Henry David Thoreau, Ktaadn *(1848)*

Increasing the ability of floodplains and rivers to flood and function inside an increased 'freedom space'[39] represents not only a major paradigm shift from the way that they are currently managed, but also creates opportunities for 'nature-based solutions' and ecosystem services to generate socio-ecological gains that are too often undervalued today. Restricted, interventionist efforts are very expensive and rarely achieve the recovery of basic ecosystem function at the spatial scales desired, and rewilding promises to lower ongoing costs associated with management of water quality by slowing flows, reducing erosion, and flushing salt, and to provide higher-quality and better-connected habitats for plants and wildlife.[40] Crucially, returning rivers to a more autonomous state with natural flood and drought regimes may actually *help* riverscapes and their ecosystems adapt to the greater hydrological extremes expected under climate change.

Reimagining rivers as wild places – as entities unto themselves – also brings with it the opportunity to bridge the gap that has traditionally existed between Indigenous and colonial understandings of riverscapes and their place in human society.[41] Perhaps, then, it will be when the next great drought strikes, and the economic, environmental and human costs of maintaining climatically obsolete human infrastructure built during the 19th and 20th centuries become unsustainable, that our now strangled rivers rediscover their voice.

I find that an encouraging thought.

Endnotes

[1] 'Basin plan reduced to ashes as community's emotions run high'. *The Daily Advertiser* (Wagga Wagga), 14 October 2010.

[2] Murray-Darling Basin Commission (MDBC) (2005) 'Living Murray Business Plan'. MDBC Publication No. 05/05. Canberra, ACT; Crouch B (2005) 'Murray's big dreams left high and dry'. *The Sunday Mail*, 10 April 2005.

[3] Murray-Darling Basin Ministerial Council (2007) 'Review of The Living Murray Implementation 2005/06'. Report of the Independent Audit Group. MDBC, Canberra, ACT; Connell D, Grafton RQ (2011) *Water Resources Research* **47**, W00G03.

[4] Skinner D, Langford J (2013) *Water Policy* **15**, 871–894; 'Howard announces $10 billion water plan'. *The Sydney Morning Herald*, 25 January 2007.

[5] Connell D (2011) *Water Resources Management* **25**, 3993–4003; Skinner D, Langford J (2013) *Water Policy* **15**, 871–894.

[6] The Guide reported that cuts of between 3000 and 7600 GL per year of long-term average flows might be needed to meet specific environment objectives but lowered the upper limit to 4000 GL per year due to concerns over social and economic impacts on local communities. See Murray–Darling Basin Authority (2010) 'Guide to the proposed Basin Plan: Overview'. Murray–Darling Basin Authority, Canberra; Skinner D, Langford J (2013) *Water Policy* **15**, 871–894.

[7] 'Angry crowd burns copy of Murray-Darling report'. *ABC News*, 13 October 2010; Jopson D, Arup T (2010) 'Irrigators vent fury at proposed water cuts'. *The Sydney Morning Herald*, 14 October 2010.

[8] 'Basin plan reduced to ashes as community's emotions run high'. *The Daily Advertiser* (Wagga Wagga), 14 October 2010.

[9] Huntly D (2010) *The Daily Advertiser* (Wagga Wagga), 7 December 2010; Rodgers E (2010) 'Murray-Darling boss resigns'. *ABC Local*, 7 December 2010; Kenny M, Political Editor (2010) *The Advertiser* (Adelaide), 7 December 2010.

[10] Australian Green Party (2012) 'Minority report – The management of the Murray Darling Basin'. Inquiry into the management of the Murray-Darling Basin; Second interim report: The Basin plan. Senate Standing Committees on Rural and Regional Affairs and Transport, Commonwealth of Australia 2012.

[11] Chiew FHS *et al.* (2008) 'Rainfall-runoff modelling across the Murray-Darling Basin'. A report to the Australian Government from the CSIRO Murray-Darling Basin Sustainable Yields Project. CSIRO, Australia.

[12] Boening C *et al.* (2012) *Geophysical Research Letters* **39**, L19602.

[13] The original definition of resilience is the amount of disturbance that a system can withstand before changing to an alternative stable state; Holling CS (1973) *Annual Review of Ecology and Systematics* **4**, 1–23. Further discussion see Angeler DG, Allen CR (2016) *Journal of Applied Ecology* **53**, 617–624; Gunderson LH (2000) *Annual Review of Ecology and Systematics* **31**, 425–439; Folke C (2006) *Global Environmental Change* **16**, 253–267; Chambers JC, Allen CR, Cushman SA (2019) *Frontiers in Ecology and Evolution* **7**, 241, doi:10.3389/fevo.2019.00241.

[14] Colloff MJ, Baldwin DS (2010) *The Rangeland Journal* **32**, 305–314.

[15] Wedderburn SD, Barnes TC, Hillyard KA (2014) *Hydrobiologia* **730**, 179–190.

[16] Cunningham SC *et al.* (2013) 'Mapping the condition of river red gum (*Eucalyptus camaldulensis* Dehnh.) and black box (*Eucalyptus largiflorens* F.Muell.) stands in the Living Murray Icon sites. Stand Condition Report 2012'. Murray-Darling Basin Authority, Canberra.

[17] Dittmann S et al. (2015) *Estuarine, Coastal and Shelf Science* **165**, 36–51; Kingsford RT et al. (2015) *Marine and Freshwater Research* **66**, 970–980.

[18] Doody TM (2014) *Marine and Freshwater Research* **65**, 1082–1093.

[19] Ye Q et al. (2014) 'Ecological responses to flooding in the lower River Murray following an extended drought. Synthesis report of the Murray Flood Ecology Project'. Goyder Institute for Water Research Technical Report Series No. 14/6.

[20] King AJ, Tonkin Z, Lieshcke J (2012) *Marine and Freshwater Research* **63**, 576–586; Ye Q et al. (2014).

[21] C.f., Colloff MJ et al. (2015) *Marine and Freshwater Research* **66**, 957–969; Kingsford RT et al. (2015).

[22] Chen Y et al. (2020) *Marine and Freshwater Research* **72**, 601–619.

[23] Botterill L (2009) The role of agrarian sentiment in Australian rural policy. In *Tracking Rural Change*. (Eds Merlan F, Raftery D) pp. 59–78. ANU Press, Canberra; Day D (2005) *Claiming A Continent*. Harper Collins, Sydney.

[24] Downey H, Clune T (2020) *Critical Social Policy* **40**, 108–129.

[25] For example, see the *MLDRIN Echuca Declaration 2007* by the Murray Lower Darling Rivers Indigenous Nations; available at https://www.mldrin.org.au/wp-content/uploads/2018/07/Echuca-Declaration-Final-PDF.pdf; Downey, Clune (2020).

[26] Lane et al. (2004) *Australian Geographic Studies* **42**, 103–115; Lane MB, Corbett T (2006) *Journal of Environmental Policy & Planning* **7**, 141–159.

[27] Graham P (1990) *International Journal of Public Sector Management* **3**, 30–39.

[28] McKibben B (1989) *The end of nature*. Random House, New York; The roots of attitudes towards dominion over nature in Western culture are discussed in Nir B (2020) *Genealogy* **4**, 68, doi:10.3390/geneology4030068.

[29] The extreme expression of the idealised garden may be seen in the geometrical landscaping movement of the French Renaissance period.

[30] Haig-Brown RL (1974) *A river never sleeps*. Crown Publishers, New York.

[31] Marcuse H (1964) *One dimensional man*. Reprint, C. Nicholls and Company Ltd., Great Britain, 1970; Marcuse H (1992) *Capitalism Nature Socialism* **3**, 29–38; Habermas J (1971) *Toward a rational society: Student protest, science and politics*. Heinemann Educational, London.

[32] A philosophical approach that links sustainability, environmental autonomy and the freedom and self-determination of individuals is provided in Droz L (2019) *Philosophies* **4**, 42. The term *mors ontologica*, which means 'the death of being', is from Phillip K Dick's (1977) *A scanner darkly*. Doubleday & Company, Inc., Garden City, NY.

[33] Waylen KA, Blackstock KL, Holstead KL (2015) *Ecology and Society* **20**, 21.

[34] Bredahl AC (1989) *New ground: Western American narrative and the literary canon*. Chappel Hill: University of North Carolina Press.

[35] Famous examples include wolves in Yellowstone National Park. For reviews see Bakker ES, Svenning J-C (2018) *Philosophical Transactions of the Royal Society* B **373**, 20170432; Jepson P, Schepers F, Helmer W (2018) *Philosophical Transactions of the Royal Society* B **373**, 20170434.

[36] Rideout NK et al. (2021) *Marine and Freshwater Research* **72**, 1118–1124.

[37] European Union Commission (2021) *European Biodiversity Strategy for 2030*. First edition. Publications Office of the European Union, Luxembourg; Belletti B et al. (2020) *Nature* **588**, 436–441; Schiermeier Q (2018) *Nature* **557**, 290–291.

[38] Ding L et al. (2019) *Chinese Geographical Science* **29**, 1–12.

[39] Skidmore P, Wheaton J (2022) *Anthropocene* **38**, 100334.

[40] Bakker ES, Svenning J-S (2018) *Philosophical Transactions of the Royal Society B* **373**, 20170432; Keesstra S et al. (2018) *Science of the Total Environment* **610–611**, 997–1009; Rideout NK et al. (2021).

[41] Brierley G et al. (2022) *Land* **11**, 1272.

19

What lies beneath

Out of the crooked timber of humanity, no straight thing was ever made.
— *Immanuel Kant[1]*

LIFE IN A NORTHERN TOWN

In the Darling–Barwon River country (Fig. 19.1), the years that immediately preceded the disastrous drought of 2017–19 were a time of uneasy consolidation. Here, the scars left by the Millennium Drought ran less deep than they did further south, and the La Niña years of 2010 and 2011 had partly restored faith in the ability of the hydrological system to provide for the people on the river's banks and the irrigated crops upstream. The return of dry weather and serious heatwaves in 2013 and 2014, which brought drought to Queensland, caused some difficulty, but it was not until an El Niño event developed in late 2014, accompanied by dry conditions and maximum temperatures 1–2°C above normal, that the climatic situation took a more serious turn. By late 2015 this event had grown into a 'super El Niño' that eventually became one the three strongest on record (with 1997–98 and 1982–83).[2]

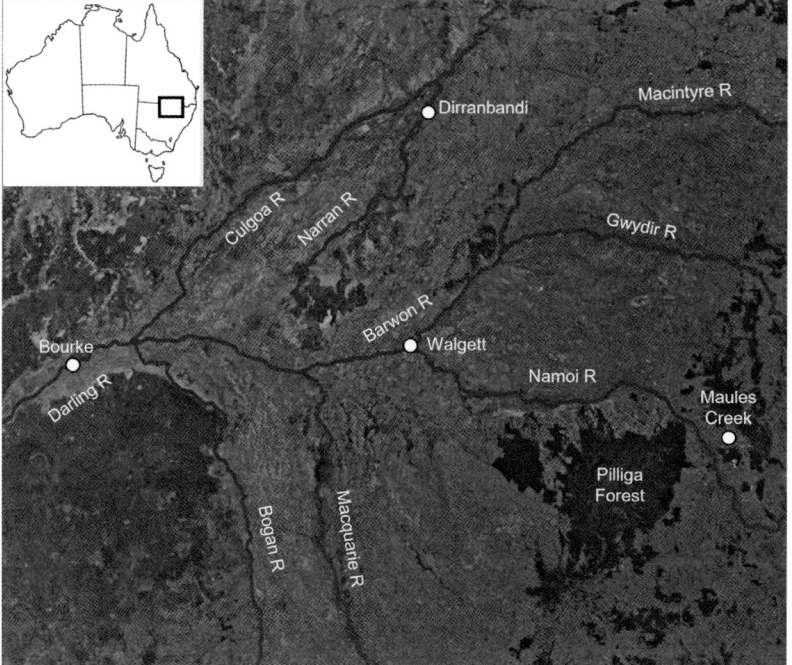

Fig. 19.1: Major rivers and some locations mentioned in the text. Satellite image courtesy Google Earth.

Again, however, the region was relatively lucky. To the north, 80% of Queensland was drought-declared following a patchy and unconvincing wet season, prompting the earmarking of more than $52 million in drought relief for primary producers.[3] But apart from a small area between Walgett and the Queensland border, the rest of New South Wales avoided severe drought, and in the winter and spring of 2016 a series of moisture-laden systems drifted across the continent, dropping record rainfall across the interior. Keepit Dam on the Namoi River and the much bigger Copeton Dam on the Gwydir captured years of water for irrigation (Fig. 19.2).[4]

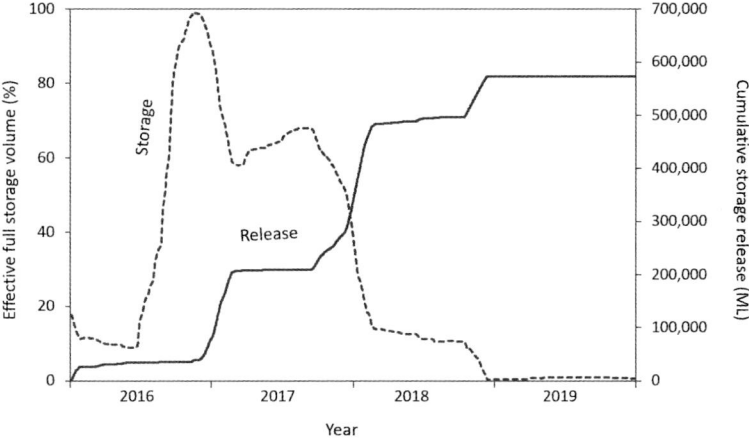

Fig. 19.2: Keepit Dam, the major storage on the Namoi River, showing the impact of the wet year of 2016 and the 2017–19 drought. Top: change in dam volume as a percentage of the storage when full (dashed line) and cumulative volume of water releases in megalitres (solid line) between 1 January 2016 and 31 December 2019. In late 2016 the dam was full but following water releases in the spring–summer irrigation seasons in 2016–17 and 2017–18 the dam was less than 20% full. After a final release in late 2018, and cumulative releases of 573 000 ML, the dam reached a volume of 1% and stayed at this level, or lower, throughout 2019. Bottom: the dam, essentially dry, in 2019. Photo courtesy of Peter Hardin.

For the first time in 4 years, drought had been banished from virtually the entire continent, the main exceptions being the northern half of the Cape York Peninsula and Arnhem Land.

But discontent bubbled just beneath the surface. In the Darling catchment the most important legacy of the Millennium Drought was the introduction in 2012 of the Murray–Darling Basin Plan, which called for a total reduction in surface water diversions of 2750 GL to improve the ailing health of water-starved rivers, floodplains and wetlands. Despite the fact that the targeted recovery from the northern Basin was revised down from 390 GL to 320 GL in 2016,[5] significant community opposition to the return of such large volumes of water for the environment remained.

> *And the land grows old and the people never*
> *Will see the worth of the Darling River*
>
> – Henry Lawson, 'The Song of the Darling River', *1899*

In Wee Waa, a small town of 2000 people dependent on the irrigated cotton industry, business leaders voiced concerns that the planned recovery of an additional 7 GL for a total of 20 GL from the Namoi River would leave the town economically ruined.[6] This was understandable, for they had witnessed the smaller town of Collarenebri, 115 km to the north-west, devastated by the recovery of around 30 GL of water from the Barwon River, roughly 66% of the 46 GL of local surface water entitlements. Much of this was purchased from the Twynam Agricultural Group, owner of the 35 600 ha Collymongle Station, in 2009, for a reported $300 million. During the late 1980s and 1990s, Collarenebri had thrived, as the cotton industry replaced wool production, but the population and local economy had begun to decline as drought and mechanisation reduced the demand for labour during the 2000s. But that single water buyback and the associated loss of irrigation and economic activity was the primary cause for the subsequent decline in the town population from 479 to 387 between 2006 and 2011.[7] Reduced to a dryland operation, Collymongle sold in 2012 for a reported $40 million.[8]

Although the impact on Collarenebri was particularly hard for locals to swallow, the economic consequences of water recovery at many other towns in the northern basin were also significant. Flow-on financial effects affected Moree, on the Gwydir, the town of Warren on the Macquarie River, where water purchases in 2009–10 targeted the recovery of 55 GL and, in Queensland, Dirranbandi and St George, both dependent on irrigation water from the Condamine–Balonne system. On the other hand, some towns benefitted from government investments in infrastructure to reduce inefficiencies and water loss, and graziers below St George and Dirranbandi looked forward to a return to increased pasture growth and carrying capacity on floodplains left high and dry by water extraction upstream.[9]

At the other end of the debate, concerns were growing that much of the vast government expenditure on the purchase of water entitlements and subsidies, which by 2015 had exceeded $5.8 billion, had been largely ineffective, since the extra cost of water acquired from infrastructure projects, particularly those run by the states, seemed greater than the cost of simply buying the

Fig. 19.3: A protest sign erected at the Menindee Lakes, which dried up in 2015, causing a water crisis in Broken Hill. The Menindee Lake System, historically a series of shallow ephemeral lakes, was modified in the mid-20th century to provide permanent water storage, supplying Broken Hill, Menindee and the lower Darling River. Photo courtesy of Peter Rae/Sydney Morning Herald.

water entitlements from willing sellers.[10] Questions were also being asked about the purchase of water from private business: the payment of $38 million by the Department of Agriculture and Water Resources in 2017 for 21 901 ML of water licences held by Webster Limited for its lower-Darling property Tandou, and a further $40 million compensation for loss of future business and the surrender of works approvals, for example, was highly controversial.[11]

Allegations of water theft and meter tampering along the Barwon and Darling rivers that aired on the ABC's Four Corners program on 24 July 2017 added further fuel to the fire.[12] These seemed particularly callous given that only 2 years previously Broken Hill had suffered a water crisis as the Menindee Lakes had dried up, and again faced the real prospect of running out of water (Fig. 19.3).[13] The allegations also drew the ire of farmers along the mid- to lower Darling, since a water-sharing plan for the Darling–Barwon enacted in 2012 had already incorporated last-minute changes that had benefitted upstream irrigators.[14] While some of the irrigators responsible for the water thefts were subsequently prosecuted,[15] ongoing concerns over maladministration, the use of water originally purchased by taxpayers for environmental purposes by irrigators for cotton operations, and the effectiveness of the Basin Plan for delivering ecological outcomes continued to bedevil water planning across the catchment.[16]

BLACK EYE

> *This is coal. Don't be afraid. Don't be scared. It won't hurt you.*
> *– Australian Treasurer Scott Morrison, House of Representatives, 9 February 2017*

Further upstream, tensions of a different kind were also reaching the boiling point. The 2010–12 La Niña had ushered in 3 years that were relatively cool and wet compared to those of the prior decade – although interestingly the Australian area-averaged mean temperature was still slightly above the 1961–90 average in both 2010 (+0.19°C) and 2012 (+0.14°C).[17] But severe heat returned in November 2012, bringing many of the hottest temperatures (some exceeding 45°C) ever recorded this early in the season to north-west Victoria, western New South Wales and northern South Australia.[18] Two months later, multiple waves of extreme heat swept across the continent, breaking all-time maximum temperature records at numerous meteorological stations. Australia set a record for the hottest day continent-wide on 7 January of 40.3°C, beating the previous hottest day on 21 December 1972 (40.17°C), while January 2013 and the December 2012 to February 2013 summer were the hottest on record for Australia.[19] Ultimately, the 12 months from 1 September 2012 to 31 August 2013 were the hottest such period on record at 1.11°C above the 1961–90 average. September 2013 was also easily the warmest on record, the +2.75°C anomaly setting a record for Australia's largest positive value for any monthly mean temperature.[20]

By 10 October 2013, unseasonably hot conditions and high fuel loads developed into extremely dangerous fire conditions across eastern New South Wales, particularly in the Hunter, Sydney Basin and Illawarra regions. Major fires erupted in the Blue Mountains, the Hunter–Central Coast region and the Southern Highlands of New South Wales, which, by the end of October, had burnt more than 100 000 ha of land, destroyed hundreds of houses and other buildings, and killed two people. The 'Red October' fires, which were the worst in New South Wales for decades, sparked significant international media coverage and heated political debate, mainly centred on the role that climate change had played in the disaster. In a notable exchange, former US Vice-President Al Gore likened the response of the Australian Prime Minister, Tony Abbott, who had stated that the fires were 'nothing to do with climate change', to 'the tobacco industry claiming smoking doesn't cause lung cancer'.[21]

In 2013, Maules Creek, 18 km north-west of Boggabri in the Namoi River catchment of northern New South Wales, found itself at the centre of this controversy. Here, beneath the ground, lie vast beds of coal formed during the early Permian, when Australia was part of the supercontinent Pangea.[22] More than 250 million years later, in 2010, a proposal was made by Aston Coal, later taken over by Whitehaven Coal, to develop and operate the Maules Creek Project, a 5-km-wide open-cut coal mine with recoverable total coal reserves estimated at over 400 million tonnes.[23] Despite mounting scientific and public opposition and a protracted mine blockade – one of the largest displays of civil disobedience in Australia's history – the project received the necessary regulatory approvals to commence construction by the Federal Government in 2013. In December 2014 the first coal was railed, destined for the port of Newcastle and then on to customers in the Asia Pacific (Fig. 19.4).[24]

At the same time, trouble was also brewing over plans by the energy company Eastern Star Gas, later acquired by Santos, to establish the Narrabri Gas Project a short distance to the north-west on the eastern edge of the Pilliga Forest. This project, which would involve construction of more than 800 coal seam gas (CSG) wells over a 95 000 ha project area with a footprint of around 1000 ha, also met with significant community opposition, including a targeted campaign by the Lock the Gate Alliance, a grassroots movement reportedly

Fig. 19.4: Changing landscapes in northern New South Wales. Top: the Maules Creek open pit mine sits atop a massive coal resource in the Namoi River catchment. Coal mines are an increasingly significant consumer of both surface water and groundwater resources and an exporter of 'virtual water' in eastern Australia, although they use far less than agricultural activities in most areas. Wikimedia Commons; photography by Dean Sewell. Bottom: a coal train near Boggabri in the Liverpool Plains of New South Wales. Photo: Holly Godfree.

supported by more than 100 000 community members and over 260 local groups opposed to CSG development.[25] Farm gates, fences and buildings across the region soon became adorned with 'Lock the Gate' signs. In an interesting fusion of feminism and environmental activism, the 'Knitting Nanas', reportedly formed out of frustration with inaction among male anti-CSG colleagues in north-east New South Wales, combined with numerous other groups and members of the public to form a highway protest spanning up to 2500 km of New South Wales and Queensland in April 2015.

Now, my primary intention here is to explain why much of the resistance to these industries is rooted in concerns over water security and the potential role of carbon emissions in driving changes to the climate. But it is also important to realise loss of amenity (essentially the pleasantness, attractiveness and desirability of a place) is an issue that galvanises fierce opposition to the development of gas fields and coal mines, especially by people living in, or who have ancestral ties to, the immediate area. The fact that the costs and benefits of mining are not borne in an equal way doesn't help: 97% of royalties collected by the New South Wales government are lumped into consolidated revenue, the bulk of which is spent in non-mining metropolitan areas, while the majority of costs are borne locally.[26] When construction of the Maules Creek mine commenced, the massive lights that were installed soon lit up the sky on the southern horizon of our farm, 20 km away. The night now provided no reprieve from the parched landscape, the crickets no relief from the dim hum of diesel earthmovers.[27]

The potential impact of both projects on water quality and availability was another bone of contention, because demand for both surface and groundwater for irrigation in the Namoi catchment had been steadily growing since the 1980s. This was driven in large part by the conversion of grazing properties to irrigated farms, which in the early 2000s sprawled over around 112 000 ha of land, 71% of which was under cotton. By 2004–05 surface water was already a 'highly developed' resource, with 359 GL out of a total of 965 GL per year (37%) extracted annually, and licences for these resources had become fiercely contested. A further 255 GL of groundwater was also being extracted annually, mainly from the Upper and Lower Namoi alluvia (39% and 35% of the total available respectively).[28] The most intense pressure occurred during periods of low river flow, such as 1982–83, 1994–95 and the early years of the Millennium Drought.[29]

Unfortunately, this level of groundwater extraction is at or above the long-term ability of aquifers to recharge themselves, and since ground and surface waters are closely connected, the whole hydrological system can be affected. Indeed, excessive extraction of water from bores can cause movement of water into aquifers from adjacent rivers, draining the surface water system. In 2004–05, for example, streamflow losses associated with groundwater extraction were estimated to be 99 GL per year, a staggering 28% of all surface water diversions.[30] This dynamic is currently playing out in the Maules Creek alluvial aquifer. After steadily increasing since the 1980s, water extraction for irrigation became unsustainable in 1994, when the Namoi River switched from gaining water *from* the aquifer to losing it *to* the aquifer, decreasing the hydraulic head of groundwater and causing smaller tributaries to experience longer periods of low or no flow.[31] A similar scenario has played out along the Namoi Valley, and for locals the impacts are obvious:

shallow wells that were once productive no longer produce, permanent or semi-permanent pools on the lower reaches of tributaries have gone dry, and the health of native trees has suffered.[32]

It is not difficult, then, to see why concerns were raised over the potential impact of these projects on water availability, especially since planning approaches in New South Wales do not have a good track record of capturing the cumulative impact of such developments on water resources.[33] Although standardised data on water use by coal mines are not publicly available, recent research indicates that a typical 'water footprint' for New South Wales coal mines is 653 L per tonne of coal produced with the largest mines using 5000 ML or more of water per year.[34] About 80% of this is sourced from natural watercourses or waterbodies, extraction of groundwater, and interception of rainfall or run-off.[35] The Maules Creek coal mine obtains water from 3000 ML high-security licence from the Namoi River, groundwater inflows of around 200 to 600 ML into the mine itself, and interception of rainfall and catchment run-off. Importantly, the demand on river water increases in times of drought: for example, 3104 ML were pumped in the drought year of 2018 compared to 269 ML in the wetter year of 2020.[36]

Compared with the water demand from irrigation and other major water users, these quantities are relatively small. However, there are now so many coal mines that their combined water use is no longer inconsequential. There are more than 80 operating black coal mines in New South Wales and Queensland, and in 2018 these used an estimated 225 000 ML of water, more than the entire population of Greater Sydney.[37] In the Namoi catchment there are seven operating coal mines (six open-cut and one underground)[38] and many more new mines and extensions are in planning stages or have been approved. If all of these become operational, significant groundwater drawdown (0.2 m or more) is very likely to occur up to 5 km from each mine (a total area of 156 km^2), although very unlikely more than 20 km away (an area of 2299 km^2). Close to mines, drawdown is likely to exceed 5 m.[39] Although less likely, surface water characteristics in unregulated systems could also be affected many kilometres downstream.

The Narrabri Gas Project is also expected to have some impact on local groundwater levels.[40] But for most, the greater concern is the potential for CSG water, which contains salt and toxins, to contaminate alluvial aquifers, thus threatening the viability of industries dependent on these resources.[41] A full review of this issue would require more attention than I can provide here, but in September 2014, Mary O'Kane, the New South Wales Chief Scientist and Engineer, reported that this risk can be managed, provided the geology and hydrogeology of the target area are sufficiently understood, and that engineering and scientific solutions are available to deal effectively with produced water and salts.[42] At the time of writing, Resources Minister Madeleine King expressed support for the Narrabri Gas Project, and calls are being made to fast-track the remaining approvals.[43]

Collectively, since ground and surface water resources are already fully (or over-) exploited, the development of the fossil fuel industry in the Namoi Valley reflects a very significant reprioritisation of the way that water is used at the catchment level. Due to the lifespan of new coal mines and CSG projects, these changes appear to be locked in for at least several decades to come, and there is little to suggest that this process will not continue. Virtually all campaigns to block coal mine and CSG developments in New South Wales and Queensland have failed, and

Australia's aspirational status as a 'global energy superpower' based on the export of fossil fuels (mainly to Asia), which was clearly articulated in the Australian Government's 2015 *Energy White Paper*, has since received bipartisan political support.[44] It would seem that we are rapidly entering a period that has been fittingly called the 'second era of coal'.[45]

The elephant in the room, of course, is that the security of water resources in Australia and in other parts of the globe where water is scarce is being placed under threat by the burning of these very same fossil fuels. Recent data show that governments of fossil-fuel-producing countries are planning to produce around 50% more fossil fuels than the amount required to keep global temperatures from rising less than 2°C.[46] Australia, already the world's largest exporter of coal and natural gas, anticipates increases of 34% and 33% for coal and gas exports between 2018 and 2030 respectively, and is therefore a major contributor to this looming 'production gap'. Recent analyses argue that, for coal, this plan is 'completely inconsistent with global implementation of the Paris Agreement' – in other words the world simply cannot absorb these exports while having any realistic chance of keeping below the 2°C target.[47] Indeed, without strengthening policies aimed at cutting global greenhouse gas emissions even further, increases of more than 3°C are on the table.[48]

Projections for the Murray–Darling Basin under such a climate make for sobering reading. More intense and severe droughts and heatwaves, along with earlier and more intense bushfire seasons, are expected and, particularly relevant to our discussion, streamflow is projected to have declined by around 20% (median projection) in 2060 relative to 1976–2005, driven by falling winter precipitation and increasing evapotranspiration.[49] This means reduced water storages and lower river levels but, at the same time, growing demand for water by crops and pastures. Inevitably, conflicts will arise over access to the dwindling supply of natural resources by an increasingly stressed agricultural sector.

But ultimately these are just words: what would it be like *actually living* in such a hot, dry and crowded world?

SOUTHERN DISCOMFORT

In fact, we do not need to look too far afield to see a region that has been dealing with gradual constriction of water availability in a situation that might be analogous to what other parts of Australia (and the world) are likely to face in the future. Beginning in the late 1960s and especially in the mid-1970s, just as the dry epoch and dust bowl conditions that afflicted central Australia was closing, south-western Western Australia underwent an unusual 'step change' to a significantly drier climate, and through the 1980s and 1990s conditions remained dry, with ca. 10–25% less rain falling during the late autumn and early winter (cf. panels in Plate 12).[50] In the mid-1970s and particularly during the extreme drought of 1975–79 this pattern began to seriously affect the Wheatbelt region, with rainfall between 1975 and 1999 running at close to 14% below the 1900–34 average.[51] Since 2000 the dry conditions have continued, and in many areas the mean annual rainfall has declined further, primarily on the back of an increased number of exceptionally dry years. In the Wheatbelt, for example, rainfall in the period 2000–16 fell by a further 14% below the 1975–99 average.

While these may not seem like major declines, they have had certainly placed food security and agricultural production in the Wheatbelt under serious pressure. Since south-western Western Australia experiences a predominately Mediterranean climate in which three-quarters or more of the annual precipitation falls between April and October, the decline in rainfall occurred at precisely the time of year when staple winter crops are being grown across the Wheatbelt. Much of this region, particularly in the east, was already marginal for cropping, and the observed reduction in rainfall shifted the potential wheat yield isoline (a line joining areas of equal yield potential) to the south-west by an average of 70 km between 1900 and 1934, which was a wetter period, and 2000–16.[52]

Fortunately, thanks to improvements in crop genetics and management and rising atmospheric CO_2 concentrations, which increase plant water use efficiency, actual wheat yield has either stayed largely stable or only slightly declined in parts of the eastern Wheatbelt.[53] This is undoubtedly an agricultural success story in which science and technological development have mitigated what would certainly have been an agricultural drought catastrophe. Whether this remains true will depend on whether future improvements can keep up with climate change, which is expected to place further pressure on water availability during the winter growing season.

The challenge posed to water security in the far south-west was far more significant. One of the unfortunate features of hydrological systems in these (and most other) forested catchments is that the amount of run-off generated per unit of rainfall – known as the run-off coefficient – can be influenced by many factors that change over time and hence often respond in a non-linear way to changing precipitation patterns. In this case, although rainfall declined by around 15–20%, water run-off into Perth's drinking water catchments (the Integrated Water Supply System) declined two-thirds from 338 GL per year in 1911–74 to 114 GL in 1997–2005.[54] Streams that were once perennial became ephemeral, often ceasing to flow even during the winter. Following another step change in 2000 inflows declined to 82 GL in 2001–06 and then to just 46 GL in 2010–17.[55] Even more concerning is that the more recent declines in run-off observed in catchments across south-west Western Australia[56] occurred in the absence of major changes in rainfall. The conclusion is inescapable: a new hydrological regime has developed in the region, one that is likely only to revert following many years of above-average rainfall.[57]

In the 1970s the decline in rainfall was thought to be part of natural variability, but by the 1980s the first connections were being made to climate change, with predictions of further declines in rainfall and run-off.[58] And, while there is likely an anthropogenic component, recent research has shown that the recent dry conditions are not unusual in a longer context and, indeed, much more severe multi-decadal megadroughts have occurred in the past. At least one has occurred in virtually every century since the 1300s, the most recent being in 1755–85, 1828–59, 1878–88 and 1889–1908, and conditions from 1908 until 1974 have, if anything, been unusually wet compared to those experienced in past centuries.[59] Perhaps the foundation of Perth's water supply was built on sand in both a metaphorical and a physical sense after all.

Whatever the cause, additional pressure was being applied by population growth. In 1950 Perth's population was tiny: just 312 000 people with an annual water demand of just 59 GL/year,[60] a volume that could easily be supplied by watersheds, even under the new rainfall regime. But by the 1975 the population had more than doubled to 770 000, as had drinking water demand, most of which was supplied still supplied by catchment run-off. Today the Perth population is 2.1 million with households consuming around 300 GL annually (with an additional ca. 100 GL used by commercial operators and for parks and gardens), and in 2050 it will have grown to around 3 million with an annual household consumption of over 450 GL per year. Surface run-off now provides just a small fraction of Perth's water, the majority being supplied by groundwater and desalination.[61] New options are being considered to accommodate further growth, including further groundwater replenishment with treated wastewater, expansion of seawater desalination, and expansion of local groundwater resources, although the latter too are under pressure from historically low recharge rates.[62]

The diversification of Perth's water supply for a more crowded and drier future is a remarkable success story that continues to unfold. As always, however, there have been unintended consequences of these changes. One of interest to ecologists is that the previously strong links that existed between forest catchment management and urban water supply have uncoupled as surface run-off has been replaced by alternative water sources. Ironically, as forest management diminishes as a driver of urban water availability, the risk is that fewer financial resources will be available for maintaining the health and biodiversity value of jarrah (*Eucalyptus marginata*) forests in the region.[63] This is particularly concerning given the broad-scale mortality in these and other forests across south-western WA due to drought over the past decade.[64] New approaches will therefore be needed to maintain the health and conservation value of these forests in the future drying and warming climate.

In the future, water availability elsewhere in southern Australia will probably come to resemble that of south-western Australia, because the pattern of lower rainfall in the south but wetter conditions in the north (Plate 12) is very likely a fundamental symptom of anthropogenic climate change.[65] In the south-east we have already seen the decline in rainfall that took place first in the late 1990s and then more severely during the Millennium Drought, and the consequences of the associated decline in streamflow across the southern Murray–Darling Basin. Recent research has highlighted that seasonal and annual surface water availability and streamflow have declined more generally over the past three decades in New South Wales, Victoria, Tasmania, south-east Queensland and South Australia, with only far northern catchments from the Kimberley to Arnhem Land (which have significantly increased) and northern Queensland (little or no change) bucking this trend,[66] mainly due to increased rainfall.

Among the southern states Tasmania has been arguably least affected, although there has been a decline in mean annual rainfall after 1975, which has manifested as a conspicuous lack of very wet years and several severe droughts. The worst of these occurred in 1979–83, 2006–08 (the driest on record over much of the state), 2014–15 and 2019, each of which caused significant impacts on forest ecosystems, including tree dieback and increased bushfire activity. Streamflow has also declined in some areas, and some climate models project it to decline further in coming decades, especially in the central mountainous part of the state.[67] Given

Tasmania's heavy dependence on hydroelectricity production, this would have significant consequences for energy security.

It is often said that the Earth's hydroclimate has played a central role in the history of human settlement patterns across the Earth, and perhaps nowhere is this more evident than in Australia. Here, most of the population clings to a narrow belt of temperate country around the southern and eastern fringe of the continent, benefiting from the cool and wet conditions and relatively secure water supplies. But I think that these case studies have shown that due to a combination of rising temperatures, declining rainfall and streamflow, increasingly severe hydrological drought and a rapidly growing human population, managing water resources across the south is set to dominate discourse for decades to come.

Endnotes

[1] From *Idea for a general history with a cosmopolitan purpose* (1784). Many variants exist of this quote, which was translated from the original German.

[2] Ren H–L et al. (2017) *Journal of Meteorological Research* **31**, 278–294.

[3] McConchie R (2015) 'Queensland drought spreads to record 80 per cent of the state'. *ABC Rural*, 13 May; Queensland Government (2015) 'Budget to continue to tackle widespread drought'. *The Queensland Cabinet and Ministerial Directory*, 14 July.

[4] The data are available at https://realtimedata.waternsw.com.au/.

[5] Murray Darling Basin Authority (MDBA) (2016) 'Northern Basin Review: Understanding the economic, social and environmental outcomes from water recovery in the northern Basin'. MDBA publication no: 39/16. Source: Creative Commons Attribution 4.0 Licence.

[6] Timms P (2017) 'Show us the science': Wee Waa demands to see MDBA's evidence behind request for more water. *ABC News*, 15 July.

[7] MDBA (2016) 'Collarenebri. Understanding community conditions'. MDBA publication no: 30/16; ISBN (online): 978-1-925221-52-7.

[8] Chancellor J (2012) 'Kahlbetzer family's Tynam Group sells Collymongle in northern NSW'. *urban.com.au*, 29 January.

[9] MDBA (2016) 'Northern Basin Review: Understanding the economic, social and environmental outcomes from water recovery in the northern Basin'. MDBA publication no: 39/16; Bartley K (2017) *The Daily Liberal* (Dubbo), 18 January.

[10] Grafton RQ (2017) *Water Economics and Policy* **3**, 1702001.

[11] Slattery M, Campbell R (2018) 'I'll have what they're having. A step-by-step guide to valuing compensation in the Lower-Darling'. The Australia Institute, Canberra.

[12] Australian Broadcasting Corporation (2017) 'Pumped'. *Four Corners*, 24 July.

[13] Stein G (2015) 'Broken Hill faces another water crisis as drought lingers and Menindee Lakes dry up'. *ABC Landline*, 20 March.

[14] Davies A (2018) 'NSW Minister altered Barwon-Darling water-sharing plan to favour irrigators'. *The Guardian*, 8 February.

[15] Brown B (2019) 'NSW cotton farmer Anthony Barlow fined $190,000 for breaching Water Management Act'. *ABC News*, 22 March; Newsroom (2020) 'Northern NSW irrigators Peter and Jane Harris found guilty of water theft from Barwon River'. *Moree Champion*, 20 March.

[16] Besser L, Fallon M, Carter L (2017) 'Murray-Darling Basin Plan: Taxpayer-purchased water intended for rivers harvested by irrigators', *ABC Four Corners*, 24 July; also Slattery M, Campbell R (2018) 'The Basin Files. Maladministration of the Murray-Darling Basin Plan: Volume I'. *The Australia Institute*, Canberra, June 2018.

[17] Each of the years 2010, 2011 and 2012 was much cooler than the decadal mean temperature anomaly of +0.48°C experienced between 2000 and 2009.

[18] Australian Bureau of Meteorology (2012) 'Extreme November heat in eastern Australia'. Special Climate Statement 41.

[19] The previous hottest day occurred on 21 December 1972 (40.17°C).

[20] Australian Bureau of Meteorology (2013) 'Australia's warmest 12-month period on record'. Climate update issued September 2013; Australian Bureau of Meteorology (2013) 'Australia's warmest 12-month-period on record, again'. Climate update issued October 2013.

[21] Australian Associated Press (2013) 'Al Gore attacks Tony Abbott's refusal to link bushfires with climate change'. *The Guardian*, 24 October.

[22] The paleogeography of Australia during the Permian is discussed in Scotese CR, Langford RP (1995) Pangea and the Paleogeography of the Permian. In *The Permian of Northern Pangea*. (Eds PA Scholle, TM Peryt, DS Ulmer-Scholle). Springer, Berlin, Heidelberg.

[23] The Maules Creek Coal Mine is a joint venture between Whitehaven Coal (75%), ICRA MC Pty Ltd (15%) and J-Power Australia Pty Ltd (10%). See Whitehaven Coal (2021) 'Annual Report 2021'. Whitehaven Coal, George Street, Sydney.

[24] Conditional approvals were obtained from the NSW Planning Assessment Commission on 23 October 2012 and the then Commonwealth Minister for Sustainability, Environment, Water, Population and Communities on 11 February 2013.

[25] The Lock the Gate Alliance was formed in 2010 in response to the rapid expansion of coal and coal seam gas development in NSW and Queensland.

[26] Drew J, Dollery BE, Blackwell BD (2018) *Resources Policy* **55**, 113–122.

[27] Apologies to T.S. Eliot; cf. 'And the dead tree gives no shelter, the cricket no relief' from '*The Wasteland*', which itself references Ecclesiastes 12: 5.

[28] Figures 2–4 in CSIRO (2007) 'Water availability in the Namoi. A report to the Australian Government from the CSIRO Murray-Darling Basin Sustainable Yields Project'. CSIRO, Australia.

[29] CSIRO (2007), Figures 2–4.

[30] CSIRO (2007); Australian Government National Water Commission (2007) 'Australian water resources 2005. A baseline assessment of water resources for the National Water Initiative. Key findings of the Level 2 Assessment: Summary results'. National Water Commission, Canberra.

[31] Giambastiani BMS *et al.* (2012) *Hydrogeology Journal* **20**, 1027–1044; CSIRO (2007).

[32] Reid N *et al.* (2007) 'Causes of eucalypt tree decline in the Namoi Valley, NSW'. Final Report to Land and Water Australia on Project UNE 42.

[33] Tan P-L, George D, Comino M (2015) *International Journal of Water Resources Development* **31**, 682–700.

[34] Black coal production uses significant quantities of water for processing, dust suppression, on-site facilities, among other uses; Overton IC (2020) 'Water for coal: Coal mining and coal-fired power generation impacts on water availability and quality in New South Wales and Queensland'. Report prepared by Natural Economy for the Australian Conservation Foundation.

[35] Overton (2020).

[36] Maules Creek Coal Mine Annual Review 2018, Table 14 (p. 62), and in Maules Creek Coal Mine Annual Review 2020, Table 19 (p. 74).

[37] Overton (2020).

[38] Namely Maules Creek, Boggabri, Tarrawonga, RocGlen, Sunnyside, Werris Creek, Narrabri.

[39] Post DA *et al.* (2020) *Journal of Hydrology* **591**, 125281.

[40] Post *et al.* (2020), Figure 5.

[41] Exposure pathways for CSG water are discussed in Navi M *et al.* (2015) *International Journal of Environmental Health Research* **25**, 162–183.

[42] NSW Government Chief Scientists and Engineer (2014) 'Final Report – Independent Review of Coal Seam Gas Activities in NSW'; Post *et al.* (2020).

[43] Crowe D, Foley M (2022) 'NSW will need Narrabri gas, federal resources minister says.' *Brisbane Times*, 15 June.

[44] See Department of Industry and Science (2015) 'Energy White Paper'. Australian Government. Since then

support for fossil fuel industries has been expressed on many occasions; see Australian Government (2019) 'National Resources Statement'; Henderson A (2020) 'Anthony Albanese says coal mining could continue in a net zero emissions world', *ABC News*, 23 February; Foley M, Chung L (2022) 'Labour digs in on support for coal to negate damaging climate debate'. *The Sydney Morning Herald*, 18 April.

[45] Connor LH (2016) *Energy Policy* **99**, 233–241.

[46] SEI, IISD, ODI, Climate Analytics, CICERO & UNEP (2019) 'The production gap: The discrepancy between countries' planned fossil fuel production and global production levels consistent with limiting warming to 1.5°C or 2°C'. Available at http://productiongap.org/.

[47] Parra PY *et al.* (2019) 'Evaluating the significance of Australia's global fossil fuel carbon footprint'. Report prepared by Climate Analytics for the Australian Conservation Foundation (ACF).

[48] IPCC 2022: Summary for Policymakers. In: *Climate Change 2022: Mitigation of Climate Change. Contribution of Working Group III to the Sixth Assessment Report of the Intergovernmental Panel on Climate Change.* (Eds PR Shukla *et al.*). Cambridge University Press, Cambridge, UK and New York, NY, USA.

[49] Prosser IP, Chiew FHS, Stafford Smith M (2021) *Water* **13**, 2504.

[50] Estimates vary depending on the region and time of year. See Timbal C, Arblaster JM, Power S (2006) *Journal of Climate* **19**, 2046–2062; Raut BA, Jakob C, Reeder MJ (2014) *Journal of Climate* **27**, 5801–5814.

[51] Fletcher AL *et al.* (2020) *Climatic Change* **159**, 347–364.

[52] Fletcher *et al.* (2020).

[53] Scanlon TT, Doncon G (2020) *Crop and Pasture Science* **71**, 128–133.

[54] Bates BC *et al.* (2008) *Climatic Change* **89**, 339–354.

[55] Liu N *et al.* (2019) *Journal of Hydrology* **572**, 761–770.

[56] Liu *et al.* (2019).

[57] Petrone KC *et al.* (2010) *Geophysical Research Letters* **37**, L11401.

[58] Sadler BS, Mauger GW, Stokes RA (1988) The water resource implications of a drying climate in south-west Western Australia. In: *Greenhouse: planning for climate change*. (Ed. GI Pearman) pp. 296–311. E.J. Brill, Leiden; McFarlane D *et al.* (2020) *Journal of the Royal Society of Western Australia* **103**, 9–27.

[59] O'Donnell AJ *et al.* (2021) *Climate Dynamics* **57**, 1817–1831.

[60] Petrone *et al.* (2010) *Geophysical Research Letters* **37**, L11401.

[61] Department of Water (2016) 'Water for growth: Urban. Western Australia's water supply and demand outlook to 2050'. St Georges Terrace, Perth.

[62] Priestley SC *et al.* (2023) *Communications Earth and Environment* **4**, 206.

[63] Harper R *et al.* (2019) *Annals of Forest Science* **76**, 95.

[64] Matusick G *et al.* (2013) *European Journal of Forest Research* **132**, 497–510.

[65] Dey R (2019) *WIREs Climate Change* **10**, e577; Timbal C, Arblaster JM, Power S (2006) *Journal of Climate* **19**, 2046–2062.

[66] Amirthanathan GE *et al.* (2023) *Hydrology and Earth System Sciences* **27**, 229–254; Zhang XS *et al.* (2016) *Hydrology and Earth System Sciences* **20**, 3947–3965.

[67] Bennett JC *et al.* (2012) *Hydrology and Earth System Sciences* **16**, 1287–1303.

20

Day zero

This is ... an environmental disaster.
— Niall Blair, NSW Primary Industries Minister, January 2019[1]

IT IS DIFFICULT TO CONVEY THE DESPERATION ONE FEELS WHEN FACED WITH a natural disaster that is so hostile, so extreme and on such a massive scale that resistance seems futile in the face of it. When witness to a phenomenon so powerful that it *will* unfold despite your wishes, your helplessness impacts you on a visceral level, and the security afforded by civilisation seems far away. To fully appreciate how this feels you must experience it first-hand,[2] as generations of people who depended on the land for a living have done in the past. But as a series of catastrophes ravaged the most populated parts of country during the 2019–20 'Black Summer', Australians from all walks of society were forced to confront the frightening prospect that the climate had permanently turned against them.

For me, that journey began on 3 January 2014, when the temperature at Narrabri climbed to 47.8°C, and at Walgett, 165 km away, 49.1°C. No temperature this great had been experienced in the north-west since the 1939 heatwave, and I had never seen anything like it. The strong north-westerly winds blowing off the western plains felt like a furnace. All around were signs of stress: birds flocked to the dams and stood submerged to their backs – magpies, hawks and honeyeaters bonded by a primal need to stay cool. Others drowned in buckets and troughs when seeking water, while chickens expired in their pens, dogs collapsed while working and kangaroos lay prostrate under shrubs. Grasses and forbs withered and turned brown and leaves of the hardy motherumbah (*Acacia cheelii*) turned yellow, curled up and fell to the forest floor. Everything cooked impartially in that great earthen griddle.[3] As the heatwave moved into Queensland, more than 100 000 flying foxes died of heatstroke across the south-east, the corpses piling up under their roosts.[4]

Then, 3 years later, it happened again. In December 2016 and January 2017, multiple, back-to-back heatwaves swept across eastern Australia, setting many temperature records. The average maximum temperature at Narrabri during January was 37.8°C (100°F), and then, following 12 straight days of 37°C or more, it reached 46.5°C on February 12. Walgett sweltered at 47.9°C. Moree, 100 km to the north, suffered through 54 straight days of 35°C or above, the longest such streak on record in NSW.[5] There was nothing you could do except watch the last vestiges of moisture that had fallen in 2016 bake away as if the very air itself begrudged the land any relief.

The return of drought then began to teach an amnesic country a tough lesson about limits. By March 2017, despite above-average rainfall over the previous year, most of Queensland (87%) was drought-declared[6], with central and south-eastern parts of the state worst-affected. At Biloela, cattle properties were completely destocked for the first time in 60 years. From Bundaberg south, crops including peanuts and sugarcane suffered severe stress and yield declines, and pastures failed.[7] The extreme temperatures that afflicted the entire continent during the 2016–17 summer made the situation worse, as rising evapotranspiration rates increased plant water stress. Since most of the water taken up by plants is vaporised in the air, keeping leaf tissue cool, water stress can cause plants to overheat and suffer yield declines.[8] Evidence is emerging that increasingly severe drought and heatwaves pose a global threat to crop yields[9] and, although within certain limits these can be mitigated by irrigation,[10] this presupposes highly efficient water distribution systems and, more importantly, adequate access to water.

Further south in New South Wales, this situation was becoming extremely tenuous by early 2018 as another dry summer and autumn intensified rainfall deficiencies that had commenced in 2017. Major water storages, which had begun 2017 at healthy levels, dropped dramatically over the summer 2017–18 irrigation season, and by August intense drought had enveloped the tablelands, slopes and plains north of the Sydney Basin. By mid-2018 virtually the whole state was drought-declared.[11] With inflows drying up, Copeton Dam fell to less than 30% capacity, Burrendong and Windamere dams (Macquarie catchment) to less than 40%, and Keepit Dam to around 10%, after having released around 230 000 ML (= 230 GL) of water over the previous spring and summer (Fig. 19.2).[12] As none of these releases could be replenished, supplies remained for just one more irrigation season at reduced allocations.

Irrigated crops bore the immediate brunt of these water limitations, and in October 2018 Cotton Australia predicted that the cotton crop in the eastern cropping belt of New South Wales and Queensland would shrink to 2.2 million bales, down more than 50% from 2017–18.[13] The following year (2019–20) would prove to be a disaster: with major water storages either dry or critically low, cotton production plunged to a 37-year low of 589 656 bales, 87% down from 2017–18.[14] Other crops across the state suffered the same fate.[15] In Narrabri, with its heavy reliance on irrigation-based agriculture, the economic desperation was palpable: the flow of people into shops slowed to a trickle, business lurched to a standstill, and property prices fell.

By late 2018, news began to flow into the media that many inland towns were running out of water. Downstream, at the town of Wilcannia, the Darling River had for months been reduced to toxic stagnant pools dotted with dead animals, forcing more than a hundred protesters, mainly local Barkindji people, to blockade the town's bridge to draw attention to the destruction of the river.[16] Walgett followed soon after; the Namoi ceased to flow upstream of the town in April 2018, and the Barwon 3 months later.[17] According to local elders, it was the first time in their lifetime that it was possible to drive across the beds of both rivers.[18] With the local weir dry, the town was forced to use bore water from the Great Artesian Basin, but

> *Oh, the Darling River. We are the people of the river. The Barkindji people of the river. The river is our home.*
>
> – *Barkindji children at Wilcannia protesting the state of the Darling River*[20]

concerns were raised that the high salt content may have exposed the Aboriginal population, who have a high incidence of chronic disease, to additional health problems. From Walgett to the Murray, people had become united in their voice to save the dying river.[19]

But the worst was yet to come. In December 2018 a local property owner discovered hundreds of dead fish in the Darling River near Menindee, the stagnant water reduced to a 'dirty rotten green'.[21] The next day the true extent was revealed – 10 000 Murray cod, silver perch, golden perch and bony bream had been killed over a 50-km stretch of the river. The immediate cause was attributed to low oxygen saturation due to an outbreak of blue–green algae (actually cyanobacteria), which thrive under warm temperatures, low river flow and high nutrient levels (usually generated by agricultural run-off), although later independent investigations revealed that the sudden mortality was exacerbated by rapid mixing of oxygen-depleted bottom water that had become highly stratified due to excessively high temperatures.[22] It got worse: hundreds of thousands more died in early January 2019,[23] and then a third time later in the month (Fig. 20.1).[24] In total, up to a million fish died, truckloads of which were buried in mass graves near Menindee.[25] In a major embarrassment to the nation, news of the disaster rapidly spread around the world in outlets as varied as *The New York Times*, *New Scientist*, *The World Economic Forum* and social media.

The images of thousands of decaying fish floating in the Darling River starkly revealed not only the true state of aquatic ecosystems in the Murray–Darling Basin, but also the failure of state and federal governments to ultimately come to grips with the problem despite having spent billions of dollars and decades of work on water management and legislation. NSW Primary Industries Minister Niall Blair faced the full brunt of community anger, even reportedly receiving threats to himself, his staff and his family.[26] I doubt anyone who saw decades-old Murray cod gasping along the riverbank for their last breath could have remained unaffected

Fig. 20.1: Dead fish in the Menindee Weir Pool, 29 January 2019. AAP/Graeme McCrabb.

Box 20.1: Water footprint and virtual water

The concept of virtual water, which refers to the volume of water that is used in the production of a commodity or service, is one of the more interesting ideas to have emerged in the study of global sustainability and resource use in recent decades. By directly relating water inputs to industrial outputs, virtual water can be used to compare the intensity of water use across different industries and regions and as an accounting method to quantify the geographical exchange of water via traded goods, which has been called the *virtual water trade*. Crucially, when a country exports goods overseas that have required domestic water to produce, it is also virtually exporting the water that is 'embedded' in its production. Since agricultural production is the main user of water globally, with crops alone consuming between 5000 and 10 000 km^3 (5–10 million GL) of fresh water annually, trade in agricultural products is a major source of virtual water redistribution around the planet.

The Middle East and North Africa are major importers of agricultural products, and it is argued that the global virtual water trade has essentially allowed these regions to avoid conflict over what is an increasingly scarce local resource. On the other hand, countries that export large quantities of food are also effectively exporting water, which has been appropriated from groundwater or streamflow resources (blue water), the soil (green water), or recycled or reused sources (grey water), in return for revenue. It is estimated that in 2014 Australia's 'hidden' net export of virtual water totalled 35 km^3 (35 000 GL), with irrigated crops like cotton, which has a large water footprint (roughly 4–5 ML of water per tonne of production), playing a significant role in this imbalance. If we also account for the value of commodities produced, the most efficient agricultural products in terms of 'water value' are sheep meat and wine grapes and the least being cotton, rice and sugar.

The idea of virtual water has been criticised as an indicator of sustainability for not taking into consideration local water scarcity, and a range of other economic and practical considerations. Nevertheless, it has also stimulated significant discussion in policy circles about the most efficient way to use water for maximum economic gain, particularly as climate change, increasingly severe droughts and population growth drive greater water scarcity in the future.

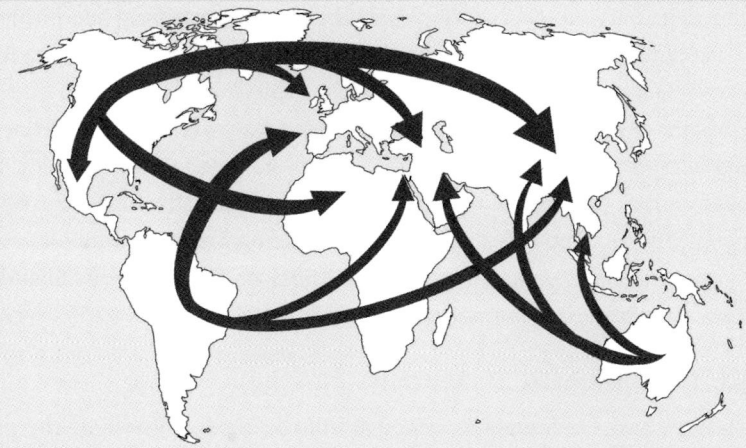

Some major export routes for virtual water across the globe.

by the situation, and Blair's obviously emotional interviews near Menindee at the height of the crisis make for painful viewing. Two months later he stepped down from the ministry and in 2020 left the National Party.[27]

Fish kills soon spread to smaller catchments upstream, and then to other major river systems. In December 2018 and January 2019 hundreds to thousands of fish were seen dead in the upper Namoi, Macquarie, Macintyre, Murrumbidgee, Dumaresq and Barwon rivers, and the following spring and summer similar events occurred in the Mehi, Border, Gwydir, Peel, Macquarie, Murray, Severn, Mooki, Cudgegong, Barwon, Darling and Lachlan rivers.[28] Most fish kills in 2018–19 were also caused by low oxygen levels associated with very low flows and high temperatures, although in some cases refugial waterholes containing stranded fish simply dried up. Others in February and March 2020 occurred as flows returned to drought-affected rivers, with either large volumes of organic material resulting in a rapid reduction in dissolved oxygen levels, or ash and silt washing into rivers from bushfire-affected landscapes.

As these events unfolded, the backlash against irrigators in the upper catchments, and particularly the major cotton interests in northern New South Wales and southern Queensland, was severe. To many the Murray–Darling Basin Plan lacked transparency, oversight and direction, and that water allocations to the irrigation sector were subject to heavy political influence by the cotton industry – the fish, as they say, rots from the head down – with decisions ultimately made by 'captive politicians'.[29] Some political figures called for an emergency embargo on water extraction for cotton irrigation,[30] while the concept of 'virtual water' was used by the South Australia-based Centre Alliance Party to argue for a ban on growing the crop entirely (Box 20.1).

Although the cotton industry resented being made a 'whipping boy' for the disaster,[31] and argued that cotton – as they put it, a 'desert plant that is water-efficient' – simply represented the best return per megalitre of water purchased by farmers,[32] the fact remains that the industry consumed in the order of 2500 GL of water in 2017–18, much of it in the Darling River catchment. The Barwon–Darling has always been prone to drying out during multi-year periods of extreme drought,[33] but the extraction of such a large volume of water for irrigation from fully or over-allocated tributaries meant that this water was simply unavailable to wetlands and other ecosystems downstream, and the decision to release water from storages for irrigation and other purposes in 2017 and 2018 spelt doom for the entire system. The Darling River's unspoiled aquatic ecosystems, survivors of countless drought and witnesses to the lives of Barkindji people, the exploration parties of Sturt and Mitchell, the paddle steamer captains with their mighty cargoes of wool, and countless others who had born, lived and died on its banks, were now no more than a memory.[34]

CODE RED: THE 'BLACK SUMMER'

Although the shock and suddenness of the fish kills of 2018–19 brought the plight of the Darling to national and international attention, above the banks, ecological decline had also been occurring in a slower, but no less devastating, way. Here, the lack of rainfall and ceaseless attentions of hungry livestock had reduced the ground to dust: every skerrick of water and grass had been blasted from the landscape, leaving the earth naked against the blazing heat. Furnace-like winds whipped up mighty storms and immense walls of choking dust that reached hundreds of kilometres across the landscape and deep into the sky, reawakening 'dust bowl' landscapes on a scale not seen for over 70 years (Plate 21).[35] Only goats seemed to prosper, returning the Cobar Peneplain country to a state of degradation resembling that caused by the rabbit plagues of the 1890s–1940s (Plate 16).

With no feed and little water, graziers destocked or lost their animals to starvation; sheep numbers fell to under 66 million, fewer than at any time since the end of the Federation Drought.[36] Beef cattle numbers declined to the lowest point since at least the early 1990s.[37] Crop production fared no better: Australian wheat production in the 2019–20 season fell to 14.48 million tonnes, the lowest since 2007, at the height of the Millennium Drought, and down 55% on 2016 (31.82 million tonnes).[38] The shortage in high-protein wheat required imports to be secured from Canada, the first foreign-owned grain to enter Australia since 2007.[39]

Native animals died in droves. At Lake Cawndilla, near Menindee, thousands of emus flocked to the last vestiges of water, collapsing in the mud from hunger (Fig. 20.2). Kangaroos invaded small towns like Angledool, in northern New South Wales, as the last of the water gave out, then littered the ground with their desiccated corpses. Even populations in more favoured

Fig. 20.2: Top left: starving emus bogged in the drying Lake Cawndilla, 20 km south-west of Menindee. Top right: eastern grey kangaroos that have died of thirst, near Mt Kaputar, northern NSW. Bottom left: extensive dieback of coolabah (*Eucalyptus coolabah*) on floodplains near Angledool, NSW. Bottom right: dead river she-oak (*Casuarina cunninghamiana*) on the banks of the Namoi River at Warrabah National Park. Photos: Robert Godfree.

areas suffered: on our property, grey kangaroos, euros and red-neck wallabies, which had been engaged in a grim battle against starvation for months, faced a desperate situation when the last dam dried up completely. Some became bogged in the mud, others dug up the roots of white box trees and tore off the bark to get at the water within. They sought solace in shade, dozens lying dead and dying in the hay and machinery sheds and under buildings (Fig. 20.2). I euthanised those still alive and buried the bodies with a loader.

In the eastern ranges, a vast belt of remnant forests and woodlands that clothe the rugged and inhospitable country lying between the coastal fringe and the tablelands was also fast drying out. With agricultural production collapsing and towns running out of water, this change had so far received very little attention, but when I travelled from Brisbane to northern New South Wales in July 2019, I was surprised to see scattered patches of trees turning brown from the southern Darling Downs and Granite Belt of Queensland to the Northern Tablelands of New South Wales. Even rainforests in the Border Ranges, which usually avoid severe drought, were wilting and dropping their leaves to the forest floor.

At the time, I was conducting radio interviews about a paper that I and my colleagues had published in the scientific journal *PNAS* on the impact of the 1895–1903 Federation Drought on plants and animals across the continent.[40] When asked about the potential for widespread forest death to occur during the present drought, I suggested that, based on our work, ecosystems in this region appeared to suffer this fate after 3 successive years in which rainfall is less than half of the average. With record dry conditions gripping northern New South Wales and drought predicted to continue, I expected to see the widespread death of canopy trees in this region beginning towards the end of 2019, just as they had in 1902.

During the first 6 months of 2019, still no significant rain had fallen and with maximum temperatures running at 2–3°C above normal soil moisture dropped to critically low levels and leaf litter on the forest floor lay as if it had been kiln-dried. Much of eastern Queensland was now facing an elevated threat of bushfires, a highly combustible situation that had been building for several years.[41] This raised alarm bells for Queenslanders, since hundreds of fires had raged across the state in the previous spring and summer, and although the responses of emergency management services and the broader community to these events were seen in a favourable light – 'a climate for good neighbours'[42] – the fires had deeply affected many communities, and preparation for the coming season began ahead of schedule.

By the end of the winter, following week after week of dry weather, the situation had become dire. The Australian Seasonal Bushfire Outlook for August 2019 made sobering reading, with a vast region from Rockhampton, Qld to south-east Victoria, as well as parts of eastern Tasmania and southern South Australia and Western Australia, all facing above-normal fire potential and the strong possibility of active bushfires.[43] In north-east and south-east New South Wales the bushfire season officially began on 1 August, 2 months earlier than normal, and by the middle of the month serious fires had broken out in the North Coast region, with multiple houses lost.[44] By November people across northern New South Wales were waking up to ash and dust[45] – fires that many described as 'apocalyptic' were raging from the Border Ranges to the Oxley Wild Rivers and inland to the biodiverse forests of the Torrington, Hanging Rock and Mount

Kaputar regions. The flames lit up the sky for weeks, the smoke filtering down from the peaks and settling on the farms below (Plate 21).

The rest is history. By February 2020, more than 10 million ha of forest, woodland and other native vegetation in south-eastern Australia had been incinerated, killing millions of animals, destroying thousands of homes and leaving 31 people dead. In total, 11 fires exceeded 100 000 ha in area, the usual standard for a megafire, but the scale of the catastrophe was so great that seven fires exceeded 1 million ha in size, making them among the largest ever seen on Earth.[46] No fire season in eastern Australia within the past 200 years has been even close to this magnitude.[47] Apart from severely impacting upon air quality across the east coast[48] and releasing significantly more CO_2 than Australia's normal total annual fossil-fuel emissions,[49] the event triggered phytoplankton blooms in the Southern Ocean[50] and a smoke cloud that reached 35 km altitude, blocking solar radiation on a planetary scale and altering the composition of the stratosphere.[51]

I was fortunate to be involved in CSIRO's early research into the impact of the fires on native Australia plant species, and what we found was eye-opening. Because much of the habitat burnt in the fires lies within a global biodiversity hotspot (an area of exceptional taxonomic diversity), and because many species in this region have quite small ranges (under 500 km across), my colleagues and I determined that more than 800 plant species had experienced fire over half or more of their ranges. Most of these species are well adapted to fire, possessing the ability to regenerate from epicormic buds, from lignotubers or from a persistent seedbank, and so are already well on the path to recovery. However, a small number of species that lack these traits, or are particularly sensitive to fire, such as epiphytic orchids that are forest canopy specialists, and certain rainforest taxa, have suffered very significant and probably persistent declines on their ranges.[52] Research into the consequences of the Black Summer fires for impacted ecosystems is ongoing.[53]

One interesting question is whether the extreme drought conditions exacerbated the impact of the fires on plant populations. We know that drought–fire interactions are increasingly restructuring ecosystems around the globe – a extreme example is the Amazon Basin, where high-intensity fires combined with extreme drought have caused widespread tree mortality, canopy decline, loss of above-ground biomass and invasion by flammable grasses.[54] Recent work conducted in the western and northern Sydney Basin, which reported that severe drought reduced juvenile tree survival and the replacement of dead individuals in two different forest types after fire, show that these concerns are well founded here, too.[55] The resprouting species that dominate many Australian forests are probably more resilient to such broad-scale changes, but it is also possible that other disturbances and biotic stressors such as grazing by feral animals may increase mortality and reduce recruitment after fire, especially if the frequency of burning is increasing under climate change.[56] At least we were very fortunate that the Black Summer fires were followed by more than 2 years of well-above-average rainfall.

On the western slopes and plains, however, the rain came too late to save vast stands of trees that simply died due to the severity of the drought. In the drier ranges, whole stands of black and white pine, red ironbark (*Eucalyptus crebra*) and white box were killed, along with rare species like ooline (*Cadellia pentastylis*). Along the creeklines vast numbers of river she-oak (*Casuarina cunninghamiana*), rough-barked apple (*Angophora floribunda*) and river red gum stood dead, leaving the banks exposed to erosion (Fig. 20.2). Tree dieback was not limited to the north-west: the New England Tablelands, central and far west New South Wales and southern Queensland were all affected. Without a doubt, the worst I have seen was in the Angledool–Lightning Ridge region of north-west New South Wales, where extensive patches of forest exist in which where virtually every individual tree and shrub was killed, young and old alike, and even the hardiest species like budda (*Eremophila mitchellii*) and wilga (*Geijera parviflora*).

When travelling through this region to document the dieback, it seemed as though a bomb had been dropped on the landscape (Fig. 20.2). Rory Treweeke, owner of Angledool Station, thinks that many woodlands and forests here will take decades to regenerate, and there are concerns that some species, like white cypress pine, may struggle to recover at all, since in many places even the older seed trees were killed. Perhaps we are witnessing a transition of these systems to a new state; if droughts like this become more frequent or severe in the future, I think that major change will be inevitable.

From a biogeographical perspective, then, changes in the composition and structure of ecosystems caused by the Black Summer fires and the 2017–19 drought should not be viewed in isolation – they are drivers of change that have left both overlapping and discrete footprints on the landscape. The fact that so many forests and woodlands that remained unburnt by the fires suffered extensive dieback due solely to the severity of the drought is a major concern for conservation biologists, since in prior times they would have served as refugia for plant and animal populations. Furthermore, the landscape has changed drastically in the last 200 years: forests have been extensively cleared and fragmented, riverine systems have been degraded by water extraction and diversion, and invasive species are spreading into remaining habitat. We simply cannot be certain that these systems retain the resilience that they once had. Ultimately, I think, satellite images tell this story better than any words can (Plate 22).

THE LEGACY

Let us conclude by focusing on the question that is now on everyone's mind: whether long-term changes to the climate contributed to the intensity and magnitude of the Black Summer fires and drought. Even during the early stages of the disaster, the Climate Council stated unambiguously that climate change caused by the burning of coal, oil and gas was 'supercharging' extreme weather events in Australia, and that the 'terrible trifecta' of bushfire, heatwaves and drought that was unfolding during late 2019 was not 'normal'.[57] This perspective is consistent with both the increase in fire danger observed across Australia since the 1970s and predictions made in several earlier studies, including the Garnaut Climate Change Review of 2008, which concluded that rising atmospheric temperatures and shifts in rainfall patterns should cause fire seasons in Australia to lengthen and become more intense, changes that should be observable by 2020.[58]

In the past couple of years, several studies have attempted to quantify the extent to which the 2019–20 Australia fire season might have been exacerbated by anthropogenic climatic change.

Unsurprisingly, such 'attribution' studies are extremely difficult because fire danger depends on many factors, including the dryness of the fuel, weather conditions, and whether an ignition source is available.[59] In Australia, fire weather conditions are modelled using McArthur's Forest Fire Danger Index (FFDI), which is based on the daily maximum temperature, wind speed, relative humidity and a 'drought factor' – a proxy for fuel dryness.[60] Models investigating the impacts of CO_2 emissions and climate change on fire risk must therefore incorporate these factors and, ideally, the amount of fuel (plant material) available. At the risk of doing these complex studies an injustice, I will attempt to extract the main messages that I think emerge from their data.

Generally, empirical data and state-of-the-art models indicate that fire weather in Australia has increased significantly in recent decades, due in large part to a rise in Australia's mean temperature by 1.4°C and increased prevalence of extreme heat events since 1910. These changes have been at least partly attributed to anthropogenic warming.[61] Although some studies have been unable to detect an anthropogenic signal in the rainfall deficiencies that occurred during 2017–19, a decrease in rainfall has been experienced across southern and eastern parts of the continent, consistent with some climate model projections.[62] Both changes are responsible for increasing fire weather severity.

Rising atmospheric CO_2 is also almost certainly causing an increase in fuel load due to carbon fertilisation, a process that results in an increased rate of photosynthesis but reduced evapotranspiration through the stomata. Indeed, in CO_2-rich environments plants may stay greener (and hence productive) longer during periods of drought, and this process is now detectable even at the planetary scale – vegetation across much of the globe has greened significantly in recent decades.[63] Climate–land-surface-model simulations suggest that the increase in annual fine leaf litter accumulation in Australia temperate forests could be 1.5 tonnes per hectare or more, adding significantly to fuel load and fire danger.[64] Collectively, I think that the balance of evidence shows that anthropogenic climate change associated with CO_2 emissions combined with carbon fertilisation has increased the probability of dangerous fire weather like the conditions experienced during 2019, and also contributed to the growing scale and intensity of fires in recent decades.[65]

It is important to note, however, that some uncertainty still exists about how fire weather across Australia might change in the future, particularly at the regional scale. One problem is that climate models frequently fail to adequately replicate important aspects of global and regional climates and often differ significantly in climate projections. Although great strides are being made, the reliability of probabilistic climate change projections based on a collection of such models ('ensembles'), and attribution studies that use them, is uncertain.[66] Compounding this problem is that future regional fire scenarios are influenced by many factors including the magnitude of atmospheric warming, changes in rainfall, wind speed and humidity, and the degree to which nutrient limitations restrict plant growth; these are not expected to uniformly respond to CO_2 forcing in the same way. Because of these and other complications, climate models still do not agree on the magnitude of FFDI increases that we might see over time in different parts of Australia, and some even predict slight reductions.[67]

Ultimately, we probably don't want to test the global climate system further and find out. I think the famous pioneering climatologist and geophysicist Wallace Smith Broecker put it best: 'the climate system is an angry beast and we are poking it with sticks.'

Endnotes

[1] This quote is taken from an interview provided by Niall Blair to 7NEWS on 9 January 2019 while visiting a fish kill near Menindee. The full quote is: 'Look this is nothing more than an environmental disaster that has come about from a range of factors including the drought.'. To be clear, Blair was unambiguously stating that the extent of the fish kill was an environmental disaster.

[2] Cf. Immanuel Kant's *The critique of pure reason*: 'That all our knowledge begins with experience there can be no doubt.'

[3] Cormac McCarthy, *Blood meridian* (1985). This line references a similar one spoken by Judge Holden near the wells of Alamo Mucho (= Mocho) well in the Sonoran Desert west of Yuma, although the 'griddle' he described was siliceous – a dune field.

[4] Bavas J (2014) 'About 100 000 bats dead after heatwave in southern Queensland.' *ABC News*, 8 January.

[5] Breaking the records of 50 days at Bourke Airport; see Australian Bureau of Meteorology (2017) 'Special Climate Statement 61'.

[6] 'Drought declared in more southern Queensland after hot, dry summer'. *ABC News*, 13 March, 2017.

[7] Buchanan K (2017) 'Drought declared as parts of Queensland suffer one of the hottest and driest summers on record'. *Qld Country Hour, ABC Rural*, 6 March; 'Wet season arrives too late for central Queensland farmer after record dry summer'. *ABC News*, 21 March 2017.

[8] Dodds PE, Meyer WS, Barton A (2005) 'A review of methods to estimate irrigated reference crop evapotranspiration'. CRC for Irrigation Futures Technical Report No. 04/05.

[9] The following three papers are relevant: Brás TA *et al.* (2021) *Environmental Research Letters* **16**, 065012; Asseng S *et al.* (2017) *Global Change Biology* **23**, 2464–2472; Zampieri M *et al.* (2017) *Environmental Research Letters* **12**, 064008.

[10] Schauberger B *et al.* (2017) *Nature Communications* **8**, 13931; Troy TJ, Kipgen C, Pal I (2015) *Environmental Research Letters* **10**, 054013.

[11] Eads L (2018) 'State of New South Wales now 100% in drought'. *The Drinks Business*, 8 August.

[12] Historic data for these and other NSW storages can be accessed at https://realtimedata.waternsw.com.au/water.stm.

[13] Lyon N (2018) 'Water shortage cuts cotton potential in half as sowing gains momentum'. *Grain Central*, 31 October.

[14] Australian Bureau of Statistics (2021) 'Australia's cotton production at a 37-year low'. 14 May.

[15] Lyon N (2018) 'Reduced corn plantings compete for scarce water'. *Grain Central*, 27 November.

[16] Wainwright S, Volkofsky A (2018) 'Drying up of the Darling River angers community, sparks second protest'. *ABC Broken Hill*, 1 April 2018.

[17] The Barwon River ceased flowing on 22 July 2018 and then remained that way through the end of 2019, apart from a brief pulse in June-July 2019. The Namoi had a minor flow in April 2019; see https://www.waternsw.com.au/.

[18] Carbonell R, Davies J, Bonica D (2018) 'Health expert warns residents are at risk from high sodium in water in drought-stricken NSW town of Walgett'. *ABC Rural*, 11 December.

[19] Gooch D (2018) 'Menindee locals regroup to ensure their town survives amid Murray-Darling water fight'. *ABC Broken Hill*, 23 April; Hannam P (2018) "Outrageous': Farmers, Broken Hill mayor call for $467 million pipeline halt.' *The Sydney Morning Herald*, 6 February.

[20] Wainwright S, Volkofsky A (2018) 'Drying up of the Darling River angers community, sparks second protest'. *ABC Broken Hill*, 1 April.

[21] Tomevska S, McKinnell J (2018) 'Dead perch found in far-western NSW river sparks concern about water quality'. *ABC News*, 16 December.

22. Tomevska S (2018) 'Darling River water quality declines with 10 000 native fish found dead from blue-green algae bloom'. *ABC Broken Hill*, 20 December; Montoya D, Ravlich E, NSW Parliamentary Research Service (2019) 'Murray Darling Basin: Fish kills and current conditions'. *NSW Parliament issues backgrounder*, No. 2, July.
23. Davies A (2019) 'Hundreds of thousands of native fish dead in second Murray-Darling incident'. *The Guardian*, 7 January.
24. Davies A, Martin L, AAP (2019) 'Menindee fish kill: Another mass death on Darling River "worse than last time"'. *The Guardian*, 28 January.
25. Nguyen K (2019) 'Experts concerned after carp survives Menindee's mass fish kills'. *ABC News*, 30 January.
26. Woodburn J, Condon M & Shields M (2019) 'NSW Water Minister bows out following "threats" and "aggression"'. NSW Country Hour, *ABC News*, 25 March.
27. 'Niall Blair resigns as minister over 'continued campaign of threats' against him and his family'. *NSW Country Hour, ABC*, 25 March 2019; Smith A (2020) 'Liberals accept defecting Nationals MP after two others quit the party'. *The Sydney Morning Herald*, 21 September.
28. Based on data provided in Montoya D & Ravlich E, NSW Parliamentary Research Service (2019) 'Murray Darling Basin: Fish kills and current conditions'. *NSW Parliament issues backgrounder*, No. 2, July.
29. Pini M (2019) 'Editorial: Something stinks in the Coalition and it's not just dead fish'. *Independent Australia*, 17 January; Grafton Q (2020) 'While towns run dry, cotton extracts 5 Sydney Harbours' worth of Murray Darling water a year. It's time to reset the balance'. *The Conversation*, 14 April; Stevens B (2019) 'Dead fish in Menindee are about mismanagement not drought'. *Sydney Morning Herald*, 14 January.
30. 'Greens call for emergency water measures in wake of unprecedented fish kill'. *The Greens website*, 9 January 2019.
31. Cox L (2019) 'Darling River fish kill: Cotton industry says it won't be 'the whipping boy' for disaster'. *The Guardian*, 10 January.
32. Kay A (2019) 'Cotton farmers did not kill the fish – it's time to hear the facts'. *The Sydney Morning Herald*, 17 January.
33. Charles Sturt saw the Darling–Barwon reduced to a series of saline pools in 1828–29, and long stretches of the river were dry during the Federation Drought; cf. 'The Darling River is simply not supposed to dry out, even in drought'; Sheldon F (2019) *The Conversation*, 16 January.
34. Murphy S (2021) 'Darling River ecology 'extinct' and Murray cod 'in real trouble', warns expert Dr Stuart Rowland'. *ABC Landline*, 13 March.
35. Australian Associated Press (2018) 'Dust descends on NSW drought-hit farmers'. *news.com.au*, 8 November.
36. Data sourced from Australian Bureau of Statistics (2020) Agricultural Commodities, Australia, 2018–2019 financial year.
37. The Australian Bureau of Statistics reported the number of beef cattle on 30 June 2020 at 21 million head, although ABS data consistently underestimate the size of the national beef cattle herd; Fordyce G *et al.* (2023). *Animal Production Science* **63**, 410–421.
38. Australia's wheat production has been increasing since the 1960s, when annual production was around 8 million tonnes per hectare.
39. Sullivan K (2019) 'Australia approves foreign grain imports for the first time in over a decade'. *ABC Rural*, 15 May.
40. Godfree RC *et al.* (2019) *PNAS* **116**, 15580–15589.
41. Withey A (2019) 'Bushfire season starts early across northern Australia due to ongoing hot, dry conditions.' *ABC News*, 27 June.
42. Office of the Inspector-General Emergency Management (2019) The 2018 Queensland Bushfires Review Report 2: 2018-2019. State of Queensland (Inspector-General Emergency Management).
43. Bushfire and Natural Hazards CRC (2019) 'Australian Seasonal Bushfire Outlook: August 2019'. Hazard Note Issue 63.
44. Noyes J (2019) 'It's still winter, but parts of NSW are now officially in bushfire season'. *The Sydney Morning Herald*, 1 August; Shoebridge J, Rubbo L, Siossian E (2019) 'It was like a Christmas tree, all lit up': Bushfires in northern NSW have already taken four homes this winter'. *ABC Mid North Coast*, 14 August.

[45] With apologies to the American pop band, Imagine Dragons.
[46] Godfree RC *et al.* (2021) *Nature Communications* **12**, 1–13.
[47] Collins L *et al.* (2021) *Environmental Research Letters* **16**, 044029.
[48] Nguyen HD *et al.* (2021) *International Journal of Environmental Research and Public Health* **18**, 3538.
[49] van der Velde IR *et al.* (2021) *Nature* **597**, 366–369.
[50] Tang W *et al.* (2021) *Nature* **597**, 370–375.
[51] Khaykin S *et al.* (2020) *Communications Earth & Environment* **1**, 1–12.
[52] Godfree *et al.* (2021).
[53] E.g. Gibson RK, Hislop S (2022) *International Journal of Wildland Fire* **31**, 545–557; Legge S *et al.* (2022) *Global Ecology and Biogeography* **31**, 2085–2104; Keith DA *et al.* (2022) *Global Ecology and Biogeography* **31**, 2070–2084; Gallagher RV *et al.* (2021) *Diversity & Distributions* **27**, 1166–1179.
[54] Brando PM (2014) *PNAS* **111**, 6347–6352.
[55] Bendall ER *et al.* (2022) *Plant Ecology* **223**, 907–923.
[56] A phenomenon known as interval squeeze. See Nolan RH *et al.* (2021) *Plant, Cell and Environment* **44**, 3471–3489.
[57] In other words, they would not be expected to occur in the absence of anthropogenic carbon emissions. See Steffen W *et al.* (2019) 'Dangerous summer: Escalating bushfire, heat and drought risk'. Climate Council of Australia.
[58] Clarke H, Lucas C, Smith P (2013) *International Journal of Climatology* **33**, 931–944; Garnaut R (2008) 'The Garnaut Climate Change Review: Final Report'. Cambridge University Press.
[59] Bradstock RA (2010) *Global Ecology and Biogeography* **19**, 145–158.
[60] Noble IR, Barry GAV, Gill AM (1980) *Australian Journal of Ecology* **5**, 201–203.
[61] Van Oldenborgh *et al.* (2021) *Natural Hazards and Earth System Sciences* **21**, 941–960.
[62] Timbal B, Drosdowsky W (2013) *International Journal of Climatology* **33**, 1021–1034.
[63] A recent study indicates that 70% of the greening trend in global vegetation since 1982 has been caused by CO_2 fertilisation; Zhu Z *et al.* (2016) *Nature Climate Change* **6**, 791–795.
[64] Clarke H *et al.* (2016) *Climatic Change* **139**, 591–605.
[65] Richardson D *et al.* (2021) *Weather and Climate Extremes* **34**, 100397; Canadell JG *et al.* (2021) *Nature Communications* **12**, 1–11.
[66] Nissan H *et al.* (2019) *WIREs Climate Change* **10**, e579; Stott PA *et al.* (2016) WIREs *Climate Change* **7**, 23–41.
[67] Clarke H, Evans JP (2019) *Theoretical and Applied Climatology* **136**, 513–527.

21
Hung out to dry

The wine of life is spilt upon the sand,
My heart is as some famine-murdered land

– Oscar Wilde, 'E Tenebris'[1]

IT IS TIME FOR US TO FINALLY CONFRONT AN UNCOMFORTABLE ISSUE THAT WE have for some time been skirting around – the decline in mental health and wellbeing that afflicts many people on the land during times of severe drought, and the consequences for families and the broader community. We have already seen that droughts have taken a terrible toll on the fortunes of squatters and farmers since the earliest days of European settlement. While the focus of the press during these events was squarely on the economic and agricultural dimensions of drought, historical newspaper articles also reveal that during these disasters farming men and women in Australia have always suffered great psychological stress and not infrequently take their own lives.[2]

Given the gravity of this topic, you might expect that efforts to better understand, and therefore treat, drought-related stress in farming communities would have been undertaken since at least Federation, when the continent was in the grips of one of the great droughts of the post-colonial era. But, surprisingly, you would be wrong: the development of a concerted and systematic sociological investigation of farmer mental health dates only to the 1980s and 1990s.[3] Prior to this, sociologists and the political class had focused their attention to more trendy cosmopolitan concerns, a pattern that may be related to broader development of an 'age of agricultural ignorance' that accompanied the demographic shift of people from farms to urban areas. History, it is said, focuses on battlefields, not ploughed fields.[4]

This began to change with the publication of early investigations into the 'readjustment' of agricultural systems in response to economic, technical and institutional drivers.[5] Farming commodities had always suffered from price volatility, but beginning in the early 1970s agricultural protections were gradually removed by successive Australian governments, a process that left farmers increasingly exposed to market forces. While this has led to increased farm productivity, it has also had severe consequences for the wellbeing of the many who have succumbed to mounting economic pressures. Early studies documented these impacts in the dairy and sugar industries,[6] and later during the South Australian farm crisis of the late 1980s to early 1990s: a period of high interest rates, financial deregulation, rapid farm turnover and expansion, and the abandonment of the Wool Reserve Price Scheme in 1991.[7] During this period, farmers experienced very high levels of physical and psychological ill-health and high levels of stress.[8] Farm exit pressures and associated economic stresses have continued, in many cases exacerbated by periodic increases in input costs and subsidies and commodity dumping by other foreign producers (especially the United States).[9]

Recent data also confirm that, even under ideal economic conditions, farming remains a very difficult occupation that regularly tests peoples' physical and psychological limits.[10] For example, in 2020, despite recent safety improvements, the agriculture, forestry or fishing sector still had a fatality rate of 13.1 per 100 000 (overwhelmingly male) workers, more than quadruple that of any other sector apart from transport, postal and warehousing (7.8 per 100 000 workers). Non-fatal injury rates are similarly high.[11] The risk of death and injury is exacerbated by extended work hours during peak agricultural seasons, difficult weather conditions and the use of old equipment.[12] In addition to these problems, farmers face incremental government regulation of their activities, isolation, reduced access to health care and blurred separation of work and home life; they take fewer holidays and are less likely to retire than the general population (Fig. 21.1).[13]

Unsurprisingly, stress, depression and anxiety also exist at high levels among Australian farmers, and chronic pain, obesity, alcohol misuse and risk of hypertension and diabetes are prevalent in the rural population generally.[14] Whether mental health differs significantly between rural and urban residents is an open question,[15] but its most tragic manifestation – suicide – is demonstrably higher in rural areas: recent data show that suicide is one of the top 10 leading causes of death in regional and remote areas, and the risk increases with remoteness.[16] Those involved in farming are a high-risk group: data from the period 1988–97 show that farm managers had a suicide rate more than double that of the comparative national rate, while the rate for farm labourers was almost as high.[17] Males comprised 97% of the 921 reported farming suicides. I can tell you from personal experience that these incidents devastate families and local communities – wounded and weary faces do not lie.[18]

Apart from the obvious societal impacts, this collection of challenges suggests that the psychological health of farmers and farming communities might also be particularly exposed to drought and other natural disasters. Once again, however, beyond the obvious economic pain caused by the failure of agricultural concerns little interest has been shown in this field of research until relatively recently. The field of 'disaster research' itself did not begin to take formal shape until the publication of the sociologist Samuel Henry Prince's *Catastrophe and social change: Based upon a sociological study of the Halifax disaster* in 1920,[19] and it took decades for authors such as John Dewey and Gilbert White to gradually expand the investigation of natural disasters into domain of *human ecology* – the study of human interactions with their environment. It was not until the 1980s and 1990s that studies investigating the psychological reactions of societies to disasters along gender, racial and socio-economic lines began to proliferate.[20]

But I do not want to leave you with the impression that the public were oblivious to these issues during major drought episodes of the past – this is false. For example, James K Chisholm of Camden Park clearly described in 1864 how drought, flood and disease of the local farming population left many burdened with overwhelming despair, misery and little hope for the future.[21] Noting the contribution of agriculture to the wellbeing and happiness of the broader population, he also called on Christian charity and benevolence to meet those in need. As we

Fig. 21.1: The burden of the farmer has changed little over the past two centuries. This iconic image was created in 1892 at the height of the rabbit plague. Melbourne: McCarron, Bird & Co., 1892; courtesy National Library of Australia, nla.obj-135861338.

Figure 21.2: Sheep slaughter, Runnymede, Vic, 26 September 1994. Farming can be a brutal business. During 1994 scenes like this took place across drought-stricken regions. Photo courtesy of Bruce Postle and National Library of Australia.

have seen, the Federation Drought shook the entire country to its core, and the 'stress and worry' and 'terrible strain' that it caused were thought by some to explain the rise in insanity and asylum committals seen at the time.[22] The need for widespread organisation and nursing services to be extended to remote rural areas was recognised as early as 1909[23], although later the impacts of the WWI and WWII droughts understandably took a back seat during the social upheaval taking place at that time.

> *There is often a noble feeling of pride and reserve in the character of our countrymen which leads them to conceal from the knowledge of the public the real nature and extent of their suffering ... It is when the minister of religion pays his sympathising visits that the aching heart of many a poor creature unburdens its load of misery, and pours into his ear many a tale of want and destitution which, alas! he can only pity, but not alleviate.*
>
> *– James K Chisholm, Camden Park, 11 February 1864*

Despite this history, sociological research initially neglected droughts as a driver of psychological morbidity, perhaps because, as 'slow catastrophes',[24] they lack the immediacy and sense of urgency of other natural disasters. Indeed, most of the earliest work involved not drought but bushfires, including the 'Ash Wednesday' fires that affected Victoria and South Australia on 16 February 1983.[25] These fires, which occurred at the culmination of the 1979–83 drought that brought record low rainfall to large parts of eastern Australia and the Western Australia wheatbelt, killed 75 people, destroyed thousands of buildings and killed hundreds of thousands of livestock. One of the more interesting findings of this work was the discovery that strategic withdrawal and disengagement as an adaptation to stress and trauma could limit the effectiveness of post-disaster mental health services.[26]

The tide, however, was beginning to turn. Interest in rural social work surged in the late 1980s,[27] and in 1994 a very serious drought that been developing in eastern Queensland and coastal New South Wales during 1991–93 spread across most of the country, bringing below to very much below average rainfall to central and western New South Wales (Plate 2). Sheep and wool prices were low at the time, and I remember visiting farmers in the Narrabri region who were forced to slaughter and bury hundreds or thousands of starving sheep (Fig. 21.2). These 'wide brown worries' resulted in the NSW Government pledging a $2 million package to combat stress and suicide in rural areas, and the adoption of a National Drought Policy to increase the self-reliance of farmers during severe drought.[28] To overcome the reluctance of some farmers to attend meetings in public, funding was provided for home visits by local counsellors, while separate mental health support services were provided to adolescents and the Aboriginal community. While only in their infancy, these efforts clearly signified growing governmental recognition of the magnitude of the problem and a willingness to target public funds to assist those affected.

SKELETONS

You know, farming looks mighty easy when your plow is a pencil and you're a thousand miles from the corn field.

— *Dwight D Eisenhower*[29]

It was during the Millennium Drought, however, that the plight of drought-affected landholders as victims of natural disasters burst into the public consciousness. Many factors were involved, but I think that the confluence of two proved instrumental. First, the Millennium Drought was the first quasi-decadal drought that most Australians had ever seen. As the landscape dried out, urban populations (where most of the voting power lies) were exposed to extreme heat and bushfires, and, for the first in more than 50 years, faced a serious threat to their drinking water supplies.[30] As we have seen, it was also the first to severely impact upon iconic Australian ecosystems like the river red gum forests and floodplains of the Murray–Darling Basin, and the pubic witnessed a new phenomenon – the emergence of pitched battles between farmers, environmentalists and even states over access to what was obviously now a finite and dwindling water supply.

Second, these events coincided with the phenomenal publicity that surrounded the release of former US Vice-President Al Gore's *An inconvenient truth* in 2006. Although concerns over the 'greenhouse effect' and later 'global warming' had received mainstream attention as early as the 1980s – when I was just a high school student in Canada in 1988 a guest speaker from the University of British Columbia spoke to our Environmental Science class about it – I think Gore's documentary opened many Australians to the idea that, instead of being the 'lucky country', they might now be at the pointy end of a developing global disaster.[31] Who could forget the iconic moment when Gore used a scissor lift to illustrate the rise of atmospheric CO_2 concentration from under 300 ppm by volume (ppmv) in pre-industrial times to a projected 600 ppmv less than 50 years into the future? This simple but effective messaging revealed the emerging power of popular culture to effectively disseminate news about environmental concerns to the public, and even the ability to change the political landscape.[32]

Voices from the drought-stricken bush, which previously had been largely quarantined from the national discourse, also began to flood the media. Some of these were so highly personal and distressing that they were impossible to ignore. I could pick many examples, but two have stayed with me ever since. The first was a phone conversation between a local radio host and a farming woman from near Mount Hope in south-central New South Wales in 2006, which at the time was little more than a rock-strewn, cauterised waste.[33] Kept awake night after night by the lowing of hungry cattle until driven nearly mad by the noise, she and her husband were considering the Government's offer of a $150 000 exit grant to walk away from their property.[34] The second was an interview with another farming woman, this time near Boorowa, 100 km north-west of Canberra. Her husband was so affected by years of battling the drought, she felt that she could not let him go out around their property alone for fear that he might not come back alive.

Figure 21.3: Fruit trees can be placed on 'life support' during drought by skeletonising the canopy to reduce water use. After severe pruning the trunks and branches are often whitewashed to reduce sunburn. Photo: courtesy of Mark Skewes.

Elsewhere, farmers were simultaneously facing erosion of income, landscape degradation, animal welfare concerns and issues of family cohesion, and sometimes were even forced to make the heartbreaking decision to sell or destroy economic assets that had taken many years and multiple generations to develop but that could no longer be supported. Along the Murray River, decades-old orchards were left to die as water allocations became insufficient to keep trees alive.[35] Even less severe options such as keeping trees alive by heavy pruning, fruit removal or even 'skeletonisation' of the canopy[36] inevitably came at the cost of future earning potential for several years (Fig. 21.3). The equivalent, in cattle country, was the sale of the highest quality breeding cows that had been selected across multiple generations for favourable traits such as growth rate, longevity, fertility and workability.[37] Such animals were only sold in desperation, and in Narrabri I knew graziers who could not muster the willpower to turn up at the saleyards when the last of their herds were being sold.

These accounts show the complex nexus of interacting agricultural, hydrological, economic and social impacts that arise during the most severe and prolonged Australian droughts.[38] And, as researchers began to take the study of drought and mental health more seriously, studies began to emerge that quantified these patterns and their impacts on rural wellbeing. For example, crucial studies published in 2006 and 2012 showed that suicide rates are 8–15 per cent more likely to occur during drought and indicated that between 1970 and 2007, drought was responsible for around 9 per cent of suicides among rural males aged 30–49 in New South Wales, which confirmed what many in the agricultural sector had long thought.[39] Since then, Australia has become a global leader in drought–mental health research, with numerous studies documenting the impacts of the Millennium Drought on the mental and physical health of rural populations, flow-on effects in the non-farming community, stress placed on irrigators in the wake of water reforms and many other issues.[40]

The past decade has seen further expansion of health care services to rural areas and targeted programs to mitigate the impacts of natural disasters. In New South Wales, for example, in addition to increased provision of medical and nursing professionals and virtual health care services to remote areas, the devastation caused by the 2017–19 drought and the 'Black Summer' disasters led to the establishment of an Emergency Drought Relief Mental Health Package at a cost of $26.5 million.[41] Similar initiatives have been established around the country in response to recent droughts, such as the Queensland's Royal Flying Doctor's Drought Wellbeing Service and Tackling Adversity in Regional Drought and Disaster Communities through Integrating Health Services scheme (established in 2015), targeted funding under Victoria's *Suicide Prevention Framework 2016–2025*, and many others. While the ultimate success of these efforts is yet to be determined,[42] we have clearly entered an era in which state and federal governments have assumed greater responsibility for the health and wellbeing of drought-affected Australians.

Unfortunately, a large and swelling body of evidence suggests that demand on these services is likely to grow as climate change increases the frequency and severity of droughts and other extreme weather events. On a global scale, flow-on effects to water availability, air quality, food availability, hygiene and sanitation, wildfire, heat and even disease will disproportionately impact upon people with geographical, economic or cultural disadvantages.[43] Australia will not be immune, and among the many people who will be affected there is particular concern that Aboriginal populations living in remote areas, who already suffer low health and wellbeing compared to the non-Indigenous population, could see the situation worsen.[44] Indeed, Anangu people from the Uluru region, whose ancestors have for generations survived in central Australia through a myriad of different climate regimes, have expressed to me concern for themselves and their country, caused in large part by excessively high summer temperatures and highly variable rainfall in recent years. For a people who play no meaningful role in anthropogenic CO_2 emissions, this seems particularly tragic.

Returning, however, to the plight of farmers, I would like to conclude on a note of optimism. Despite the obvious health disparities between rural and urban people, many farmers are members of close-knit communities with relationships that have been forged over generations. Until very recently, the full weight of drought has fallen almost exclusively on rural societies,[45] and there has been a genuine need for personal mechanisms to cope with the daily struggle of agricultural existence since the time of European settlement. Life on the land has been hard won, and although self-reliance and stoicism – which have unsurprisingly become central to the mythos of the Australian farmer – are often seen as liabilities in modern society, in challenging times they may also prove to be virtues.[46] If, as predicted, the scale and severity of natural disasters increase in coming decades, and societal destabilisation begins to seriously impact upon the lives of people living in urban areas, it remains to be seen who will be best positioned to weather the storm.

Endnotes

[1] '*E Tenebris*' translates as 'out of darkness', perhaps inspired by John 1:5 '*et lux in tenebris lucet et tenebrae eam non conprehenderunt*'.

[2] E.g. *The Narracan Shire Advocate*, 15 Jan 1898, p. 4; *The Sydney and Station Journal*, 31 May 1901, p. 3; *Evening News* (Sydney), 20 June 1902, p. 4; *The Register* (Adelaide), 18 June 1919, p. 8; *Western Mail* (Perth), 3 March 1906, p. 11; *Casino and Kyogle Courier and North Coast Advertiser*, 18 September 1926, p. 1.

[3] Fraser CE *et al.* (2005) *International Journal of Social Psychiatry* **51**, 340–349.

[4] Disinterest in agricultural concerns also grew in other western nations during the same period; Evans S (2019) *Agricultural History* **93**, 4–34. The full quote from which this is abridged is provided on page 291 in Fabre JH (1928) *The wonders of instinct. Chapters in the psychology of insects*. Duckworth, London.

[5] Agricultural, rural or farm adjustment refers to the exit of farmers from agriculture, or undertaking steps to avoid doing so; Gow J, Stayner R (1995) *Review of Marketing & Agricultural Economics* **63**, 272–283.

[6] Cary JW (1980) *New Zealand Agricultural Science* **14**, 18–26; Fray P (1986) *National Farmer* **6**, 28.

[7] Argent N (1999) *Journal of Rural Studies* **15**, 1–15; Bardsley P (1994) *The Economic Journal* **104**, 1087–1105.

[8] Slee PT (1988) *Health Education Research* **3**, 331–335.

[9] Tens of thousands of farmers have left the industry in recent decades; Peel D, Berry HL, Schirmer J (2016) *Journal of Rural Studies* **47**, 41–51. Commodity dumping is the export of agricultural products at below the cost of production; Murphy S, Hansen-Kuhn K (2020) *Renewable Agriculture and Food Systems* **35**, 376–390.

[10] Fennell KM *et al.* (2016) *Journal of Rural Studies* **46**, 102–110.

[11] Safe Work Australia (2021) 'Key work health and safety statistics, Australia 2021'; Lower T, Rolfe M, Monaghan N (2017) *Journal of Agricultural Safety and Health* **23**, 139–151.

[12] Safe Work Australia (2020) 'Work-related traumatic injury fatalities Australia 2020'; Byard RW (2017) *Forensic Science Medicine and Pathology* **13**, 1–3; Lower T, Herde E (2012) *NSW Public Health Bulletin* **23**, 21–26; Lower T, Rolfe M, Monaghan N (2017) *Journal of Agricultural Safety and Health* **23**, 139–151. In 2020 the overall worker fatality rate among men (2.8 per 100 000 workers) in Australia was 28 times that of women (0.1 per 100 000 workers).

[13] Fennell KM, *et al.* (2016) *Journal of Rural Studies* **46**, 102–110; Fraser CL *et al.* (2005) *International Journal of Social Psychiatry* **51**, 340–349; Brew B *et al.* (2016) *BMC Public Health* **16**, 988; Monk A (2000) *Rural Society* **10**, 393–403.

[14] Kennedy A *et al.* (2014) *Rural and Remote Health* **14**, 2517.

[15] Kaukiainen A, Kõlves K (2020) *Rural and Remote Health* **20**, 5399; Fraser CE *et al.* (2005) *International Journal of Social Psychiatry* **51**, 340–349; Brew B *et al.* (2016) *BMC Public Health* **16**, 988.

[16] Australian Institute of Health and Welfare (2019) 'Rural & Remote Health'. Catalogue No. PHE 255. Australian Institute of Health and Welfare, Canberra.

[17] Page AN, Fragar LJ (2002) *Australian and New Zealand Journal of Psychiatry* **36**, 81–85.

[18] Apologies to Oscar Wilde, '*E Tenebris*'.

[19] Prince SH (1920) *Catastrophe and social change: Based upon a sociological study of the Halifax disaster*. New York: Columbia University Press; Scanlon TJ (1988) *International Journal of Mass Emergencies and Disasters* **6**, 213–232.

[20] Peek LA, Mileti DS (2002) The history and future of disaster research. In *Handbook of Environmental Psychology*. (Eds RB Bechtel, A Churchman) pp. 511–524. John Wiley & Sons, New York.

[21] *Sydney Morning Herald*, 19 February 1864, p. 2.

[22] *The Australasian* (Melbourne), 1 August 1903, p. 276.

[23] *The Sydney Morning Herald*, 25 August 1909, p. 5.

[24] Jones R (2017) *Slow catastrophes: Living with drought in Australia*. Monash University Publishing, Clayton, Victoria.

[25] McFarlane AC (1987) *British Journal of Psychiatry* **151**, 362–367; McFarlane AC (1990) *International Journal of Mental Health* **19**, 36–47.

[26] McFarlane AC (1989) *Stress Medicine* **5**, 29–36.

[27] Cheers B (1990) *Australian Social Work* **43**, 5–13.

[28] *Canberra Times*, 13 November 1994, p. 3; 'Wide brown worries'. *National Outlook*, October 1994; White DH, O'Meagher B (1995) *Drought Network News*, June 1995.

[29] Spoken by Dwight D Eisenhower during and address at Bradley University, Peoria, Illinois, on 25 September 1956.

[30] *Daily Advertiser* (Wagga Wagga), 14 April 1938, p.7; *The Sun* (Sydney), 10 June 1941, p. 7; *The Sydney Morning Herald*, 2 April 1940, p. 8; *The Courier-Mail* (Brisbane), 7 November 1946, p. 6; 7 December 1950, p. 1; *The Canberra Times*, 26 July 1977; 6 May 1977.

[31] Although the link between carbon emissions and global warming had been known for many decades, scientific research into the 'greenhouse effect' expanded rapidly through the 1980s.

[32] The shift in public awareness of climate change that took place at this time is discussed in Anderson D (2009) *Rural Society* **19**, 340–352. The mean atmospheric CO_2 concentration was close to 380 ppmv when *An inconvenient truth* was released; at the time of writing in August 2022 it has risen to about 418 ppmv.

[33] Apologies to Cormac McCarthy for borrowing this description of the Sonoran Desert in *Blood meridian*.

[34] AAP (2007) 'Drought aid's $1b boost'. *The Sydney Morning Herald*, 26 September 2007.

[35] Harden B (2009) The *Washington Post*, 9 December.

[36] Lopez G *et al.* (2006) *Tree Physiology* **26**, 469–477; Ye M *et al.* (2021) *Journal of Hydrology* **594**, 125651; Goodwin I, O'Connell MG (2017) *Acta Horticulturae* **1150**, 219–232; Falivene S, Giddings J, Skewes M (2018) 'Primefact 427 Fourth edition: Managing citrus orchards with less water'. NSW Department of Primary Industries 2018.

[37] E.g. Meat and Livestock Australia (2006) 'Managing the breeder herd: Practical steps to breeding livestock in northern Australia'. Meat and Livestock Australia Ltd., North Sydney; Foran BD, Stafford Smith DM (1991) *Journal of Environmental Management* **33**, 17–33; Bowen MK, Chudleigh F (2021) *The Rangeland Journal* **43**, 67–76; NSW Department of Primary Industries (2018) 'Managing and preparing for drought 2018'. Eighth Edition. Published by NSW Department of Primary Industries, a part of NSW Department of Industry, Skills and Regional Development.

[38] Edwards B, Gray M, Hunter B (2015) *Social Indicators Research* **121**, 177–194.

[39] Nicholls N, Butler CD, Hanigan I (2005) *International Journal of Biometeorology* **50**, 139–143; Hanigan IC *et al.* (2012) *Proceedings of the National Academy of Sciences of the USA* **109**, 13950–13955.

[40] Obrien LV *et al.* (2014) *Environmental Research* **131**, 181–182; Fennel KM *et al.* (2016) *Journal of Rural Studies* **46**, 102–110; Horton G, Hanna L, Kelly B (2010) *Australasian Journal on Ageing* **29**, 2–7; Austin EK *et al.* (2018) *The Medical Journal of Australia* **209**, 159–165; Edwards B, Gray M, Hunter B (2015) *Social Indicators Research* **12**, 177–194, Sartore G (2005) *Australian Journal of Rural Health* **13**, 315–320; Wheeler SA, Zuo A, Loch A (2018) *Journal of Rural Studies* **62**, 183–194.

[41] NSW Ministry of Health (2022) 'Rural Health Plan: Towards 2021'. Final Progress Review.

[42] Suicide rates have risen since the latter stages of the Millennium Drought, and the disparity between remote and metropolitan areas remains stubbornly high; see Australian Bureau of Statistics (2021) Causes of death, Australia; Biddle N *et al.* (2020) 'Suicide mortality in Australia: Estimating and projecting monthly variation and trends from 2007 to 2018 and beyond'. Australian National University Centre for Social Research and Methods.

[43] Palinkas LA, Wong M (2020) *Current Opinion in Psychology* **32**, 12–16; Hrabok M, Delorme A, Agyapong VIO (2020) *Journal of Anxiety Disorders* **76**, 102295; Ebi KL, Bowen K (2016) *Weather and Climate Extremes* **11**, 95–102; Bryan K *et al.* (2020) *Climatic Change* **163**, 2073–2095.

[44] Green D, Minchin L (2014) *EcoHealth* **11**, 263–272.

[45] For example, recent research has shown that the Millennium Drought increased distress for rural but not urban populations. See OBrien LV *et al.* (2014) *Environmental Research* **131**, 181–187.

[46] In modern usage, a 'stoic' is one who endures hardship patiently or represses their feelings. An interesting critique of the use of the term 'stoicism' in the contemporary medical and health literature is provided in Moore A *et al.* (2012) *Health* **17**, 159–173.

Obscured horizons

The optimist proclaims that we live in the best of all possible worlds; and the pessimist fears this is true.

– James Branch Cabell, The silver stallion: A comedy of redemption, *1926*

THERE IS MUCH IN THE HISTORY OF AUSTRALIAN DROUGHT TO NOURISH both optimists and pessimists. Again and again, we have seen human societies and the landscapes that they have created enveloped in a cloud of dust and death, only to witness them rise again, phoenix-like, with a baked in resilience more fitting of the land. If there is truth in the view that the influence of people's actions is greatest when the environment is at its most extreme,[1] then Australians have surely had many opportunities to be influential. But, having ploughed through three centuries of failure and excess, you will be relieved to find that I will not subject you a blow-by-blow development of a managerial-style list of 'lessons learned' through which we might chart the future. Even if that was possible, our path will probably equally resemble one of divination than of reason. Instead, I would like to leave you with some reflections that have revealed themselves as I have put this book together and wrestled with understanding the motivations of people throughout time.

Perhaps the most confounding aspect of Australia's drought history is that it is popular to attribute mistakes made in the past primarily to ignorance, and especially to a lack of scientific understanding. But I find this uncompelling, because many Europeans, especially those among the educated classes, quickly understood the nature of the country that they had set roots in. Some knowledge was passed on by Aboriginal people, who were intimately familiar with the vicissitudes of the antipodean climate and its limitations, although obviously much was overlooked. However, the poverty of certain soils and lack of permanent water was realised immediately, and before 1800 they knew of the hydrophobic earth and the salt that lay below, and that the brutal heatwaves so intense that bats and birds fell dead from the sky originated in the continental interior.

Within the first two decades of the 1800s the decline in soil fertility and porosity caused by intensive livestock grazing was evident, as was the need to plan stocking rates and pastures for the inevitable periods of severe drought. Within two more decades they knew of the great aridity of the interior, a land subject to great cycles of drought so severe that lakes dried up and forests died. By the 1860s the sensitivity of the semi-arid and arid rangeland plant and animal assemblages to grazing and the hard limits placed on traditional farming systems in these biomes were fully understood and even identified and addressed by government. It was also known that cattle and sheep grazing had the capacity to completely undermine the traditional lifestyles of Indigenous people. Yet they continued.

By 1900 it was clear that irrigation was affecting flows in the Murray River, that 'dust bowl' conditions could form in sandy country across much of the country if it was pushed too hard, that grazing was altering the composition and age structure of rangeland vegetation, and that near-permanent soil loss and scalding were occurring on a massive scale. This knowledge did not prevent these problems from reaching crisis levels over the next half century. A lucky streak of wet weather intervened, but by the 1980s it was clear that the Murray River was in trouble. The Millennium Drought shattered the illusion of catchment health, yet just a decade later, and despite a deluge of scientific papers and models, an arguably worse crisis struck in 2017–20 as megafires ravaged the east coast, rivers disappeared, towns ran out of water, and wildlife numbers collapsed in a series of massive die-offs.

THE CASSANDRA PROBLEM

Without overlooking the obvious advances that have been made in each of these areas, we should ask why Australian societies, like so many before, were unable to avoid the drought-induced catastrophes discussed throughout this book, especially given that they were clearly foreseen by some individuals. The knowledge required to avoid these events was contained within the collective human consciousness but could not alter their progression, and we need not invoke Laplacian determinism to feel a sense of inevitability about the deficit of societal agency that has emerged during such times.

The obvious explanation is that humans frequently do not cooperate to overcome a common problem when narrower interests are at stake. This phenomenon that has attracted an extensive body of research, but I think is most brilliantly brought to life in Liu Cixin's *The three body problem*, which can be seen as thinly veiled metaphor for the major environmental problems that we face today. The great difficulty of addressing their root causes, especially carbon emissions, has led an increasing number of planners and policy makers to view them as virtually intractable or *wicked problems*. Extreme anthropogenic droughts fit neatly into this category, since they involve a series of complex and interdependent problems, that can, in a real sense, be seen as symptoms of each other.[2] These have been exacerbated by the panoply of problems that we have investigated throughout this book, including shifting baselines and evolving technological, social, agricultural and climate systems, not to mention the tendency of Australian ecosystems to damage quickly and recover slowly.

Climate projections for Australia make sobering reading. Most state-of-the-art models indicate that much of the continent will suffer longer and more intense meteorological and hydrological droughts due to declining rainfall and increasing evapotranspiration in the future. The south-west and east of the continent are likely to be worst affected and the north comparatively unscathed.[3] There remains major uncertainty in rainfall projections, with a minority of models even indicating a reduction in drought duration and intensity everywhere except south-western Western Australia. But the worst projections suggest a doubling of time in extreme drought by 2070 compared to 1995, combined with a 50% increase in drought severity. By one estimate, Australia already ranks fifth in the world in terms of total economic damage and 15th in number of people affected by drought,[4] and so these projections hardly bear thinking about.

Yet, as we have seen, Australia is now the world's third biggest exporter and fifth biggest miner of fossil fuels (by CO_2 potential), the burning of which poses a very significant, and at worst existential, threat to national security, and has a long-term plan to increase these further.[5] At the same time, the Australian Government's *Intergenerational report 2023* indicates that the national population in 2062–63 may have risen by 53% to 40.5 million, further burdening natural resources that were stretched to the breaking point in 2019–20, and will also age significantly, creating an additional strain on revenue. Interestingly, there are currently calls to tax the fossil fuel industry more heavily to pay for rising health, aged care, disability, defence and interest costs,[6] although how this will reduce dependency on these industries remains unclear. Whatever the outcome, tackling this wicked problem is going to require both national and international cooperation, and it will need to happen against a backdrop of ever-increasing, global energy-related CO_2 emissions, which set a new record of 36.8 gigatonnes in 2022.

But let us end on a different note. Many Australians in the past have had to stare down drought alone, suffering starvation, impoverishment and other hardships every bit as harsh as anything we are likely to face in the future. But society has emerged stronger, as collective advances in agriculture, environmental management, science and both human and animal welfare have diminished the worst of these impacts over time. And, despite the growing tendency of the population to focus on the admittedly essential role of both national and international governments to solve these challenges, there remains a central role for the individual to navigate these changing waters. Among the pages of this three-century history there is surely much to guide us.

Endnotes

[1] Harrington GN, Wilson AD, Young MD (1984) Management of rangeland ecosystems. In *Management of Australia's Rangelands*. (Eds GN Harrington, AD Wilson, MD Young) pp. 3–13. CSIRO, Melbourne.

[2] Lönngren J, van Poeck K (2021) *International Journal of Sustainable Development & World Ecology* **28**, 481–502.

[3] Kirono DGC *et al.* (2020) *Weather and Climate Extremes* **20**, 100280.

[4] Tanago IG (2016) *Natural Hazards* **80**, 951–973.

[5] Swann T (2019) 'High carbon from a land down under: Quantifying CO_2 from Australia's fossil fuel mining and exports'. The Australia Institute, Canberra.

[6] Rollins A (2023) 'Pocock flags opposing govt changes, calls for super profits tax on fossil fuel'. *The Canberra Times*, 30 August.

Index

Abbott, Tony 242
Aboriginal 'highways' 82
Aboriginal missions 204
Aboriginal people 10
    conflict with settlers 32, 37, 63, 133, 144–5, 202, 204, 205
    famine 3, 103, 202, 204–6, 208
    health 204, 206, 208
    impact of droughts 11, 13–14, 68, 202–8
    lighting of fires 35, 45
    smallpox epidemic 14, 37, 92, 93, 100–3
    water sources 95, 201–2, 206
above-ground net primary productivity 111–12
Adnyamathanha people 144–5
agricultural drought xiv
ailuru 205
algae, blue–green 254
Allyn River 12, 13
Anangu people 200, 205–6, 272
Annean Station 199
Anthropocene 214
anthropogenic drought xiv
Antill, Henry 84
apple, rough-barked 260
aquifer recharge 244
arid zone, droughts 200
Arrernte people 207
Aston Coal 242
Austin, Thomas 155
Australian Agricultural Co. 104–6

Bagundji people 102
Ballandool River 225
bandicoot
    desert 208
    pig-footed 136–7
Barcoo River 142
Barkindji people 253
barley 32, 33, 80
Barmah Forest 220–3
Barrow Creek Pastoral Co. 198, 200
Barrow Creek 198, 199, 202
Barwon Park 155

Barwon River 94, 95, 240, 253, 256
Bass, George 7, 50, 68–71, 82
Bathurst Plains 83
beech, Antarctic 12
Belabula River 84, 85
Bell River 95
Bennett, Samuel 7, 36
bettong
    brush-tailed 137
    burrowing 53, 137, 138, 144, 156, 208
Big Dry (2017–19) xvi, 152, 214
Bigambul people 215
Bigge, John Thomas 19, 20
bilby 208
biophysical thresholds 128
birds
    heat stress 99
    waterbirds 224, 225–6
Birrbay people 13
Birrie River 225
Black Drought 189, 202
Black Summer 252–60
Black, Tom 190
Blackmore, Don 222
blackwater event 233
Blair, Niall 252, 254, 256
Blaxland, Gregory 50, 74, 78, 82
Blaxland, John 74, 78
bloodwood, red 32
Blue Mountains, crossing 82–3
bluebush, black 133
Bogan River 59, 95, 98
Bokhara River 225
Bolger, J 75
Border River 256
Bowman, RG 183
box
    apple 20
    black 220, 233
    white 259
bream, bony 254
Breeza Plains 118
Briarie River 225

brigalow 142
Broken Hill, dust storm 58
Broken River 216
Brown, David 156–7, 159
Brown, John 8
Brown, Robert 5
budda 260
Buddah Lake 96–8
bullock teams 113–14, 119, 169, 198
burr, Bathurst 134
Burrendong Dam 253
bushfire outlook (2019) 258
bushfires 8, 62, 64–5, 128, 242, 258–60, 269

Cabell, James Branch 275
Caley, George 82
calicivirus 155
Calvert, S 138, 173
Calwell, Arthur 182
Campbells River 51, 84–5, 132
Camyr Allyn 13
Canning Downs Station 169
Carnegie, David 201, 202
cassowary, southern 128
Castlereagh River 95, 118
catastrophes 128
catastrophism 127
cats 137–9, 208
cattle stations 198, 215
cattle, numbers 63, 74–7, 80, 81, 171–2, 198, 257
Chisholm, James K 266, 269
Chowilla Floodplain 220, 221
Cicero 214
Cixin, Liu 276
Clarke, George 100, 101
climate myth-building 3
clover, Darling 133
coal mining 242–6
coal seam gas 242, 244
cod, Murray 132, 233, 254
Colbee 33
Collarenebri 240
Collins, David 34, 36–7, 42, 45, 63, 65–6
Collymongle Station 240
colonial expansion 110–15
commodity prices 120, 121
community protests 244–5
continental-scale, quasi-decadal droughts 47

cooba, river 225
Cook, James 26–7
coolabah 141–3, 200, 225–6, 257
Cooral 70, 71
Coorong, The 60, 223–4, 233
Copeton Dam 226, 239, 253
cotton farming 225–6, 244, 253, 256
Cotton, Charles 201
cottonbush 133
Cowpastures 64, 75, 76, 79–82, 119
Cox, William 83, 84
Coxs River 82–4, 131
Crick, William Patrick 158, 177, 179
Crucifixion 55, 182
Cubbie Station 225
Cudgegong River 256
Culgoa River 225
Cumberland Plain 83–4, 86
cumulative rainfall departure 151–2
Cunningham, Alan 11, 93, 97, 104
Cuvier, Georges 127

da Vinci, Leonardo 215
Dangar, Henry 11
Daringha 33
Darling River 92, 95, 98, 114, 115, 116, 133, 160, 241, 253–4, 256
Darling, Ralph 93, 94
Darling-Barwon country 238–41
Darug people 66
Dawes, William 31, 42, 43
Dawson, Robert 105
'dead heart' 196
decadal droughts 47–8
depression, economic (1840s) 119–23
dieback, tree 118, 200, 220–2, 257, 260
dingoes 113, 156, 184
Dirranbandi 240
discharge, Murray River 215, 219, 222, 224, 225
diversions, Murray-Darling Basin 216–18, 220, 224
doubah 206
Dowling, Wally 208
Doyle, JT 113
drought
  (c. 1760) xv, 3–9, 11
  (1797–99) xv, 64–7
  (1802–05) xv, 75–7
  (1811) 78–9

(1813–15) xv, 84–8
(1824–29) xv, 91–2, 94–5, 97–9, 103–4
(1835–42) xv, 116–21
(1846–50) xv, 133
(1858–59) xv, 135, 137, 142
(1864–68) xv, 136–7, 144
(1875–77) xv, 156
(1890–92) xvi, 171, 197–9
(1895–1903) xvi, 2, 78, 85, 169–71, 174–7, 179–80
(1911–15) xvi, 204, 206
(1922–29) xvi, 205, 206
(1931–38) xvi, 206
(1937–46) xvi, 207–8
(1951–54) xvi, 208
(1958–65) xvi, 151
(1964–67) xvi, 151
(1975–80) xvi, 246
(1979–83) xvi, 220, 248, 269
(1991–96) xvi, 269
(2001–09) xvi, 214, 220–6, 230–6, 270
(2017–19) xvi, 214, 233, 238–42, 253, 260
Black 189, 202
Federation xvi, 2, 78, 85, 169–71, 174–7, 179–80
Millennium xvi, 214, 220–6, 230–6, 270
Settlement 30–7
droughts
  major Australian xv–xvii
  types of xiv, 47–8
dry sheep equivalent 80–1
dryland farming 112
dryland salinity 65
Drysdale, Russell 55, 182
Duguid, Charles 204
Dumaresq River 256
Dunlop Station 160
Dunlop, James 116, 130–1
dust bowl 183–4
dust storms 58, 131–2, 143, 182–3, 188–90

eagle, wedge-tailed 156, 158
eaglehawks 156, 158
Eastern Star Gas 242
Ebsworth, Alfred Martin 175
ecological resilience 233
economic collapse (1840s) 119–23
ecosystem transformation 128
Eisenhower, Dwight D 270
El Niño 36, 115, 117, 232, 238

Eliot, TS vii, 188
Elliott, Frederick 199
Energy White Paper 246
ENSO 28, 115
environmental flows 231–2
ephemeral lakes 96
erosion, soil 56, 131–2, 182–8, 193
Eucalyptus brookeriana 7, 8
Eucalyptus viminalis 7, 20
euro 144, 258
European settlement, first 27–37
Evans, George William 83, 84
evapotranspiration 98
extreme temperatures 40–5, 64, 98–9, 140, 190–3, 242, 252–3

famine 32–4, 71, 103, 202, 204–6, 208
  Aboriginal people 3, 103, 202, 204–6, 208
  water 161
farmers, health 266, 269–72
Federation Drought xvi, 2, 78, 85, 169–71, 174–7, 179–80
feed distribution 175
FFDI 261
Field, Barron 103
Finlayson, HH 136, 196
fires 8, 62, 64–5, 128, 242, 258–60, 269
  lit by Aboriginal people 35, 45
  'Red October' 242
First Fleet 27–32
fish kills 98, 132, 233, 254, 256
Fish River 84–5
fish, decline 224
Flemming, James 5, 7–8
Flinders Ranges 136, 144
Flinders, Matthew 7, 8, 69
flood-dependent plants 223
floods 17, 20, 36, 65–6, 74, 77, 115, 160, 168, 220, 225, 233
flying-foxes 43–5, 252
fold catastrophe framework 128
food shortages 28, 32–4, 77, 103
Forrest, John 197
fossil fuels 245–6, 277
foxes 136, 138, 139, 208
freshwater fish, decline 224

Gamilaraay people *see* Kamilaroi people

Georges River 26, 74, 77
Giles, Ernest 197
Gillet 201
Gipps, George 122
Glenbawn Dam 12
Gore, Al 242, 270
Gosse, William C 197
Goulburn River, NSW 12, 13
Goulburn River, Vic 216
Goulburn Weir 216
Goyder, George Woodrofe 141
Goyder's Line 59, 141
gradualism 127
grasslands, deterioration 130
grazing 78, 80–1, 104, 142–3
    impact of 128, 130–2
Great Artesian Basin 253
'great drought' 63, 152, 197–9
Green Hills Farm 169
Gregory, Augustus Charles 142
Grimes, George 5, 8
Grose River 77
groundwater extraction 244, 245
groundwater salinity 95
gum
    manna 7, 20
    river red 17, 18, 160, 220–3, 225, 226, 233, 260
    snow 12
    Tasmanian blue 7–8
Gunter, Tom viii, 155
Guringay people 13
Gweagal people 27
Gwydir River 226, 256

Hamilton, George 110
hardyhead, Murray 224, 233
Harte, Bret 156
Hawkesbury River 36, 65–6, 74, 77, 79, 83, 115
health
    Aboriginal people 204, 206, 208
    farmers 266, 269–72
    mental 265, 266, 269, 271–2
heatwaves 40–5, 64, 98–9, 140, 190–3, 242, 252–3
Henderson, John 12, 118–19
Hermannsburg Mission 200, 204, 206
historical droughts 3
Hoddle, Robert 20, 21
hop bush 32

Howe, John 11
Hudson, RA 225
Hughes, John Bristol 144
Hume River 118
Hunter River 10–14, 49, 97, 118, 132
Hunter Valley 10–14
Hunter, John 63, 64–6
Hurley, Frank 184
Hut Lake 16
Hutton, James 127
hydrological drought xiv
hydrophobicity 66

ibis 226
Ilyatjari, Nganyintja 205, 208
invasional meltdown 134
ironbark, red 259
irrigation 216–17, 230, 240, 244, 253, 256, 276

Jemison's Beach 69
Jevons, William 36, 116

Kallara Station 135, 156–7, 159
Kamilaroi people 100, 101–2
kangaroo grass 78
kangaroo rats 144
kangaroos 81, 128, 130, 144, 257
Kant, Immanuel 238
Kayawaykal people 13
Kaytetye people 202
Keepit Dam 239, 253
Kempe, Hermann 197
Kidman, Sidney 177–8
King Island 5–8
King, Madeleine 245
King, Phillip Gidley 75
King, Phillip Parker 93
Knitting Nanas 244
Knowles, Craig 231
Kooma people 215
Krefft, Gerard 136–7
Kukatja-Luritja people 206, 207

La Niña 28, 232, 241
Lachlan River 17, 84, 85, 86, 116, 118, 133, 226, 256
Lake Albacutya 18
Lake Albert 223–4
Lake Alexandrina 60, 217, 223–4

Lake Bathurst 19, 20, 116, 117
Lake Cargelligo 17, 118
Lake Cawndilla 257
Lake Cowal 96, 97
Lake Etamunbanie 141
Lake Eyre 18, 196
Lake George 18–23, 83, 116, 117
Lake Lady Blanche 141
Lake Lipson 141
Lake Tuross 69
Lawson, Henry 240
Lawson, William 82
Lewin, John 51, 84
Lindsay, David 197
Liverpool Plains 93, 94, 105, 130
livestock
    numbers 63, 74–8, 80–1, 87, 88, 135, 142–3, 171–2, 177, 198, 257
    trampling 32, 104, 130–1
    weight 81, 130
Lock the Gate Alliance 242, 244
Loddon River 116, 216
long paddock 176
Lower Balonne River 225
Lyell, Charles 127

MacFarlane, J 175
Macintyre River 256
Macquarie Marshes 59, 92, 97, 98, 99, 226
Macquarie River 84, 85, 92, 94, 95, 97, 98, 118, 256
Macquarie, Lachlan 19, 74, 78, 80, 83, 84
Mahroot 100
Mair, John 100, 102
maize 32, 33, 35, 64, 74, 80
mammals, decline 136–8, 144–5, 208
Mandandanji people 215
Manning John Edye 121, 123
Marra Creek 97
Marthaguy Creek 95
Mason, HGB 198
Mathew, Felton 11
Matra, James 26
Maules Creek Project 242–5
McArthur's Fire Danger Index 261
McCarthy, Cormac vii
McKibben, Bill 234
Meehan, James 19
megadroughts 133, 151, 247

Mehi River 256
Menindee Lakes 241
Menindee Weir 254
mental health 265, 266, 269, 271–2
Meredith, Charles 109, 110, 121, 123
Meredith, Louisa Ann 109–10, 113, 119, 121, 123
Merri Merri Creek 95
Meston, Archbald 10, 168–9
meteorological drought xiv
meter tampering 241
Michelet, Jules xii
Millennium Drought xvi, 214, 220–6, 230–6, 270
Mills, William Whitfield 197
Mitchell, Thomas 11, 13, 17, 19–20, 102, 116, 137, 142
MLDRIN Echuca Declaration 234
moira grass 223
Monaro Plains 116
Mooki River 118, 132, 256
Moorilyanna Sheep Station 206
Moree 240, 252
Morris, Henry T 142
Morrison, Scott 241
motherumbah 252
mulga 57, 164, 198, 200, 201
Murray mouth 60, 223–4, 233
Murray River 114–15, 116, 168, 215–20, 222–6, 232, 276
    barrages 223, 224
    discharge 215, 219, 222, 224, 225
    water extraction 216–17
Murray River system, inflows 217, 219, 220, 222
Murray–Darling Basin 215–20, 230
Murray–Darling Basin Authority 230, 231
Murray–Darling basin plan 230, 231–4, 240, 256
Murrumbidgee River 114, 116, 118, 132, 215, 256
myxoma virus 155

Namoi River 95, 118, 132, 218, 239, 240, 244, 245, 253, 256
Narrabri Gas Project 242, 245
Narran Lakes 225
Narran River 225
Narrandera, dust storm 58, 189
National Plan for Water Security 230
natural disasters, psychological effect 266, 269
negative water balance 159
Nepean River 77, 115, 118
'Niobe of Australia' 169

Ngarrindjeri people 215
Ngiyampaa people 95
numbat 208

O'Kane, Mary 245
oak
   desert 200, 205, 206
     swamp 32
oats 32, 80
Officer, Charles 156
Officer, Suetonius 156
ooline 260
orange, wild 163, 164
Ovens River 118, 131
overstocking 87, 130, 143
Oxley, John 17, 92, 93

paleohydrology 16
Palinkana 207
parakeelya 206
Parker, John 34
Parker, Mary Ann 34
Parramatta River 32
Parramatta 28, 29, 32, 33, 45, 116
pastoral expansion 110–14, 197–8
Paterson River 13
Peel River 118, 256
perch
   golden 233, 254
   silver 233, 254
   southern pygmy 224, 233
   Yarra pygmy 224, 233
permanent water sources 11, 135–6
Phillip, Arthur 26, 27, 28, 32, 34, 35
pine
   black 259
   white 259
Pintupi people 207
Pitjantjatjara people 205–8
Poolamacca Station 159, 161
Potato Point 69, 70
Potato Point lagoons 50, 69
pre-colonisation droughts 3–9, 11
predator control 156, 158

quasi-decadal droughts 47
quolls 137, 144, 155, 156, 158, 208
   western 137, 144, 208

rabbits 53, 81, 136, 138, 154–5, 171, 203, 208
railway network 174
rainbow fern 32
rainfall decline, WA 246–7
rainfall
   long-term 54–5, 150–2
   Sydney dataset 31–2
   Tasmania 248
rat, greater stick-nest 53, 137
rat kangaroo, desert 136, 144
rations, food 32–6, 204
'Red October' fires 242
Regent's Lake 17
River Lett 82–4
River Torrens 117–18, 131
riverine plain 52, 92, 98
Robbins, Charles 5
Rose Hill *see* Parramatta
rush, giant 221, 223
Russell, Henry Chamberlain 3–5, 10–11, 13–14, 16, 20, 36, 168

salinity 13, 65, 69, 95, 224
saltbush plains 132–6
saltbush 133, 135, 136
   oldman 52
saltwater intrusion, Murray mouth 217
Sand Drift Act 1923 188
sand drift 183–8
sandalwood 163, 164
sandplain country 133
Sandy Hook 131
sawsedge, tall 71
scalded country 57
Schwarz, Wilhelm 197
scurvy 206
sea tassel, tuberous 224
Second Fleet 28, 30
sedimentation 49, 66, 131
selfish herd theory 164
Settlement Drought (1790–1793) 30–7
Severn River 256
Sharley, Tony 222
sheep
   numbers 63, 74–7, 80, 88, 135, 142–3, 171, 172, 177, 257
   slaughter 268
sheep-boiling mania 123

Shelley, Percy Bysshe 168
she-oak
   black 32
   river 257, 260
shifting baseline syndrome 217–18
smallpox 14, 37, 92, 93, 100–3
socioeconomic drought xiv
soil compaction 104, 130, 131
soil drought xiv
soil erosion 56, 131–2, 182–8, 193
soil moisture 258
Spencer, Walter Baldwin 202
sphagnum bogs 12
spinifex 200, 201, 206
squatters 110–13, 130, 140, 143
squatters, conflict with Aboriginal people 144–5
St George 240
Stevenson screen 41
Stevenson, Thomas 41
stilt, banded 224
stock movement 114, 176–7
stock routes 143, 176–7, 179
stocking rates 80–1
Storm Boy 224
Strehlow, TGH 207
Sturt expedition 91–2, 94–5, 97–8
Sturt, Charles 17, 91, 94–5, 97–100, 103, 114, 115, 140–3
subterranean water 135
sugar, melting 99
suicide, farmers 266, 271–2
surface water
   diversions 240, 244
   extraction 225, 244–5
Sussex Station 160
Sydmouth Valley 51
Sydney Cove 27–9, 31–2, 36
Sydney
   dust storm 58
   heatwaves 40–5
   rainfall dataset 31–2

Talbragar River 95
Tandou 241
Tank Stream 27, 29, 32, 34, 78–9
Taylor, Mike 231
Tench, Watkin 31, 32–3, 40, 42, 45
thistle, Scotch 134

time, distortion of 127
tipping points 128, 129
Tjuintjara 208
Torrens River 117–18, 131
Traill, Donald 30
transportation 113–14
tree deaths 5, 7–8, 141, 142, 225, 226
Trollope, Anthony 136
tubeworm, Australian 224–5
Tuross River estuary 69, 70
turtle
   eastern long-necked 224
   southern river 224
turtles, decline 224
Twynam Agricultural Group 240

uniformitarianism 127

Vierordt's law 127
virtual water 255, 256
Vivian, Lyndsey 223

wages, farm workers 113, 123
Walgett 252, 253
wallaby
   crescent nailtail 137, 144
   red-neck 258
Wallis Plains 103
Warlpiri people 205
Warren 240
Warrimay people 13
Water Act 2007 230–2
water allocation 222, 230–1, 256
water availability 226, 244–6, 248
water balance, negative 159
water buyback 240
water demand, WA 247–8
water entitlements 234, 240–1
water famine 161
water footprint 255
water harvesting 225
water repellence, soil 66
water security 244, 247–8
water theft 241
water, selling of 104
waterbirds 224, 225–6
Wayilwan people 95
Webster Limited 241

Wee Waa 240
weeds, infestation 130, 134, 193
Wells, Lawrence A 196
Wentworth, William 82, 85
wetland disconnection 223
wheat 28, 32–6, 63–5, 74, 79–80, 83, 168, 247, 257
Wheatbelt, WA 246–7, 269
Wilde, Oscar 265
wilga 260
Wilhelm, Friedrich 206
Wilson, John 82
Winburndale Riverlet 84, 85

Windamere Dam 253
Wiradjuri people 18, 83, 97, 100, 101
witjuti grubs 206
wombat, hairy-nosed 156
Wonnarua people 10, 13
wool boom 110–11, 113
Woolloomooloo Farm 75
workers, injuries 266

Yuin people 70–1
Yumu people 206
Yuwaalaraay people 101